Recent Progress in Biotechnology

Recent Progress in Biotechnology

Edited by Sansa Gilbert

SYRAWOOD
PUBLISHING HOUSE

New York

Published by Syrawood Publishing House,
750 Third Avenue, 9th Floor,
New York, NY 10017, USA
www.syrawoodpublishinghouse.com

Recent Progress in Biotechnology
Edited by Sansa Gilbert

International Standard Book Number: 978-1-64740-088-0 (Hardback)

Cataloging-in-Publication Data

Recent progress in biotechnology / edited by Sansa Gilbert.
    p. cm.
Includes bibliographical references and index.
ISBN 978-1-64740-088-0
1. Biotechnology. 2. Genetic engineering. I. Gilbert, Sansa.
TP248.2 .R43 2022
660.6--dc23

TABLE OF CONTENTS

PREFACE

This book has been an outcome of determined endeavour from a group of educationists in the field. The primary objective was to involve a broad spectrum of professionals from diverse cultural background involved in the field for developing new researches. The book not only targets students but also scholars pursuing higher research for further enhancement of the theoretical and practical applications of the subject.

Biotechnology is an area of biology, which is concerned with the use of living systems, organisms or their derivatives for the development of products and processes for human benefit. Biotechnology has played a crucial role in advancing medicine, food production and agriculture. Practices such as the cultivation of plants, domestication of animals and improvements in either of these through designed breeding programs have been enabled by the use of biotechnology. In the modern day, microorganisms are being used for the manufacture of organic products, for bioleaching by the mining industry, for the treatment and recycling of wastes, and production of biological weapons. Modern biotechnology involves genetic engineering, cell and tissue culture technologies, applied immunology, recombinant gene techniques, etc. Biopharmaceuticals, genetic diagnosis of inherited diseases, genetically modified foods and crops, are some of the significant applications of the field. This book presents the complex subject of biotechnology in the most comprehensible and easy to understand language. The various advancements in this field are glanced at and their applications as well as ramifications are looked at in detail. This book includes contributions of experts and scientists which will provide innovative insights into this discipline.

It was an honour to edit such a profound book and also a challenging task to compile and examine all the relevant data for accuracy and originality. I wish to acknowledge the efforts of the contributors for submitting such brilliant and diverse chapters in the field and for endlessly working for the completion of the book. Last, but not the least; I thank my family for being a constant source of support in all my research endeavours.

Editor

PCR based Random Mutagenesis Approach for a Defined DNA Sequence using the Mutagenic Potential of Oxidized Nucleotide Products

Utpal Mohan[#], Shubhangi Kaushik[#] and Uttam Chand Banerjee*

Biocatalysis and Protein Engineering Group, Department of Pharmaceutical Technology, National Institute of Pharmaceutical Education and Research, Sector 67, S.A.S. Nagar-160062, Punjab, India

Abstract: Oxidizing conditions have not been explored well for the *in vitro* random mutagenesis in directed evolution. The mutagenic potential of diverse range of oxidized products is well reported in literature. Incorporation of errors during PCR in the presence of oxidized nucleotides can be a very effective alternative to error prone PCR as the transversion mutation frequency is higher in the former case. Earlier reports used a single purified oxidized nucleotide for introducing mutations during polymerase chain reaction. This could be further improved using the entire range of oxidized nucleotides to widen the mutation spectrum. The highlight of the present work lies in the fact that the oxidized nucleotides used in this study were generated by incubating the mixture of all the four nucleotides (dATP, dCTP, dTTP and dGTP) with an oxidizing agent, ferrous sulphate. This oxidized nucleotide mixture was then directly used without purification in polymerase chain reaction to introduce random mutations. The 100 μM oxidized nucleotides mixture treated with 5 mM FeSO$_4$ for 10 minutes along with 200 μM nucleotides are the optimized parameters for PCR amplification of a desired gene. The effect of manganese and magnesium ions over the incorporation of oxidized nucleotides was also investigated. An optimized PCR based approach which can be an efficient alternative to error-prone PCR for introducing random mutations in a defined gene sequence has been successfully developed.

Keywords: Random mutagenesis, oxidized nucleotides, error prone PCR.

1. INTRODUCTION

Oxidative stress has been shown to be involved in biological processes such as mutagenesis, carcinogenesis and ageing [1-3]. Reactive oxygen species (ROS) produced in cells react with DNA and its precursors, and the oxidative DNA lesions formed causes mutational events. One of the oxidative DNA lesions is 8-hydroxydeoxyguanosine (8-OH-dG; 7,8-dihydro-8- oxodeoxyguanosine) [4-7] and it pairs with dA as well as dC in *in vitro* DNA synthesis and induces mainly G → T transversions in cells [8-16]. Moreover, 2 hydroxydeoxyadenosine (2- OH-dA) and 5-hydroxydeoxy-cytidine (5-OH-dC), which are produced by ROS, are miscoding and mutagenic in nature [17-21]. Major products reported are 2-hydroxydeoxyadenosine (2-OH-dA), 8, 5'-cyclodeoxyadenosine (cyclo-dA), 5-hydroxydeoxycytidine (5-OH-dC), 8-hydroxydeoxyguanosine (8-OH-dG), 5-formyl-deoxyuridine (5-CHO-dU) and glyoxal [22]. These results suggested that triphosphates of 2-OH-dA, cyclo-dA, 8-OH-dA, cyclo-dG, 5-CHOdU, 5-OH-dC, and glyoxal-dG as well as 8-OH-dG may be produced in cells with different ratio by various types of oxidative stress and involved in mutagenesis and carcinogenesis. Glyoxal is a major product of DNA oxidation in which Fenton-type oxygen free radical-forming systems are involved. It had been reported that the yield of glyoxal was much higher (17-fold) than that of

8- hydroxydeoxyguanosine (8-OH-dG). Moreover, the formation of glyoxal was estimated to be 13- fold more than that of 8-OH-dG when mixtures of deoxynucleosides were treated [22]. Glyoxal is known to be mutagenic in *Salmonella typhimurium* strains TA100, TA102 and TA104 [23-25]. It was further reported that glyoxal induces mutations at G:C base pairs, in a study using a set of seven *S. typhimurium* strains (TA7001–TA7006 and TA98) [26]. Moreover, glyoxal induces mutations mainly at G:C base pairs in wild-type *Escherichia coli* [27]. It was found that glyoxal induced predominantly G:C→T:A transversions, followed by G:C→C:G, A:T→T:A and G:C→A:T mutations. Oxidation of the methyl group of thymine produced 5-hydroxymethyl uracil (5-hmU) and as major products. One of the attractive approaches for random mutagenesis is the addition of a mutagenic nucleotide analog during PCR, to enhance the mutation frequency. 6-(2-Deoxyβ-D-ribofuranosyl)-3,4-dihydro-8H-pyrimido-[4,5-C][1,2]oxazin-7-one-5-triphosphate (dPTP) and 8-oxo-2'-deoxyguanosine triphosphate (8-oxodGTP) were previously used to create mutations. The former induces A:T→G:C and G:C→A:T transitions, and the latter elicits A:T→C:G transversions [28]. The 2-hydroxyadenine base in DNA induces A:T→G:C and A:T→C:G mutations in living cells and 2-OH-dATP has the potential to elicit G:C→A:T and G:C→T:A mutations [29]. It was reported that a 2- substituted purine nucleotide analog, 2-hydroxy-2-deoxyadenosine 5-triphosphate (2-OHdATP), was used for the random PCR mutagenesis [30]. It was also reported by them that PCR with 8-OH-dGTP, after error-prone PCR with Mn^{2+} induced A:T→G:C and G:C→A:T transitions and A:T→T:A and A:T→C:G transversions with similar frequencies. These results indicated

*Address correspondence to this author at the Biocatalysis and Protein Engineering Group, Department of Pharmaceutical Technology, National Institute of Pharmaceutical Education and Research, Sector 67, S.A.S. Nagar-160062, Punjab, India;
E-mail: ucbanerjee@niper.ac.in

that the combination of the Mn^{2+}-PCR and 8-OH-dGTP PCR may be useful to generate random mutant libraries of proteins or functional nucleic acids [31].

It has been very well demonstrated that all these oxidized products have diverse mutagenic capacity. Oxidizing conditions have not been explored well for *in vitro* random mutagenesis in directed evolution experiments. Herein, we report the development and optimization of a PCR based system where the entire oxidized nucleotide products were used to have a highly efficient random chemical mutagenesis approach. This may be an effective alternative method for the error prone PCR and could be used for inducing wide spectrum of transition and transversion mutations with higher frequencies.

2. MATERIALS AND METHOD

2.1. Random Mutagenesis with Oxidized Nucleotides

Pseudomonas aeruginosa lipase encoding gene was amplified by the specific forward 5' GCCATATGATGACA-CACAAGAGGTGTGGCCCGC 3' flanked with *NdeI* restriction site and reverse oligonucleotides 5' CGGATGTCA-GAGGAGATAAATCTGTCAGTAGAC 3' flanked with *XhoI* restriction site in 1X Taq buffer with KCl, 1.5 mM $MgCl_2$, 200 µM dNTP, 0.1-0.5 µM primer (forward and reverse), 1-2 units *Taq* polymerase and 100 ng genomic DNA. DNA molecular weight marker and all the PCR components except primers were from MBI fermentas GMBH, Germany. PCR was done in Eppendorf Master cycler gradient (Eppendorf AG, Germany) under the following conditions: initial denaturation at 94°C for 10 min followed by cycling conditions, denaturation at 94°C for 1 min, annealing at 58°C for 1 min, elongation at 72°C for 1 min 30 sec (repeated for 30 cycles). Final extension at 72°C for 10 min was used to complete the reaction. The PCR amplification of lipase encoding gene was done in the presence of oxidized nucleotides (dNTP). dNTP mixture was incubated for various time intervals in the presence of 5-10 mM $FeSO_4$. Mannitol (0.5 M) was used to stop the oxidation reaction. This oxidized nucleotide mixture was added to the normal polymerase chain reaction mixture and the PCR was run. The oxidized nucleotide mixture was used in combination with normal dNTP in the polymerase chain reaction. Except oxidized dNTP, all other components of a normal polymerase chain reaction were added before the incubation of the nucleotides with the oxidizing agent was over. Oxidized dNTP was always the last component added in the reaction mixture in all the experiments. *Taq* polymerase was added just a minute before the incubation of nucleotides was over. Care was taken to be quick in setting the reaction. All other components were kept same as in any normal polymerase chain reaction. The densitometry analysis was performed using the Quantity one software associated with the gel documentation unit (Biorad, India).

The amplification product obtained under the optimised oxidised condition was purified by gel extraction using a QIAquick PCR purification kit (Qiagen) and ligated to a linear, pDrive (T-A based) cloning vector (Qiagen). The ligation mixture was then transformed to *E.coli* DH5α cells. The sequencing of the mutants was performed at Bangalore Genie, India.

2.2. Effect of Concentration of Oxidizing Agent

To observe the effect of concentration of oxidizing agent, dNTP were incubated with various concentrations of oxidizing agent ($FeSO_4$) for 10 minutes at 37°C. All other components were kept same as in any normal polymerase chain reaction.

2.3. Effect of Incubation Time with Oxidizing Agent ($FeSO_4$)

To observe the effect of incubation time with the oxidizing agent, dNTP was added with the oxidizing agent and incubated for various time intervals. This oxidized dNTP mixture was then added to the polymerase chain reaction mixture.

2.4. Effect of Manganese Ions on PCR Fidelity

Manganese (500 µM) in a normal polymerase chain reaction results in the incorporation of errors. The effect of manganese in the incorporation of oxidized nucleotides was checked during polymerase chain reaction. It was checked by adding 500 µM $MnCl_2$ to the polymerase chain reaction mixture in the presence and absence of oxidized nucleotides.

2.5. Effect of Oxidized Nucleotide Concentration on Taq Polymerase Activity

To observe the effect of oxidized nucleotide concentration, they were added to the polymerase chain reaction mixture in varying concentrations. All other components were kept same as in any normal polymerase chain reaction.

2.6. Effect of Concentration of Nucleotides and $MgCl_2$ over the Incorporation of Oxidized Nucleotides

To observe the effect of untreated dNTPs and $MgCl_2$, untreated dNTP was varied with and without $MgCl_2$ in the reaction mixture.

2.7. Effect of $MgCl_2$ over the Incorporation of Oxidized Nucleotides

Increase in $MgCl_2$ concentration in the polymerase chain reaction results in the decreased fidelity of *Taq* polymerase. This allows misincorporation of nucleotides in the growing DNA template in the polymerase chain reaction. To observe the effect of $MgCl_2$ over the incorporation of oxidized nucleotides, 7.5 mM $MgCl_2$ was added in the polymerase reaction mixture in the presence of 50 and 100 µM oxidized nucleotides keeping all the other components constant.

3. RESULTS AND DISCUSSION

3.1. Effect of Concentration of Oxidizing Agent

Treatment of nucleotides with an oxidizing agent in *in vitro* condition and using the oxidized nucleotides for random mutagenesis is an approach which is very less explored till now. Unlike to the earlier reports which involved the use of single purified oxidised nucleotide for introducing the errors in defined DNA sequence, the present work employed the use of non-purified oxidised nucleotide mixture for incorporation of the mutations [30, 31]. Instead of using the oxidizing agent in the PCR reaction mixture, we oxidized the nucleotide mixture by incubating it with $FeSO_4$ (Fig. **1a**). The polymerase chain reaction mixture containing all the components in optimized concentration resulted in an intense

933 bp band. This PCR reaction mixture when supplemented with only the oxidized nucleotides without any untreated dNTPs, resulted in no amplification of lipase encoding gene. Treated dNTP mixture when supplemented with untreated dNTPs, resulted in the appearance of a 933 bp band. It was observed that when the dNTPs were treated with increasing concentration of FeSO$_4$, the PCR yield was decreased. Kamiya *et al.* used the 5 mM FeSO$_4$ concentration for their studies with single nucleotide oxidation [22]. In the present work, sizeable band intensity was obtained with 5 mM FeSO$_4$ while amplification was negligible with 10 mM FeSO$_4$. Densitometry analysis was also performed to compare the relative PCR product yield (Fig. **1b**). The probable explanation behind such an observation may be the increase in the concentration of oxidized nucleotide products which resulted due to the higher concentration of oxidizing agent. This higher concentration of oxidized nucleotide products seems to stall the *Taq* polymerase enzyme for which oxidized nucleotides products are not the natural substrates.

3.2. Effect of Incubation Time with Oxidizing Agent (FeSO$_4$)

The effect of increasing concentration of oxidizing agent led us next to investigate the role of incubation time with oxidizing agent (Fig. **2**). Increasing of incubation time from 10 to 20 minutes led to no amplification of 933 bp lipase gene. This observation indicated the difficulty faced by *Taq* polymerase in incorporating oxidized nucleotides to the template. The increased time of incubation had a direct proportionality with the number of oxidized nucleotides in the solution. The effects observed with the increased incubation time might be due to the number of oxidized nucleotides in the solution, oxidation state of the nucleotides and type or chemical nature of the oxidized nucleotides. Since *Taq* polymerase doesn't have a natural affinity towards oxidized nucleotides, the enzymes seems to be stalled when faced with a high concentration of oxidized nucleotides. This observation suggested that to induce mutation with oxidized nucleotides, the ideal incubation time should be the one at which the concentration of the oxidized nucleotide does not inhibit the polymerizing activity completely. The incubation time was not increased further beyond 20 minutes due to the disappearance of desired band on increasing the incubation time.

3.3. Effect of Manganese Ion on the PCR Fidelity

Introduction of manganese in the polymerase chain reaction mixture is one of the mostly used approaches in introducing mutations in a desired DNA sequence. Earlier reports establishing the role of manganese in decreasing the fidelity of *Taq* polymerase led us to investigate the cooperative role of manganese ion in inducing mutations in case of oxidized nucleotides. Mutagenic PCR involving the combination of the Mn^{2+}-PCR and 8-OH-dGTP PCR has been exploited for inducing the various mutations [31]. In the present work, introduction of manganese in the presence of oxidized dNTPs did not result in any PCR amplification of the desired gene (Fig. **3**). Though presence of manganese ions decreased the fidelity of *Taq* polymerase, the incorporation of oxidized nucleotides was not visible when these two conditions were included in a normal polymerase chain reaction mixture. The probable reason behind such an observation might be the stalling of *Taq* polymerase in the presence of excess of oxidized nucleotides.

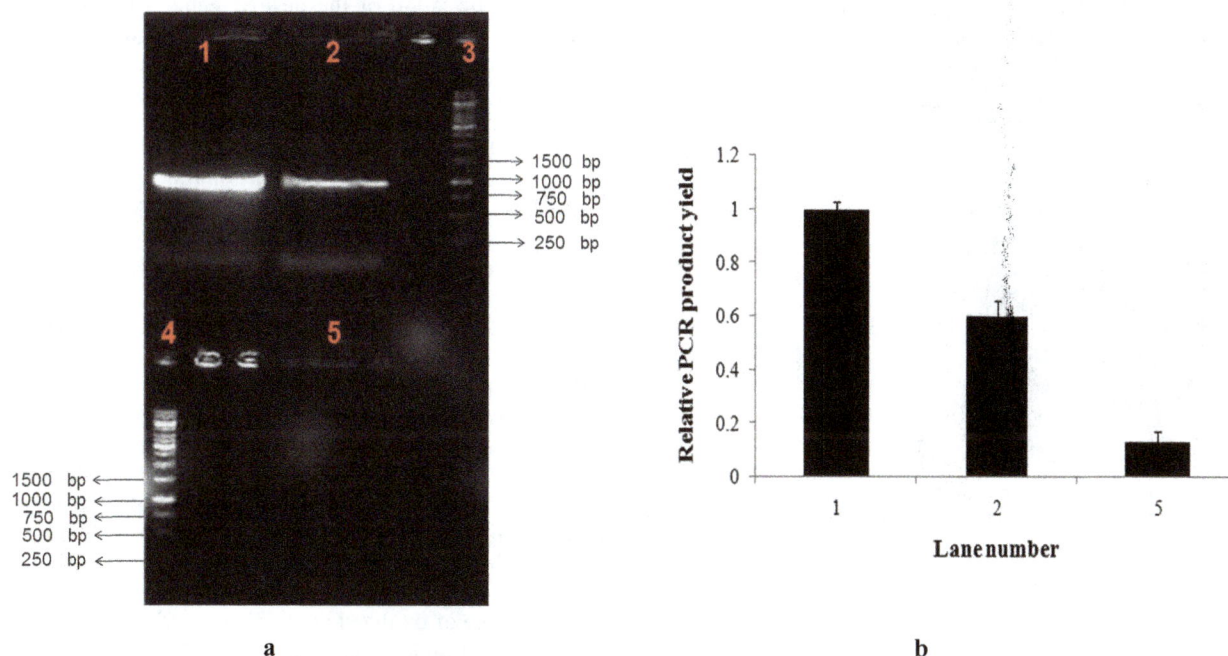

a   b

Fig. (1). a. Effect of concentration of oxidizing agent on the amplification. The lipase encoding gene was amplified in the presence of different mixtures of nucleotides treated for either 10 minutes or not: 200 µM nucleotides (lane 1); 100 µM of 5 mM FeSO$_4$ treated-nucleotides + 200 µM nucleotides (lane 2); 100 µM of 10 mM FeSO$_4$ treated-nucleotides + 200 µM nucleotides (lane 5) and visualized by agarose gel electrophoresis and ethidium bromide staining. DNA molecular weight markers are given in lanes 3 and 4.

b. Densitometry analysis to compare the relative PCR product yield. The intensity of the control reaction was designated as 1 (Lane 1) and relative intensity of other lanes were quantified against the control reaction.

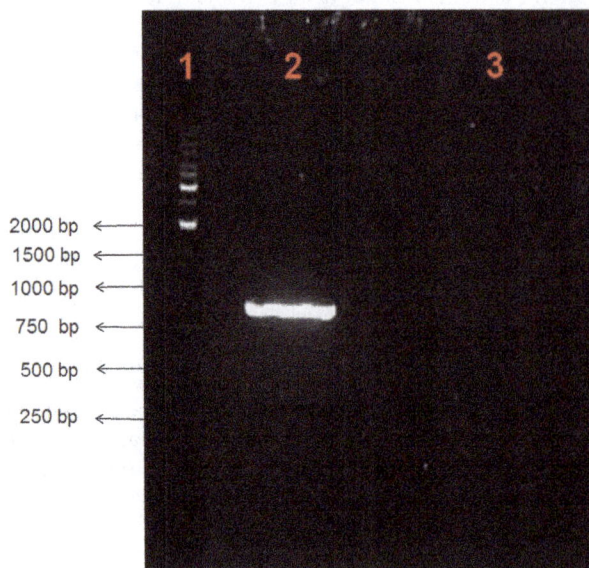

Fig. (2). Effect of incubation time with oxidizing agent on the amplification of the gene of interest. 100 μM nucleotides treated with 5 mM FeSO₄ for 10 minutes + 200 μM nucleotides (lane 2); 100 μM nucleotides treated with 5 mM FeSO₄ for 20 minutes + 200 μM nucleotides (lane 3). DNA molecular weight marker is shown in lane 1.

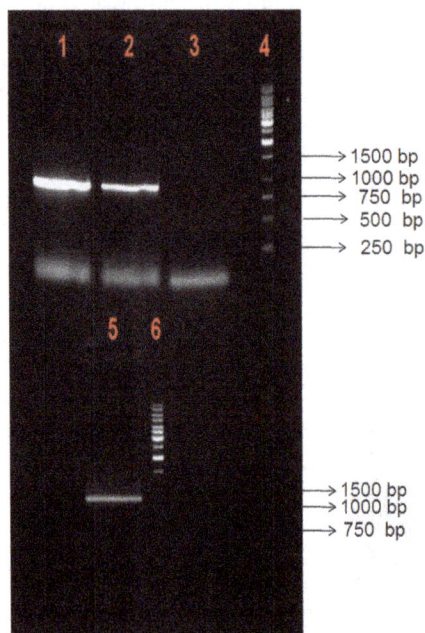

Fig. (3). Effect of manganese ions on the amplification of the gene of interest in the presence of oxidized nucleotide products. 200 μM nucleotides (lane 1); 100 μM nucleotides treated with 5 mM FeSO₄ for 10 minutes + 200 μM nucleotides (lane 2); 100 μM nucleotides treated with 5 mM FeSO₄ for 10 minutes + 200 μM nucleotides + 500 μM MnCl₂ (lane 3); 200 μM nucleotides + 500 mM MnCl₂ (lane 5). DNA molecular weight marker is shown in lane 4 and 6.

3.4. Effect of Treated Nucleotide Concentration on Taq Polymerase Activity

It was clear from the previous experiment that concentration of oxidizing agent affected PCR yield. Further, the ef-

fect of varying treated nucleotide concentration (keeping the untreated dNTP concentration constant) was checked (Fig. **4a**). It was observed that the treated dNTP concentration when decreased from 100 to 50 μM resulted in nearly the same PCR yield while the increased dNTP concentration (200 μM) resulted in no amplification of lipase encoding gene. The relative PCR product yield was also quantified for individual lane by densitometry (Fig. **4b**). From the above observation it could be concluded that the concentration of treated dNTPs in the reaction mixture is a crucial factor in polymerase chain reaction in the presence of oxidized nucleotides. The amount of oxidized nucleotides in the reaction mixture may or may not affect the *Taq* polymerase fidelity depending on whether the amount present in the reaction is enough to induce mutations or to stall the *Taq* polymerase. The excess of oxidized nucleotides in the reaction mixture leads to complete inhibition of *Taq* polymerase activity as the oxidized nucleotides are not the natural substrates of *Taq* polymerase. So, a critical concentration of oxidized nucleotides in the reaction mixture is required for polymerase chain reaction in the presence of oxidized nucleotides.

3.5. Effect of Untreated dNTPs Concentration over the Incorporation of Oxidized Nucleotides

The importance of a critical concentration of oxidized nucleotides in the reaction mixture led us to investigate the role of concentration of untreated nucleotides in the polymerase chain reaction mixture (Fig. **5**). It was observed that amplification was being done by *Taq* polymerase when 100 μM treated nucleotides were added with 200 μM untreated nucleotides. On decreasing the concentration of untreated nucleotides to 100 μM keeping the concentration of treated nucleotides at 100 μM (constant), it was observed that there was no amplification of the desired gene. This observation indicated that a proper ratio of oxidized as well as normal dNTP was required for amplification to take place. In such amplification, there were chances for oxidized dNTPs to get incorporated in the desired gene, which in further amplification cycles base pairs with normal bases and thus resulted in introduction of point mutations at random places. It is well established that increased concentration of MgCl₂ leads to the decreased substrate specificity of *Taq* polymerase. In order to investigate the effect of decreased substrate specificity of *Taq* polymerase on the incorporation of oxidized nucleotides, 7.5 mM MgCl₂ was added to the above two reaction mixture. It was found that increased concentration of MgCl₂ in the reaction mixture resulted in no amplification in both the conditions. Though increased concentration of MgCl₂ decreased the fidelity of *Taq* polymerase, the incorporation of oxidized nucleotides was not visible when these two conditions were included in a normal polymerase chain reaction mixture. The probable reason behind such an observation might be the stalling of *Taq* polymerase in the presence of excess of oxidized nucleotides which resulted due the acceptance of oxidized nucleotides by *Taq* polymerase in the presence of higher concentration of MgCl₂.

3.6. Effect of MgCl₂ over the Incorporation of Oxidized Nucleotides

The role of MgCl₂ on the fidelity of *Taq* polymerase led us to investigate the effect of MgCl₂ over the incorporation of oxidized nucleotides during polymerase chain reaction in

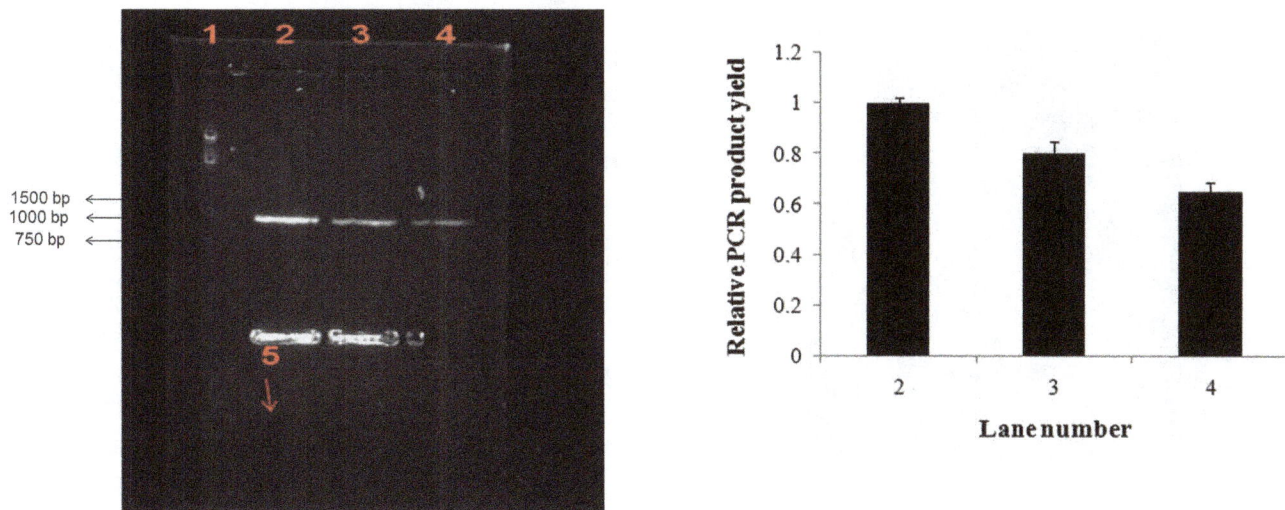

Fig. (4). a. Effect of treated nucleotide concentration on the amplification of the gene of interest. 200 µM nucleotides (lane 2); 50 µM nucleotides treated with 5 mM $FeSO_4$ for 10 minutes + 200 µM nucleotides (lane 3); 100 µM nucleotides treated with 5 mM $FeSO_4$ for 10 minutes + 200 µM nucleotides (lane 4); 200 µM nucleotides treated with 5 mM $FeSO_4$ for 10 minutes + 200 µM nucleotides (lane 5). DNA molecular weight marker is shown in lane 1.

b. Densitometry analysis to compare the relative PCR product yield. The intensity of the control reaction was designated as 1 (Lane 2) and relative intensity of other lanes were quantified against the control reaction.

the presence of oxidized nucleotide mixture (Fig. **6**). Reports are available showing the use of increased $MgCl_2$ concentration with oxidised nucleotides for the incorporation of errors in defined DNA sequence [32]. In present case, $MgCl_2$ concentration when increased to 7.5 mM keeping all other components of PCR reaction same, there was a decrease in the band intensity and amplification of non specific fragments took place (Fig. **6**, lane 3). The effect of $MgCl_2$ in the presence of oxidized dNTPs was investigated. The significant reduction in amplification of the desired gene was observed when oxidized dNTP (100 µM) was supplemented with 7.5 mM $MgCl_2$ (Fig. **6**, lane 6). There was no amplification even when the $MgCl_2$ concentration was decreased to 4.5 mM (Fig. **6**, lane 7).

3.7. Sequencing of Products of PCR Amplification with Oxidized Nucleotides

All the above results led to the conclusion that 100 µM oxidized nucleotides mixture (treated with 5 mM $FeSO_4$ for 10 minutes) along with 200 µM nucleotides are the optimized parameters for PCR amplification of a desired gene in the presence of oxidized nucleotide products. The PCR-amplified lipase encoding gene obtained using 100 µM oxidized nucleotides mixture (treated with 5 mM $FeSO_4$ for 10 minutes) and 200 µM nucleotides were cloned and sequenced. Both the transition and transversion mutations were observed in the sequenced clones (Table **1**). Sequencing of mutants resulted in the similar pattern of mutations that was earlier reported by Kamiya *et al.* [30, 31].

CONCLUSIONS

The entire studies on the oxidized nucleotides led us to conclude that to incorporate errors with oxidized dNTPs; it was necessary to adjust the ratio of the treated and untreated

dNTPs in the reaction mixture, concentration of oxidizing agent and the time of incubation. The polymerizing activity of Taq polymerase was observed to be affected in the presence of an excess oxidized nucleotides. It stalls when the

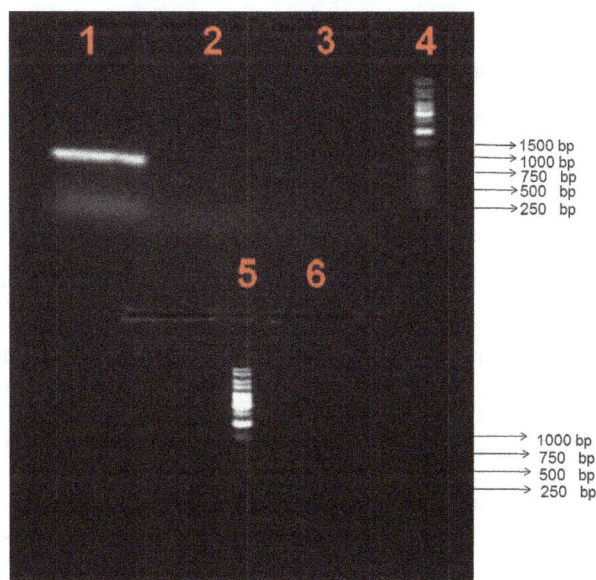

Fig. (5). Effect of untreated dNTPs concentration on the amplification of the gene of interest in the presence of oxidized nucleotide products. 100 µM nucleotides treated with 5 mM $FeSO_4$ for 10 minutes + 200 µM nucleotides (lane 1); 100 µM nucleotides treated with 5 mM $FeSO_4$ for 10 minutes + 100 µM nucleotides (lane 2); 100 µM nucleotides treated with 5 mM $FeSO_4$ for 10 minutes + 200 µM nucleotides + 7.5 mM $MgCl_2$ (lane 3); 100 µM nucleotides treated with 5 mM $FeSO_4$ for 10 minutes + 100 µM nucleotides + 7.5 mM $MgCl_2$ (lane 6). DNA molecular weight marker is shown in lane 4 and 5.

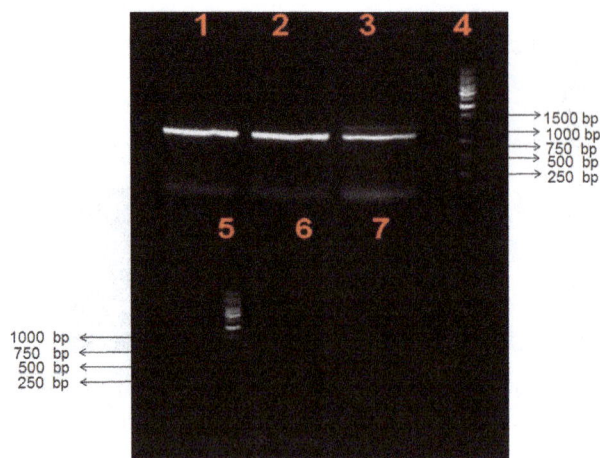

Fig. (6). Effect of MgCl$_2$ on the amplification of the gene of interest in the presence of oxidized nucleotide products. 100 µM nucleotides treated with 5 mM FeSO$_4$ for 10 minutes + 200 µM nucleotides (lane1); 50 µM nucleotides treated with 5 mM FeSO$_4$ for 10 minutes + 200 µM nucleotides (lane 2); 200 µM nucleotides + 7.5 mM MgCl$_2$ (lane 3); 50 µM nucleotides treated with 5 mM FeSO$_4$ for 10 minutes + 200 µM nucleotides + 7.5 mM MgCl$_2$ (lane 6); 50 µM nucleotides treated with 5 mM FeSO$_4$ for 10 minutes + 200 µM nucleotides + 4.5 mM MgCl$_2$ (lane 7). DNA molecular weight marker is shown in lane 4 and 5.

Table 1.   Point Mutations in the Variant Generated by PCR in the Presence of Oxidized Nucleotides

Base Pair Change	Nucleotide Position	Base Pair Change	Nucleotide Position
C → T	210	C → T	303
A → G	245	A → T	419
G → C	259	A → G	441
C → T	280	A → G	500

number of unnatural nucleotides crosses a critical concentration. Earlier reports used a single pure oxidized nucleotide for introducing mutations [30, 31] during polymerase chain reaction. This was a serious drawback as one requires tedious purification of oxidized nucleotide product before using it in the PCR reaction to induce point mutations. Moreover, the diverse range of oxidized nucleotide products in a single polymerase chain reaction has never been utilized to induce the random mutations. The highlight of the present work lies in the fact that the oxidized nucleotides used in this study were generated by incubating the mixture of all the four nucleotides (dATP, dCTP, dTTP and dGTP) with an oxidizing agent. This oxidized mixture was then used in polymerase chain reaction to introduce random mutations. The diverse range of oxidized products of these four dNTPs (which can induce a wide spectrum of mutations: both transitions and transversions) [10, 17, 21, 26, 29, 30, 31] when used together in a polymerase chain reaction could constitute an efficient alternative to error-prone PCR for introducing random mutations in a defined gene sequence.

ACKNOWLEDGEMENTS

UM acknowledges NIPER for the senior research fellowship to carry out this study. SK acknowledges DBT for the senior research fellowship to carry out this study.

REFERENCES

[1]   Harman D. The ageing process. Proc Natl Acad Sci USA 1981; 78: 7124-8.

[2]   Ames BN. Dietary carcinogens and anticarcinogens. Science 1983; 221: 1256-64.

[3]   Ozawa T. Mechanism of somatic mitochondrial DNA mutations associated with age and diseases. Biochim Biophys Acta 1995; 1271: 177-89.

[4]   Kasai H, Tanooka H, Nishimura S. Formation of 8-hydroxyguanine residues in DNA by X-irradiation. Jpn J Cancer Res 1984; 75: 1037-9.

[5]   Kasai H, Nishimura S. Hydroxylation of deoxyguanosine at the C-8 position by ascorbic acid and other reducing agents. Nucleic Acids Res 1984; 12: 2137-45.

[6]   Dizdaroglu M. Formation of an 8-hydroxyguanine moiety in deoxyribonucleic acid on γ-, irradiation in aqueous solution. Biochemistry 1985; 24: 4476-81.

[7]   Ohshima H, Iida Y, Matsuda A, Kuwabara M. Damage induced by hydroxyl radicals generated in the hydration layer of gamma-irradiated frozen aqueous solution of DNA. J Radiat Res 1996; 37: 199-207.

[8]   Wood ML, Dizdaroglu M, Gajewski E, Essigmann JM. Mechanistic studies of ionizing radiation and oxidative mutagenesis: genetic effects of a single 8-hydroxyguanine (7-hydro-8-oxoguanine) residue inserted at a unique site in a viral genome. Biochemistry 1990; 29: 7024-32.

[9]   Shibutani S, Takeshita M, Grollman AP. DNA synthesis past the oxidation-damaged base 8-oxodG. Nature 1991; 349: 431-4.

[10]   Cheng, KC, Cahill DS, Kasai H, Nishimura S, Loeb LA. 8-hydroxyguanine, an abundant form of oxidative DNA damage, causes G →T and A → C substitutions. J Biol Chem 1992; 267: 166-72.

[11]   Kamiya H, Sakaguchi T, Murata N, *et al. In vitro* replication study of modified bases in ras sequences. Chem Pharm Bull 1992; 40: 2792-5.

[12]   Kamiya H, Miura K, Ishikawa H, Inoue H, Nishimura S, Ohtsuka, E. c-Ha-ras containing 8-hydroxyguanine at codon 12 induces point mutations at the modified and adjacent positions. Cancer Res 1992; 52: 3483-5.

[13]   Moriya M. Single-stranded shuttle phagemid for mutagenesis studies in mammalian cells, 8-oxoguanine in DNA induces targeted G • C→T • A transversions in simian kidney cells. Proc Natl Acad Sci USA 1993; 90: 1122-6.

[14]   Kamiya H, Kamiya MN, Fujimuro M, *et al.* Comparison of incorporation and extension of nucleotides *in vitro* opposite 8-hydroxyguanine (7,8-dihydro-8-oxoguanine) in hot spots of the c-Ha-ras gene. Jpn J Cancer Res 1995; 86: 270-6.

[15]   Kamiya H, Kamiya MN, Koizume S, Inoue H, Nishimura S, Ohtsuka E. 8- hydroxyguanine (7,8-dihydro-8-oxoguanine) in hot spots of the c-Ha-ras gene. Carcinogenesis 1995; 16: 883-9.

[16]   Takimoto K, Tachibana A, Ayaki H, Yamamoto K. Spectrum of spontaneous mutations in the cyclic AMP receptor protein gene on chromosomal DNA of *Escherichia coli.* J Radiat Res 1997; 38: 27-36.

[17]   Purmal AA, Kow YW, Wallace SS. Major oxidative products of cytosine, 5- hydroxycytosine and 5-hydroxyuracil, exhibit sequence context-dependent mispairing *in vitro.* Nucleic Acids Res 1994; 22: 72-8.

[18]   Feig DI, Sowers LC, Loeb LA. Reverse chemical mutagenesis: identification of the mutagenic lesions resulting from reactive oxygen species-mediated damage to DNA. Proc Natl Acad Sci USA 1994; 91: 6609-13.

[19]   Kamiya H, Ueda T, Ohgi T, Matsukage A, Kasai H. Misincorporation of dAMP opposite 2-hydroxyadenine, an oxidative form of adenine. Nucleic Acids Res 1995; 23: 761- 6.

[20]   Kamiya H, Kasai H. Effects of sequence contexts on misincorporation of nucleotides opposite 2-hydroxyadenine. FEBS Lett 1996; 391: 113-6.

[21]   Kamiya H, Kasai H. Substitution and deletion mutations induced by 2-hydroxyadenine in *Escherichia coli*, effects of sequence contexts in leading and lagging strands. Nucleic Acids Res 1997; 25: 304-11.

[22]   Kamiya MN, Kamiya H, Muraoka M, Kaji H, Kasai H. Comparison of oxidation products from DNA components by γ-irradiation and Fenton-type reactions. J Radiat Res 1997; 38: 121-31.

[23]   Bjeldanes LF, Chew H. Mutagenicity of 1, 1-dicarbonyl compounds: maltol, kojic acid, diacetyl and related substances. Mutat Res 1979; 67: 367-71.

[24]   Sayato Y, Nakamuro K, Ueno H. Mutagenicity of products formed by ozonation of naphthoresorcinol in aqueous solutions. Mutat Res 1987; 189: 217-22.

[25]   Dorado L, Montoya MR, Mellado JMR. A contribution to the study of the structure mutagenicity relationship for alpha-dicarbonyl compounds using the Ames test. Mutat Res 1992; 269: 301-6.

[26]   Kamiya MN, Kaji H, Kasai H. Types of mutations induced by glyoxal, a major oxidative DNA-damage product, in *Salmonella typhimurium*. Mutat Res 1997; 377: 13-16.

[27]   Kamiya MN, Kamiya H, Kaji H, Kasai H. Mutational specificity of glyoxal, a product of DNA oxidation, in the lacI gene of wild-type *Escherichia coli* W3110. Mutat Res 1997; 377: 255-62.

[28]   Zaccolo M, Williams DM, Brown DM, Gheradi E. An approach to random mutagenesis of DNA using mixtures of triphosphate derivatives of nucleoside analogues. J Mol Biol 1996; 255: 589-603.

[29]   Kamiya H, Kasai H. Mutations induced by 2-hydroxyadenine on a shuttle vector during leading and lagging strand synthesis in mammalian cells. Biochemistry 1997; 36: 11125-30.

[30]   Kamiya H, Ito M, Harashima H. Induction of transition and transversion mutations during random mutagenesis PCR by the addition of 2-hydroxy-dATP. Biol Pharm Bull 2004; 27: 621-3.

[31]   Kamiya H, Ito M, Harashima H. Induction of various mutations during PCRs with manganese and 8-hydroxy-GTP. Biol Pharm Bull 2007; 30: 842-4.

[32]   Purmal AA, Kow YW, Wallace SS. 5-hydroxypyrimidine deoxynucleoside triphosphates are more efficiently incorporated into DNA by exonuclease free Klenow fragment than 8-oxopurine deoxynucleoside triphosphates. Nucleic Acids Res 1994; 22: 3930-5.

2

Influence of 1-butyl-3-methylimidazolium Chloride on the Ethanol Fermentation Process of *Pichia pastoris* GS115

Wenjing Huang, Yanjie Tong, Wangxiang Huang, Ke Wang, Qiming Chen, Yuanxin Wu and Shengdong Zhu[*]

Key Laboratory for Green Chemical Process of Ministry of Education, Hubei Key Laboratory of Novel Chemical Reactor and Green Chemical Technology, School of Chemical Engineering and Pharmacy, Wuhan Institute of Technology, Wuhan 430073, P.R. China

Abstract: To evaluate the influence of 1-butyl-3-methylimidazolium chloride ([Bmim]Cl) on the ethanol fermentation process of *Pichia pastoris* GS115, this paper investigated the yeast growth, ethanol formation and the fermentable sugars consumption during the ethanol fermentation process of *Pichia pastoris* GS115 at different [Bmim]Cl concentrations in the medium. The results indicated that the [Bmim]Cl had no influence on the ethanol fermentation process at its concentration less than 0.0001 g.L^{-1}. The [Bmim]Cl inhibited the yeast growth and had a negative effect on ethanol formation at its concentration higher than 0.0001 g.L^{-1}. The final biomass and ethanol concentration, and the overall ethanol yield from the fermentable sugars all decreased with its concentration increasing. The yeast growth was very slow and nearly no ethanol formed when its concentration reached 5 g.L^{-1}. Compared to *Saccharomyces cerevisiae*, the growth of *Pichia pastoris* GS115 was more sensitive to the [Bmim]Cl, and its ethanol fermentation had lower final ethanol concentration and overall ethanol yield from fermentable sugars at the same [Bmim]Cl concentration. This work provides useful information on selecting suitable strains for ethanol fermentation containing the [Bmim]Cl in the medium.

Keywords: [Bmim]Cl, ethanol fermentation, ionic liquid, *Pichia pastoris* GS115.

1. INTRODUCTION

Lignocellulosic materials are the most economical and highly renewable natural resources in the world. Lignocellulosic ethanol production has drawn much attention in recent years [1]. Ethanol is not only used as a clean and renewable energy but also as a versatile chemical. Its consumption is to keep increasing steadily. It is now one of the most widely used transport bio-fuels. In general, production of lignocellulosic ethanol needs to firstly convert the carbohydrates in lignocellulosic materials to the fermentable sugars, and then ferment the obtained fermentable sugars to ethanol. Due to the complex structure of lignin and hemicellulose with cellulose in lignocellulosic materials, the conversion of carbohydrates in lignocellulosic materials to the fermentable sugars becomes the control procedure in the lignocellulosic ethanol production. Although lots of studies have been carried out, there are still facing great challenges in converting the carbohydrates in lignocellulosic materials to the fermentable sugars in an industrial scale based on economical and environmental consideration [1, 2]. Use of ionic liquids has provided a new technical tool to convert the carbohydrates in lignocellulosic materials to the fermentable sugars for ethanol production [3]. Some studies have indicated that the carbohydrates in lignocellulosic materials can be efficiently converted to the fermentable sugars using ionic liquid technology [4]. The conversion of carbohydrates in lignocellulosic materials to the fermentable sugars using ionic liquid technology has three technical routes: ionic liquid pretreatment of lignocellulosic materials [5], enzymatic hydrolysis of lignocellulosic materials in ionic liquid medium, and chemical hydrolysis of of lignocellulosic materials in ionic liquid medium [6-8]. Whatever technical route was employed, some ionic liquids remained in the obtained fermentable sugars were inevitable. Therefore, it is extremely important to know how the residual of ionic liquids in fermentable sugars will affect the subsequent ethanol fermentation process, because the 1- butyl-3-methylimidazolium chloride ([Bmim]Cl) is one of the most widely-used and cheapest ionic liquid in conversion the carbohydrates in lignocellulosic materials to the fermentable sugars for ethanol production, it was often chosen as a model ionic liquid to study the influence of its residual in fermentable sugars on the subsequent ethanol fermentation process. Some studies have been carried out on the effects of [Bmim]Cl on the growth and ethanol fermentation of *Saccharomyces cerevisiae* in our previous work [9, 10]. Apart from *Saccharomyces cerevisiae*, *Pichia pastoris* is also often used for ethanol fermentation [11, 12]. In order to select the suitable strains for ethanol fermentation containing the [Bmim]Cl in the medium, the influence of [Bmim]Cl on the growth and ethanol fermentation of *Pichia pastoris* GS115 will be investigated in this work and compared with our previous studies.

*Address correspondence to this author at the Key Laboratory for Green Chemical Process of Ministry of Education, Hubei Key Laboratory of Novel Chemical Reactor and Green Chemical Technology, School of Chemical Engineering and Pharmacy, Wuhan Institute of Technology, Wuhan 430073, P.R. China;
E-mail: zhusd2003@21cn.com

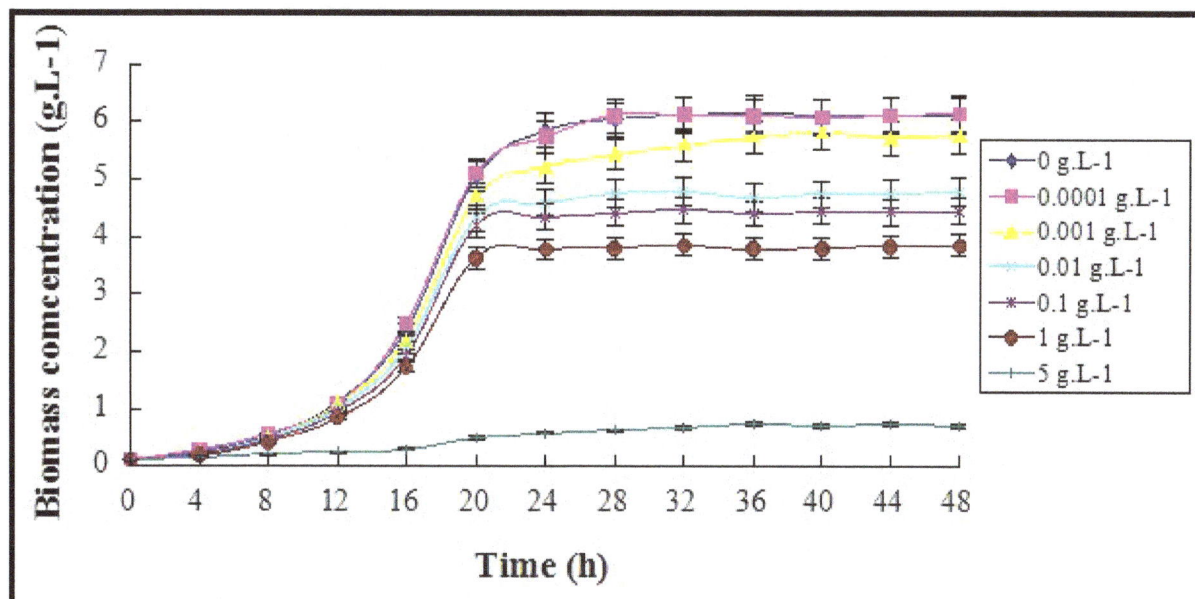

Fig. (1). Growth curves of *Pichia pastoris* GS115 for ethanol fermentation process under different [Bmim]Cl concentrations.

2. MATERIALS AND METHODS

All experiments were carried out three times, and the given numbers are the mean values, whose relative errors are within ± 5%.

2.1. Micro-Organism, Medium and Culture Conditions

The *Pichia pastoris* GS115 was used throughout this study. The medium and culture conditions were the same as our previous work for ethanol fermentation of *Saccharomyces cerevisiae* [9, 10]. During the ethanol fermentation process of *Pichia pastoris* GS115, some samples were taken at regular intervals for later analysis.

2.2. Analytical Methods

The samples taken from the ethanol fermentation process of *Pichia pastoris* GS115 were used to determine the concentration of biomass, ethanol and the fermentable sugars. Biomass concentration was determined by the dry weight method [13]. Ethanol content was determined by gas chromatography [14] and the fermentable sugars concentration was estimated using the 3,5-dinitrosalicylic acid method [15].

3. RESULTS AND DISCUSSION

3.1. Effect of the [Bmim]Cl on the Growth of *Pichia pastoris* GS115

In order to evaluate the influence of the [Bmim]Cl on the ethanol fermentation process of *Pichia pastoris* GS115, the growth of *Pichia pastoris* GS115 was first investigated at different [Bmim]Cl concentrations in the fermentation medium. Fig. (1) shows the growth curves of *Pichia pastoris* GS115 at different [Bmim]Cl concentrations form 0.0001 to 5 g.L^{-1} during the ethanol fermentation process. As indicated in Fig. (1), the [Bmim]Cl had no influence on the growth of

Pichia pastoris GS115 at its concentration less than 0.0001 g.L^{-1}. However, the [Bmim]Cl inhibited the growth of *Pichia pastoris* GS115 at its concentration higher than 0.0001 g.L^{-1}. Moreover, this inhibition became stronger with its concentration increasing. When the [Bmim]Cl in the medium reached 5 g.L^{-1}, the yeast had almost no growth and its biomass concentration increased very slowly. Compared to our previous studies [9, 10], the influence of [Bmim]Cl on the growth of *Pichia pastoris* GS115 and *Saccharomyces cerevisiae* had the same characteristic, but *Pichia pastoris* GS115 was was more sensitive to the [Bmim]Cl than *Saccharomyces cerevisiae*. Under the same [BMIM]Cl concentration, the [BMIM]Cl had more serious inhibition on the growth of *Pichia pastoris* GS115. The inhibition mechanism of [BMIM]Cl on the growth of *Pichia pastoris* GS115 might be similar with its inhibition on *Saccharomyces cerevisiae*, which comes from the interaction between [BMIM]Cl and its cytomembrane.

3.2. Effect of [Bmim]Cl on the Ethanol Fermentation Process of *Pichia pastoris* GS115

Apart from yeast growth, the [Bmim]Cl also affects the ethanol formation and fermentable sugars consumption during the ethanol fermentation process of *Pichia pastoris* GS115. The time courses of ethanol formation and fermentable sugars consumption were measured at different [Bmim]Cl concentrations in the fermentation medium and the results are shown in Figs. (2 and 3) respectively. As shown in Figs. (2 and 3), the [Bmim]Cl had no influence on the ethanol formation and fermentable sugars consumption during the ethanol fermentation process of *Pichia pastoris* GS115 at its concentration is less than 0.0001 g.L^{-1}. However, the [Bimim]Cl negatively affected the ethanol formation and fermentable sugars consumption at its concentration is higher than 0.0001 g.L^{-1}. With the [Bmim]Cl concentration

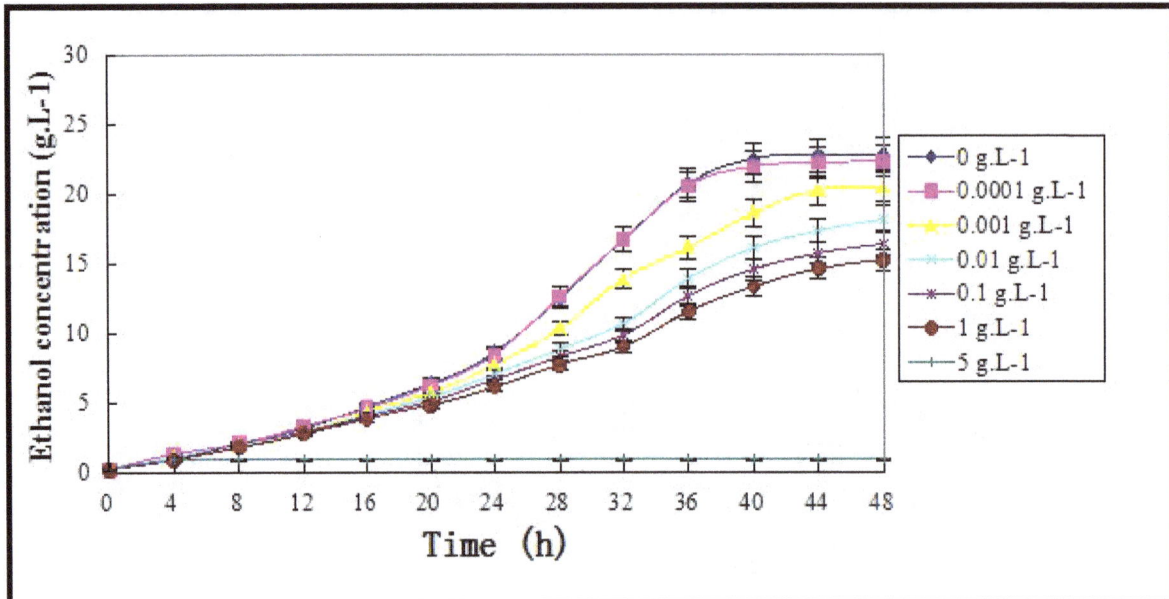

Fig. (2). Time courses of ethanol concentration for ethanol fermentation process under different [Bmim]Cl concentrations.

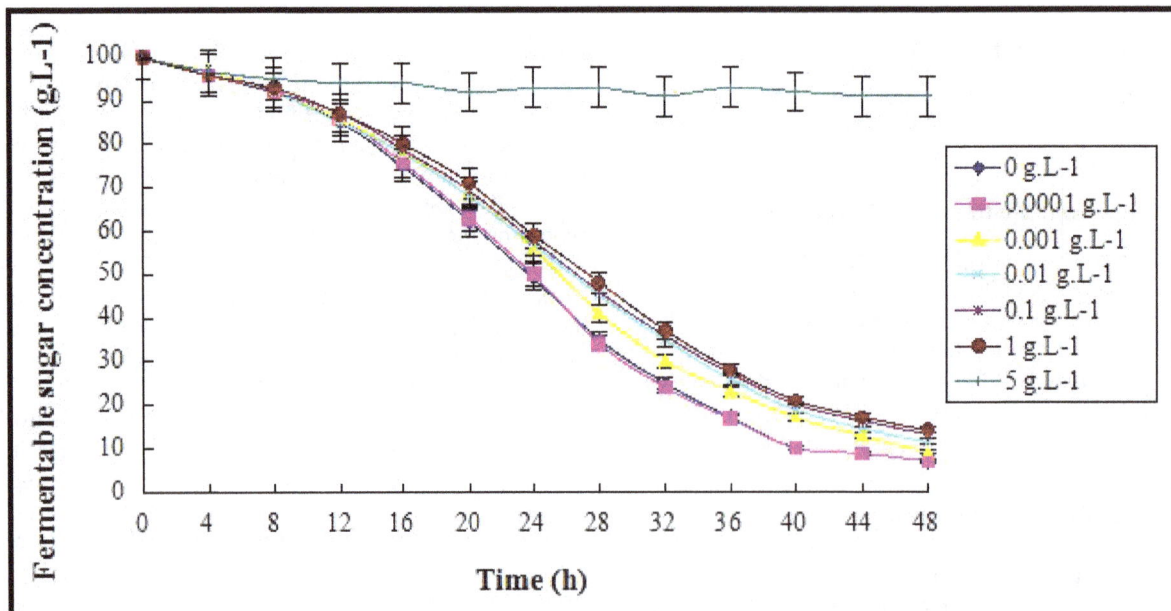

Fig. (3). Time courses of the fermentable sugars concentration for ethanol fermentation process under different [Bmim]Cl concentrations.

increasing, the ethanol formation rate and the consumption rate of fermentable sugars both decreased. When the [Bmim]Cl in the medium reached 5 g.L^{-1}, there was almost no ethanol formation and fermentable sugars consumption. Table **1** listed some important parameters during the ethanol fermentation process of *Pichia pastoris* GS115 at different [Bmim]Cl concentrations. As indicated in Table **1**, the final biomass and ethanol concentration and the overall ethanol from the fermentable sugars all decreased with the [Bmim]Cl concentration increasing, but the final remained fermentable sugars increased. It is obvious that the influence of [Bmim]Cl on the growth of *Pichia pastoris* GS115 was in good agreement with its effect on the ethanol formation and fermentable sugars consumption during the ethanol fermentation process of *Pichia pastoris* GS115, which implies that

the influence of [Bmim]Cl on the ethanol formation and fermentable sugars consumption during the ethanol fermentation process of *Pichia pastoris* GS115 came from its impact on the growth of *Pichia pastoris* GS115. This is similar with the influence of [Bmim]Cl on ethanol fermentation process of *Saccharomyces cerevisiae*. Compared to the ethanol fermentation of *Saccharomyces cerevisiae* [9, 10], the Bmim]Cl had a more serious effect on the ethanol fermentation process of *Pichia pastoris* GS115. The ethanol fermentation of *Pichia pastoris* GS115 had lower final ethanol concentration and overall ethanol yield from the fermentable sugars under the same [Bmim]Cl concentration. Comparatively speaking, *Saccharomyces cerevisiae* was more tolerant to the [Bmim]Cl and suitable for ethanol fermentation containing the [Bmim]Cl in the medium.

Table 1.   Effect of [Bmim]Cl concentration on the ethanol fermentation process parameters.

$C_i (g.L^{-1})$	0	0.0001	0.001	0.01	0.1	1	5
$C_b (g.L^{-1})$	6.12	6.15	5.75	4.78	4.45	3.85	0.72
$C_p (g.L^{-1})$	22.9	22.4	20.5	18.3	16.5	15.3	1.1
$C_s (g.L^{-1})$	6.8	7.1	9.2	11.5	13.1	14.2	91.1
Y	0.246	0.241	0.226	0.207	0.190	0.178	0.124

C_i represents the [Bmim]Cl concentration $(g.L^{-1})$, C_b represents the final biomass concentration $(g.L^{-1})$, C_p represents the final ethanol concentration $(g.L^{-1})$, C_s represents the final fermentable sugars concentration $(g.L^{-1})$, Y represents the ethanol overall yield from the fermentable sugars.

CONCLUSION

The influence of the [Bmim]Cl with different concentrations in the medium on the ethanol fermentation process of *Pichia pastoris* GS115 was investigated and the main conclusions are as follows:

1) When the [Bmim]Cl in the medium was higher than 0.0001 $g.L^{-1}$, it inhibited the yeast growth and had a negative effect on ethanol formation. The final biomass and ethanol concentration, and the overall ethanol yield from the fermentable sugars all decreased with the increase of the [Bmim]Cl concentration. When the [Bmim]Cl in the medium reached 5 $g.L^{-1}$, the yeast growth was very slow and nearly no ethanol formed.

2) Compared to the growth and ethanol fermentation of *Saccharomyces cerevisiae*, the growth of *Pichia pastoris* GS115 was more sensitive to the [Bmim]Cl, and its ethanol fermentation had lower final ethanol concentration and overall ethanol yield from the fermentable sugars under the same [Bmim]Cl concentration. This work provides useful information on selecting suitable strains for ethanol fermentation containing the [Bmim]Cl in the medium.

ACKNOWLEDGEMENTS

This work was supported by the National Natural Science Foundation of China (No. 21176196) and the Collaborative Innovation Center of Catalysis Materials of Hubei Province.

REFERENCES

[1]   Gupta A, Verma J P. Sustainable bio-ethanol production from agro-residues: A review. Renew Sust Energy Rev 2015; 41: 550-67.

[2]   Sims R E H, Mabee W, Saddler J N, Taylor M. An overview of second generation biofuel technologies. Bioresour Technol 2010; 101: 1570-80.

[3]   Wang Q, Wu Y, Zhu S. Use of ionic liquids for improvement of cellulosic ethanol production. Bioresources 2011; 6: 1-2.

[4]   Zhu S, Yu P, Lei M, et al. A mini-review on the lignocellulosic ethanol production by using ionic liquid technology. Energy Educ Sci Tech-A 2012; 30(SI-1): 95-106.

[5]   Datta S, Holmes B, Park J I, et al. Ionic liquid tolerant hyperthermophilic cellulases for biomass pretreatment and hydrolysis. Green Chem 2010; 12: 338-45.

[6]   Li C, Wan Q, Zhao Z K. Acid in ionic liquid: an efficient system for hydrolysis of lignocellulose. Green Chem 2008; 10: 177-82.

[7]   Binder J B, Raines R T. Fermentable sugars by chemical hydrolysis of biomass. Proc Natl Acad Sci 2010; 107: 4516-21.

[8]   Zhang Y, Du H, Qian X, Chen E X Y. Ionic liquid-water mixtures: enhanced Kw for efficient cellulosic biomass conversion. Energy Fuel 2010; 24: 2410-17.

[9]   Zhu S, Yu P, Tong Y, et al. Effects of the ionic liquid 1-butyl-3-methylimidazolium chloride on the growth and ethanol fermentation of *Saccharomyces cerevisiae* AY92022. Chem Biochem Eng Q 2012; 26: 105-9.

[10]   Zhu S, Yu P, Lei M, et al. Investigation of the toxicity of the ionic liquid 1-butyl-3-methylimidazolium chloride to *Saccharomyces cerevisiae* AY93161 for lignocellulosic ethanol production. Polish J Chem Technol 2013; 15: 94-8.

[11]   Margeot A, Hahn-Hagerdal B, Edlund M, Slade R, Monot F. New improvements for lignocellulosic ethanol. Curr Opin Biotechnol 2009; 20: 372-80.

[12]   Meko'o D J L, Xing Y, Shen L L. Production of ethanol from cellobiose by recombinant β-glucosidase expressing *Pichia pastoris*: submerged shake flask fermentation. Afr J Biotechnol 2012; 11: 9108-17.

[13]   Atala DIP, Costa AC, Filho MR, Mauger F. Kinetics of ethanol fermentation with high biomass concentration considering the effect of temperature. Appl Biochem Biotechnol 2001; 91-93: 353-65.

[14]   Zhu S, Wu Y, Yu Z, et al. Simultaneous saccharification and fermentation of microwave/alkali pretreated rice straw to ethanol. Biosys Eng 2005; 92: 229-35.

[15]   Miller G L. Use of dinitrosalicylic acid reagent for determination of reducing sugar. Anal Chem 1959; 31: 420-28.

3

The Uncertainty Assessment of Threonine Determination in Ginseng and Its Immune Activity

Jane yuxia Qin, Yan Chen, Dianshuai Gao and Ye Xiong*

Department of Neurobiology, Xuzhou Medical College, Jiangsu Province, 221994, China

Abstract: Ginseng is one of famous herbs, which has many medical functions, such as neuroprotective effects, and anticancer function, because ginseng contains many active substances, and threonine is one important ingredient. In this study, to establish a mathematical model of uncertainty assessment for Threonine content in Ginseng by the amino acid assay, the sources of uncertainty in the measurement process were completely concerned, the main sources of uncertainty were fully analyzed, and evaluated and calculated. The results showed, for 0.2083g sample, the Threonine determination in Ginseng showed a good linear relationship. IN conclusion, this method developed in this study is suitable for the Uncertainty factors assessment of threonine measurement in Ginseng by amino acid assay. We also did animal test. To detect the immune activity of threonine in Ginseng, we detected the CD3+, CD19+ using FACS of mice after oral administration.

Keywords: Ginseng, threonine, Immune Activity, CD3+.

1. INTRODUCTION

Ginseng is a powerful herb which can improve immunity because they contain many medicine ingredient and nutritional ingredient. Threonine is one of the important nutritional ingredients. For a long time, the main method for determination of threonine in Ginseng is amino acid assay, and the experimental procedures have been very developed. But the results of the same sample showed a certain difference, which because those experiments are performed by different labs or different people in same lab. We analyzed the causes of the difference through the determination of threonine in Genseng, evaluated the sources of this uncertainty of these quantitative results, and found that the quantification of threonine showed in a certain confidence interval, which play the role of the correction on the measurement of threonine.

2. MATERIALS AND METHODOLOGY

Ginseng samples were farmed in Jilin Province, China, dried at 60°C after mashing using organizations broken machine, filtered with 60 mesh sieve after crushing by pulverizer, and then mixed, separately stored in bottle for detection.

Threonine reference substance were purchased from the Beijing Academy of Agricultural Sciences, 14.62 mg/L, 6.0 mol/L hydrochloric acid solution, guarantee reagent, Beijing Chemical industry; water is pure water, Hangzhou Wahaha

Group Co., Ltd. MCI Buffer L-8500-PH Kit for Mitsubishi Chemical Corporationg; Coloration liquid: R1, R2 is Japan, and by Wako Pure Chemical Industries Co., Ltd. All glassware and experimental apparatus are immersed by concentrated sulfuric acid, washed with deionized water.

0.1g of Ginseng powder were weighed by 0.1 mg precision balance accurately and placed in the hydrolysis tube added 10.00 ml 6.0 mol/L hydrochloric acid, N2 flowed through for 1min, and covered with rubber plug. This hydrolysis tube was sealed after vacuumed by vacuum pump, placed in a 110°C thermostatic oven, hydrolyzed for 22 h. When the hydrolysis ended, hydrolysis solution was filtered and transferred to a 50.00 ml volumetric flask and add deionized water to a constant volume. 1.00 ml filtrate was put into the beaker, and evaporated to dryness in a vacuum dryer, and then 1-2 ml of water was dissolved and then evaporated to dryness and repeated 2 times, and finally the residue was dissolved with 1.00 ml of 0.02 mol/L hydrochloric acid, filtered with 0.22 μm polyethylene ether sulfone membrane filtration then for determination on the machine.

The temperature of the laboratory (20 ± 5) °C; volumetric flask, straw and other glass container are according to the JJG196-2006 test procedures [1] Class B equipment standards; 0.1mg division balance was used according to JJG 1036-2008 Verification regulation for Electronic balance [2] requirements; testing instruments (Agilent1200) meet JJG705-2002 test requirements for Verification regulation of liquid chromatographs [3]. Methanol and other reagents meet the analytical criteria, water was Wahaha water.

Japan's Hitachi L-8800 amino acid analyzer; electronic analytical balance: 1712mp8 oven thermostat: DG 30/14-II.

*Address correspondence to this author at Tongshan Road, Xuzhou, China. 221004; Email: yuxiaemail@126.com

Table 1. The experimental design.

Group	amount
control	25 mice
sinsenoside	25 mice
threonine	25 mice
Sinsenoside+threonine	25 mice

Immunizing dose and approaches

Chromatographic column: Elipses XDB C18 (4.6 × 250 mm, 4.6 μm); buffer: PH1, PH2, PH3 and PH4, pH 5; detection wavelength for reaction solution: R1, R2; 570nm (visible light); buffer flow rate was 4.0 ml/min, and the flow rate of reaction liquid was 4.0 ml/min; column temperature: 57°C; injection volume was 20 μl.

Immunity Activity Detection on Animal

100 Mice are divided into four groups as following (Table **1**):

5mg sinsenoside, threonine and Sinsenoside+threonine respectively was given to one Bal/c mice by oral administration per day. 8 days later, half of each group mice for facs, and half of it were forlymphocyte transformation test.

Facs approaches:

1. The mice were killed and the put in 1% benzalkonium bromide for 3-5min.

2. Sterile gauze was prepared in sterile plates, and 2 ml PBS buffer was added into plates. The left lumbar region skin of mice was cut, spleen was taken out, put on gauze, and the spleen was minced into single cells.

3. The cells were transferred into centrifuge tubes, labeled, 1000 r/min for 10min, and discarded supernatant.

4. 500μl distilled water was added to precipitate, shaked gently, leaved there for 20s, removed red blood cell by cell disruption.

5. Appropriated amount PBS buffer was added into solution, in case the white cells break.

6. 2 min later, the supernatant was transferred into another tube, and labeled.

7. The cell solution was diluted into $1×10^6$ /100μl for facs.

$CD3^+T$, $CD4^+/CD8^+$ and $CD19^+$ was detected μl FITC labeled antibody to mice $CD3^+$ put into each tubes. PE labeled antibody to mice $CD4^+T$, $CD8^+T$, and $CD19^+$, then 100μl cell supernatant solution (10^6/ml), incubated for 30min at room temperature (18°C-25°C), then detected by facs. 488nm vave length for the detection of FITC, 533 nm vave

length for the detection of PE,10000 cells were detected, and analyzed by facs software.

3. RESULTS AND DISCUSSION

Preparation of standard curve: Take exactly 0.1, 0.2, 0.3, 0.4, 0.6 ml reference substance Radix angelicae dahuricae with a 1 ml single channel pipette. Get constant volume to 1ml and place in an automatic sampling bottle. According to above "chromatographic condition" to test peak area, using the least square method with the peak area A and Radix angelicae dahuricae standard concentration C (mg/ml) were for linear fit. The Radix angelicae dahuricae graticule is:

$A = 0.2088 + 44.9649C$, $r = 0.9999$.

Determination of sample: Filtrate the sample solution with $0.22\ \mu$ m nylon membrane, according to 2 "chromatographic condition", determine on the machine. The peak area of the sample area was measured, and the results were shown in Table **2**.

Evaluation of Uncertainty

Mathematical model

Mathematical calculation formula for Threonine content measurement:

$$R = \frac{C_{standard} \times A_{sample} \times V_{load} \times V_{volume}}{A_{standard} \times m \times V_{removal} \times 10^6}$$

where R is the threonine content in the sample, %; Cstandard means reference sample concentration, mg/L; V_{volume} is the final volume of the sample, mL; A_{sample} means peak area of a sample mAu; $A_{standard}$ means the peak area of the reference substance, mAu; m is the sample weight, g; the volume of the sample V_{load} is the sample volume loaded on the machine, mL; $V_{removal}$ is the sample volume removed.

The Major Source of Measurement Uncertainty

The method of Determination of Amino Acids in Foods according to GB/T 5009.124-2003A [4], there are the following mainly sources of uncertainty: (1) the uncertainty introduced by sample weight; (2) the uncertainty introduced by sample volume; (3) sample process introduces uncertainty; (4) the uncertainty introduced by the sample

Table 2. The content of threonine in different volume injection sample (n = 6).

Sample Amount, m(g)	Peak area	Result R, (%)	Mean, (%)	RSD, (%)
0.2083	469234	0.2968		
0.2056	461412	0.2958		
0.2053	457122	0.2933		
0.2028	463257	0.3009	0.2971	0.008656
0.2015	456055	0.2982		
0.2084	471341	0.2979		

peak area; (5) the uncertainty introduced by sample measurement repeat; (6) the uncertainty introduced by the non-linear of standard curve.

Evaluation of Measurement Uncertainty

The samples weighing introduced uncertainty [u(m)]: The balance calibration introduced uncertainty: the balance test showed analytical balance error was 0.1 mg, according to rectangular distribution ($k = \sqrt{3}$), uncertainty component introduced by the balance calibration:

$$u(m) = \sqrt{2[(0.00005)^2 + (0.00005)^2]} = 0.0001g$$

Weighing variability introduced uncertainty: according to Uncertainty Evaluation Guide in chemical analysis [5], the analytical balance variability is about $0.5 \times$ the final significant figure, and the final significant figure of the analysis balance in our laboratory was 0.1 mg. So, weighing variability introduced the uncertainty components:

$$u_2(m) = 0.5 \times 0.1 mg = 0.00005 g$$

Weighing by the difference method should calculated uncertainty twice, so the uncertainty introduced by the weighing scales:

$$u(m) = \sqrt{2[(0.00005)^2 + (0.00005)^2]} = 0.0001g$$

The sample weighted m = 0.2083g, the relative standard uncertainty introduced by weighing:

$$u_{rel}(m) = 0.0001g / 0.2083g = 0.000480$$

The Final Volume of Samples Introduced Uncertainty [u(V)]

50 mL volumetric flask bring uncertainty: the uncertainty brought by the volume container includes the following four sources:

Uncertainty introduced by the calibration error: 50 mL A grade volumetric flask error allowed $\pm$ 0.05 mL [1] rectangular distribution ($k = \sqrt{3}$), 50 mL volumetric flask calibration introduced uncertainty:

$$u_1(V_{volume50}) = 0.05mL / \sqrt{3} = 0.0288mL$$

Repeatability: filled 10 times to 50 mL volumetric flask in one experiments, the standard deviation 0.02 mL can be directly as uncertainty, namely: $u_2 (V_{50}) = 0.02$ mL.

Uncertainty introduced by the temperature: the temperature range is $\pm$ 5°C, in the manual, water expansion coefficient is 2.1×10^{-4} mL °C^{-1}, then 50 mL volumetric flask volume change were:

$$\Delta V_{volume50} = 50 \times 2.1 \times 10^{-4} \times 5 = 0.0525mL$$

The confidence level of 0.95, contains the factor K = 1.96, the uncertainty introduced by the temperature change:

$$u_3(V_{volume50}) = 0.0525mL / 1.96 = 0.0267mL$$

Uncertainty components caused by reading: relative standard uncertainty introduced from 50 mL volumetric flask:

$$u_4(V_{volume50}) = 0.01 \times 50mL / \sqrt{6} = 0.204mL$$

The relative standard uncertainty brought by 50ml volumetric fask:

$$u_{rel}(V_{volume50}) = \sqrt{0.0288^2 + 0.02^2 + 0.0267^2 + 0.204^2} / 50mL = 0.00156$$

Simple Processing Introduced Uncertainty [u(rep)]

Repeatable measurement introduced uncertainty: the threonine content repeatable measurement results are shown in Table **3**, the standard deviation obtained by Bessel formula:

$$S_x = \sqrt{\frac{\sum_{n=1}^{i}(x_i - \bar{x})^2}{n-1}} = 0.000386\%$$

Repeatability uncertainty:

$$u_{rel}(rep) = u_{(rep)} \Big/ \bar{R} = \frac{0.000157\%}{0.2971\%} = 0.000528$$

$$u_{(rep)} = S_x / \sqrt{n} = 0.000386 / \sqrt{6} = 0.000157\%$$

Table 3. Peak area of threonine

Sample Injected Order	Peak Area	Mean	RSD%
1	469234		
2	461412		
3	457122	463070.1	0.01344
4	463257		
5	456055		
6	471341		

Sample Peak Area Measurement Introduced Uncertainty [u(AS)]

Generated Uncertainty of Peak Area by Repeated Measurements

Because there is only 6 data in the Table **2**, so use range method for evaluation according to Errors Analysis and Measurement Uncertainty Evaluation [6], in which $n = 6$, $C = 2.53$.

$$u_A = \frac{R_{range}}{C_{difference\ coefficient}};$$

Threonine: $u_{rel} = \frac{u_A}{A} = \frac{6041.897}{463070.1} = 0.0103$

Instrumental Data Processing System Introduced Uncertainty

According to instrument manual and the general performance of integrator, so far, the maximum error of peak area convolution procedure by the liquid chromatography is 0.2% to 1%, then the peak area relative uncertainty components: $u_A = \frac{0.01}{\sqrt{3}} = 0.00577$

Liquid chromatography used micro-injector for measurement, the injection uncertainty was 1%, then the relative uncertainty components: $u_{injection} = \frac{0.01}{\sqrt{3}} = 0.00577$

The uncertainty introduced by data-processing system:

$$u_{process} = \sqrt{u_A{}^2 + u_{load}{}^2} = 0.0082 \qquad (14)$$

So we can get the evaluation of the uncertainty introduced by sample peak area measurement:

$$u_{rel}(A_S) = \sqrt{u_{process}{}^2 + u_{rel}{}^2}$$

Threonine:

$$u_{rel}(A_S) = \sqrt{0.0103^2 + 0.0082^2}$$

$$=0.0131\%; \quad u_{rel}(A_S) = u_{(rep)} \bigg/ \overline{R} = \frac{0.0131\%}{0.2971\%} = 0.0440 \qquad (16)$$

The nonlinear standard curve introduced uncertainty [u(line)]: The preparation concentration 0, 0.1, 0.2, 0.3, 0.4, 0.6 mL five threonine standard solution, measured twice for each concentration. According to the measurement data, using the least squares method to prepare standard working curve equation [6-8]: $A = 9130c - 0.002857$, correlation coefficient $r = 1$. The standard deviation of the standard curve equation was calculated, i.e. residual standard deviation. The peak area measured values by instruments were calculated according to the linear equation (in Table **4**).

Standard curve residual standard deviation

$$S_R = \sqrt{\frac{\sum_{j=1}^{n}[A_{0j} - (a + bC_{0j})]^2}{n-2}} = 37.562\,mg/L$$

The uncertainty introduced by standard curve fitting:

$$u_{(line)} = \frac{S_R}{b}\sqrt{\frac{1}{p} + \frac{1}{n} + \frac{(\overline{C} - \overline{C_0})^2}{\sum_{j=1}^{n}(C_{0j} - \overline{C_0})^2}} = \frac{37.562}{9130}$$

$$\sqrt{\frac{1}{6} + \frac{1}{12} + \frac{(50.7160 - 26.4)^2}{2286.9}} = 0.00209\,mg/L$$

where S_R: Residual standard deviation of standard curve (residual standard deviation); b: Slope; p: Repeatable times for the measurement of samples; n: The points of standard curve; $\overline{C}$: The mean of the sample concentration; $\overline{C_0}$: The mean of the each points concentration on standard curve; C_{0j}: The concentration of each standard solution.

So the uncertainty introduce by standard curve:

$$u_{rel}(line) = \frac{u_{(line)}}{C} = \frac{0.00209\,mg/L}{50.7160\,mg/L} = 0.0000412$$

Composition Uncertainty, Expanded Uncertainty and their Results Expression

The above uncertainties are separated, so the composition relative uncertainty is:

Table 4. The results of residual calculation for standard curve

n	Concentration Coj (mg/l)	Response Value A0j	Calculated Value a+bC0j	[A0j-(a+bC0j)]2	$(Coj-\overline{C_0})2$	$(Coj-\overline{C_0})2$
1	0	0 0	-0.002857	0.000008162 0.000008162	-26.4	696.96
2	9.9	90393.4 90393.6	90386.997	40.9984 43.5996	-16.5	272.25
3	19.8	180787.2 180787.3	180773.977	174.3192 176.9698	-6.6	43.56
4	29.6	271180.4 271180.6	271160.997	376.4764 384.2776	3.3	10.89
5	39.6	361574.2 361574.2	361547.997	686.5972 686.5972	13.2	174.24
6	59.4	542361 542361.4	542321.997	1521.2340 1552.5964	33	1089
	$\overline{C_0}$ =26.4			$\sum$ 5643.6658		$\sum$ 2286.9

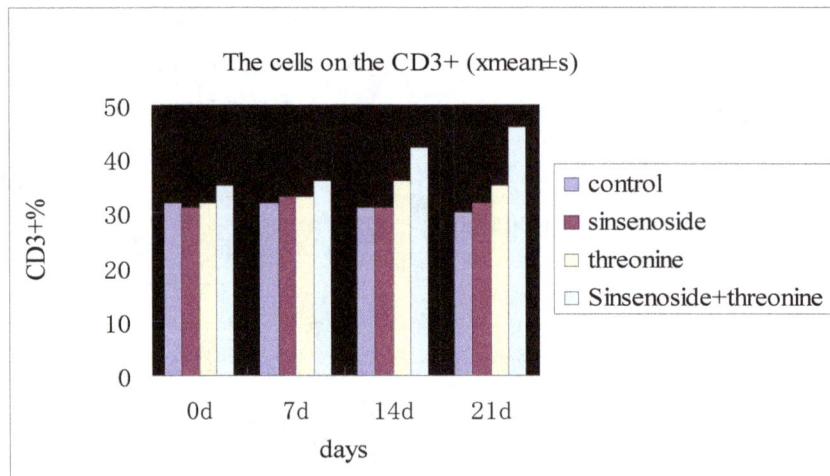

Fig. (1). Variation of CD3+T lymphocytes in spleen of mice after oral administration.

$$u_{rel}(R) = \sqrt{\begin{array}{c}\left[u_{rel}(m)\right]^2 + \left[u_{rel}(V_{volume50})\right]^2 \\ + \left[u_{rel}(rep)\right]^2 + \left[u_{rel}(As)\right]^2 + \left[u_{rel}(line)\right]^2\end{array}}$$

$$= \sqrt{\begin{array}{c}0.000480^2 + 0.00156^2 + 0.000528^2 \\ +0.0440^2 + 0.0000412^2\end{array}} = 0.0440$$

Composition uncertainty:

$$u(R) = u_{rel}(R) \times R = 0.0440 \times 0.2971\% = 0.0130\%$$

in which, inclusion factor $k = 2$; so the expended uncertainty:

$$U = ku(R) = 2 \times 0.0130\% = 0.0260\%$$

Finally, the content of threonine expressed as following:

$$R = (0.2971 \pm 0.0260)\%, k = 2$$

The Results of FACS

CD4+T, CD8+T, CD19+ cell subgroups were detected by FACS, and analyzed DBS (Figs. **1-3**).

After oral administration, the three test groups compared to the control group, the increase of CD3+T, CD4+/CD8+, and CD19+ is significant (P <0.01). The diversify of CD3+T, CD4+/CD8+, and CD19+ between groups

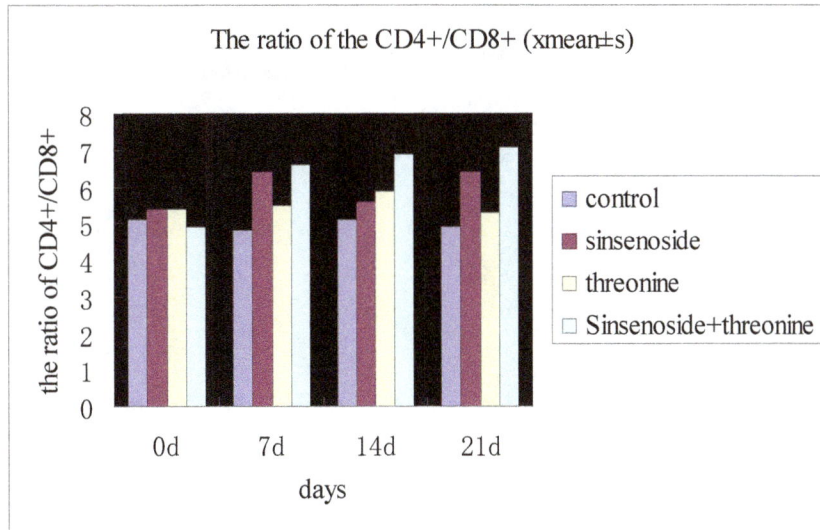

Fig. (2). Variation of C $CD4^+/CD8^+$ of lymphocytes in spleen of mice after oral administration.

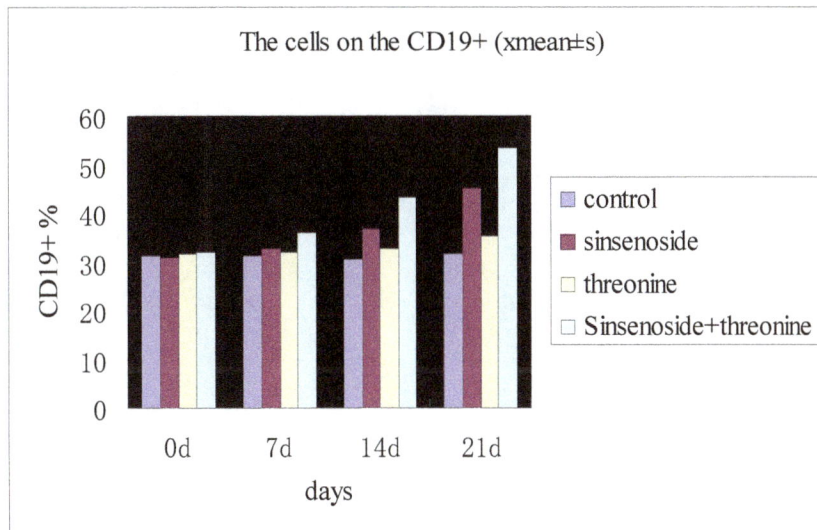

Fig. (3). Variation of CD19+T lymphocytes in spleen of mice after oral administration.

sinsenoside and threonine changed little (P >0.05). Compared to control, sinsenoside and threonine group, the sinsenoside + threonine group, all tests change significantly (P <0.01). So the threonine can improve the immunity, also this activity of threonine threonine in ginseng has synergism with ginsenoside.

CONCLUSION

From the whole evaluating process of the uncertainty, when threonine in Ginseng was determined by amino acids assay, whose uncertainty comes mainly from the sample peak area. So the uncertainty introduced by the sample peak area is the largest component. So for the determination of threonine in Ginseng, the control of the instrument sensitivity could reduce the uncertainty components. To reduce this component is the efficient way to reduce the measurement uncertainty. Also the threonine play an important role on the enhancement of immunity, and this activity of threonine threonine in ginseng has synergism with ginsenoside. This study provides an effective analysis approaches for the identification of Ginseng, which is an important contribution for the application of Ginseng in Chinese Medicine.

ACKNOWLEDGEMENTS

This study was funded by the National Natural Science Funds of China (grant no. ISIS584763SN:2810171 and 81101899). And also was sponsored by Qing Lan Project: 53041305.

REFERENCES

[1]   Du Shuli, Zhangzhiqing, Xie Junyan, et al. JJG 196-2006 People's Republic of China National Metrological Verification regulation of working glass container, China Metrology Press, Beijing, China, 2006; 2-8.

[2]   Ding Jingan, Huang Jian, Li Yong, et al. JJG 1036-2008 People's Republic of China National Metrological Verification regulation for Electronic balance, China Metrology Press, Beijing, China, 2008; 1-8.

[3]   Zhao Min, Wu Fangdi, He Yajuan. JJG 705-2002 People's Republic of China National Metrological Verification regulation of liquid chromatographs, China Metrology Press, Beijing, China, 2002; 1-15.

[4]   Jia Jianbin, Zhao Xihe. GB/T 5009.124-2003 Determination of Amino Acids in Foods, China Standard Publishing House, Beijing, China, 2003; 115-119.

[5]   Wei Hao, Qiao D. Guidance on Evaluating the Uncertainty in Chemical Analysis. China Metrology Press, Beijing, China, 2002; 100-129.

[6]   Sha Dingguo. Errors Analysis and Measurement Uncertainty Evaluation, China Metrology Publishing House, Beijing, China, 2003; 174-181.

4

Microbial, Urease Activities and Organic Matter Responses to Nitrogen Rate in Cultivated Soil

Sun Jingjing[1], Zhu Mijia[1], Yang Xiaoqia[2], Zhang Chi[1] and Yao Jun[*,1]

[1]School of Civil & Environmental Engineering and National "International Cooperation Based on Environment and Energy" and Key Laboratory of "Metal and Mine Efficiently Exploiting and Safety" Ministry of Education, University of Science and Technology Beijing, Beijing 100083, P.R. China

[2]College of Resources and Environment Science, Agricultural University of Hebei, Baoding, Hebei, 071001, P.R. China

Abstract : Nitrogen (N) fertilizer is an important field management, this paper was to investigate the responses of microbial activity measured by microcalorimetric technique, urease activity and organic matter to proper nitrogen (N) fertilizer in cultivated system in the North China Plain. The microcalorimetry results showed that microorganisms under proper N fertilizer rate got the efficient metabolism. The maximum heat production (Pmax) of microcalorimetry parameter implied that increasing the N fertilizer rate, Pmax did not increase always but firstly increased than decreased. In addition, from soil urease activity and organic matter studies, high soil urease activity due to both N fertilizer rate and growth stage and these had light influence on soil organic matter. These results suggest that only proper amount of N fertilizer rate and suitable topdressing for a good wheat-growing environment.

Keywords: Microbial activity, microcalorimetry, organic matter, urease

1. INTRODUCTION

Nitrogen (N) fertilizer application is an important measure to improve the soil fertility and crop yield [1], also it is one of the most management measure in agriculture [2]. Excess N fertilizer into soil is very common in China at present [3]. However, with increasing yield of crops and soil fertility due to N fertilizer, also making a large amount of N wasted, leading to N leach polluting ground water, ammonia volatilization and eutrophication caused pollution to environment [4]. Winter wheat (*Triticum aestivum L.*) is the major crop in China. The North China Plain is the most important grain-producing areas in China, two-thirds of China's winter wheat come from here [5]. Therefore, to study this area's soil on different growth stages of winter wheat is extremely important, also it is becoming essential to research the microbial activity, urease activity and soil organic matter under different N rates in winter wheat system. It will provide proper N availability, reduce the pollution of N on the environment play a guiding role.

Soil microorganisms are important components of soil ecosystem, leading the nutrient cycle [6] and energy flow, meanwhile, it play an important role on maintaining the ecosystem stability and sustainability. 90% of microorganism involved in the soil reaction process, could change the soil fertilization. For example, soil bacterial can enhance the availability of nutrients [7]. The proper amount of nitrogen fertilizer can improve the activity of microorganism, then improve soil fertility, also reduce the environment pollution. Nowdays, microbial technologies gain an increasing attention for agricultural and environmental problems [8]. In this paper, we use microcalorimetry technology. It is a simple and straightforward method for the study of microbial activity. It can measure the heat flow of microbial process, obtain a kinetic and thermodynamic parameters.

Soil organic matter (SOM) is a critical factor which affecting soil quality [9], decreasing soil erosion [10], improving water holding capacity [11] and keeping the sustainable development of agriculture ecosystem [12]. A part of N fertilizer may become SOM when it into soil [13]. Meanwhile, SOM is accounted for atmospheric carbon once it preservation in soils [14]. Therefore, N fertilizer use change, according to change the organic matter accumulation in soil and crop growth [15], thus, impact agricultural ecosystem. On the other hand, soil urease is one of most active hydrolytic enzymes in the soil, the hydrolysis of soil urease to urea which was applied into the soil, release ammonium that can use by crops, play an important role in soil N cycle. Soil urease is paid more and more attention because it can be used as important indexes to evaluate SOM and N application [16] and more quickly response to environment change and agricultural management [17].

The main objective of this study was to evaluate and compare soils from long-term N fertilizer experiments in the North China Plain. The experiments have the same range of phosphorous-potassium fertilizer treatments and field management, different N fertilizer treatments. We investigated the different N fertilizer on soil microbial and

*Address correspondence to this author at the School of Civil & Environmental Engineering, University of Science and Technology Beijing, Beijing, 100083, P.R. China;
E-mail: yaojun@ustb.edu.cn

urease activity, soil organic matter. We also examined what the relative effects were of underlying microcalorimetry parameters, urease activity and organic matter.

2. MATERIAL AND METHODS

2.1.Soil Sampling Sites and Sampling Collection

Soil sampling selected in The North China Plain, The area is a temperate monsoon climate, with annual average temperature 12°C, annual precipitation 550 mm. Soil samples taken down from 0 to 30 cm soil depth and were taken on trough winter stage, reviving stage, jointing stage, blossoming stage, maturing stage of winter wheat respectively. Soil samples was air dried and passed through a 1mm sieve for determination of microbial, urease activities and a 0.15mm sieve for organic matter. It has a sandy loam texture (0-20 cm) with 37% sand, 48% silt and 13% clay, pH 8.5, the surface soil (0-20 cm) contained 16.8g kg^{-1} organic matter, 0.9 g kg^{-1} total nitrogen, 16.6 mg kg^{-1} Olson-P, 99.3mg kg^{-1} Olsen-K, CEC (16.3 cmol kg^{-1}).

2.2. Experimental Design

Five N fertilizer treatments were designed:0 kg hm^{-2}(N0), 100 kg hm^{-2}(N1), 180 kg hm^{-2}(N2), 250 kg hm^{-2}(N3), 300 kg hm^{-2}(N4). N fertilizer applied as urea, P and K fertilizer applied as calcium superphosphate and potassium chloride which the rate was 120 and 120 kg hm^{-2}, ZnSO$_4$ fertilizer applied as 15 kg hm^{-2}. All the P, K and ZnSO$_4$ fertilizer were applied into soil as basic fertilizer, and N fertilizer applied as basic fertilizer (40%), topdressing at jointing (40%) and blossoming (20%) stage. From sowing in October to harvest in next June needed to irrigate 4 times. Other field management measures are the same in each treatment.

2.3. Microcalorimetry

TAMⅢ can provide a metabolic heat output production of the soil system. 4 ml of ampouls with soil (1 g) and 0.2 ml nutrition solution containing glucose (2.5 mg) and ammonium sulfate (2.5 mg). All the ampoules temperatures were set and controlled at 28°C , then the microcalorimetric data will record by a computer automatically, when the power-time curve returned to baseline, the experiment was finished.

2.4. Urease Assay

Urease activity was measured using colorimetric method [18]. 2.5 g soil sample were mixed with 0.5 mL of toluene for 15min, then added to 2.5 mL of 10% urea and 5mL citrate buffer (pH 6.7) in incubator at 38°C for 24 h. Then with 38°C distilled water diluted, filtered, then 4 mL of sodium phenate and 3 mL of sodium hypochlorite were added to 1 mL filtrate and diluted to 50 mL for 20 min, finally, it was measured at wavelengths of 578 nm using spectrophotometer.

2.5. Soil Organic Matter

Soil organic matter was determined by a standard potassium dichromate digest method [19]. 0.2500 g soil sample mixed with 5 ml of 0.8000 mol L^{-1} K$_2$Cr$_2$O$_7$ and 5 ml

of concentrated H$_2$SO$_4$ at 190°C for 5min, followed by titration of the extracts with standardized FeSO$_4$.

2.6. Statistical Analyses

Data and graphs are presented in the results was carried out using Microsoft Excel and Origin8.0. , Standard analysis of variance (AMOVA) procedures were used to calculate treatment means.

3. RESULTS

3.1. Microcalorimetry

Five stages (trough winter stage, reviving stage, jointing stage, blossoming stage, maturing stage) of winter wheat under five N fertilizer rate treatments on microbial activity were analyzed. As illustrated in (Fig. 1), the power-time curves of soil microbial growth were influenced obviously by N fertilizer rate.

In order to better interpretation of the power-time curves, we calculated some parameters from the curves (Table 1), such as Q_T: the total heat production; Pmax: the maximum heat production; Tmax: the time in the maximum heat production; t: the time course of metabolism. These are very important parameters for evaluation of microbial metabolism. Q_T released heat production of soil microorganisms from beginning to end, in blossoming stage N0, N1, N3 had the lowest values, while N2 and N4 had the lowest values in maturing stage. The values of Q_T had light difference at trough, reviving and jointing stages with increasing nitrogen fertilizer rate, but increasing nitrogen fertilizer rate that Q_T showed a trend of obvious W shape at blossoming stage and at maturing stage Q_T decreased with increasing nitrogen fertilizer rate. N3 had the highest Pmax which affected by the N fertilizer rate, with the N fertilizer rate increasing, Pmax showed increased firstly and then decreased trend in each stage, respectively. N3 all reach the highest thermal power in every stage. Tmax is the time when the soil microbial activity reached the maximum heat production. N0, N1, N2, N4 in jointing stage had the maximum time, and N3 had the maximum time in blossoming stage, N0, N1, N4 had the minimal time in blossoming stage, N2 in trough winter stage had the minimal time, N3 in reviving stage had the minimal time and each stage' s time had little change. t which indicated the time of soil microorganisms metabolism in the whole process, in trough winter and reviving stage, no influence on N rate (Table 1), jointing stage had the longest time of t, blossoming stage used greater time of t than maturing stage except for N0,N1.

3.2. Soil Urease, Organic Matter in Different Growth Stages

In the whole growth of winter wheat, soil urease activity in the jointing stage reached the peak, then it dropped to the low in maturing stage (Fig. 2). From sowing to trough winter stage, the change of urease activity depended on N rate, four N treatment increased 36.0%, 25.9%, 64.6%, 44.3% compared to CK, respectively. In reviving stage, urease activity had a slight increase. However, urease activity picked up quickly in jointing stage and reached the top of

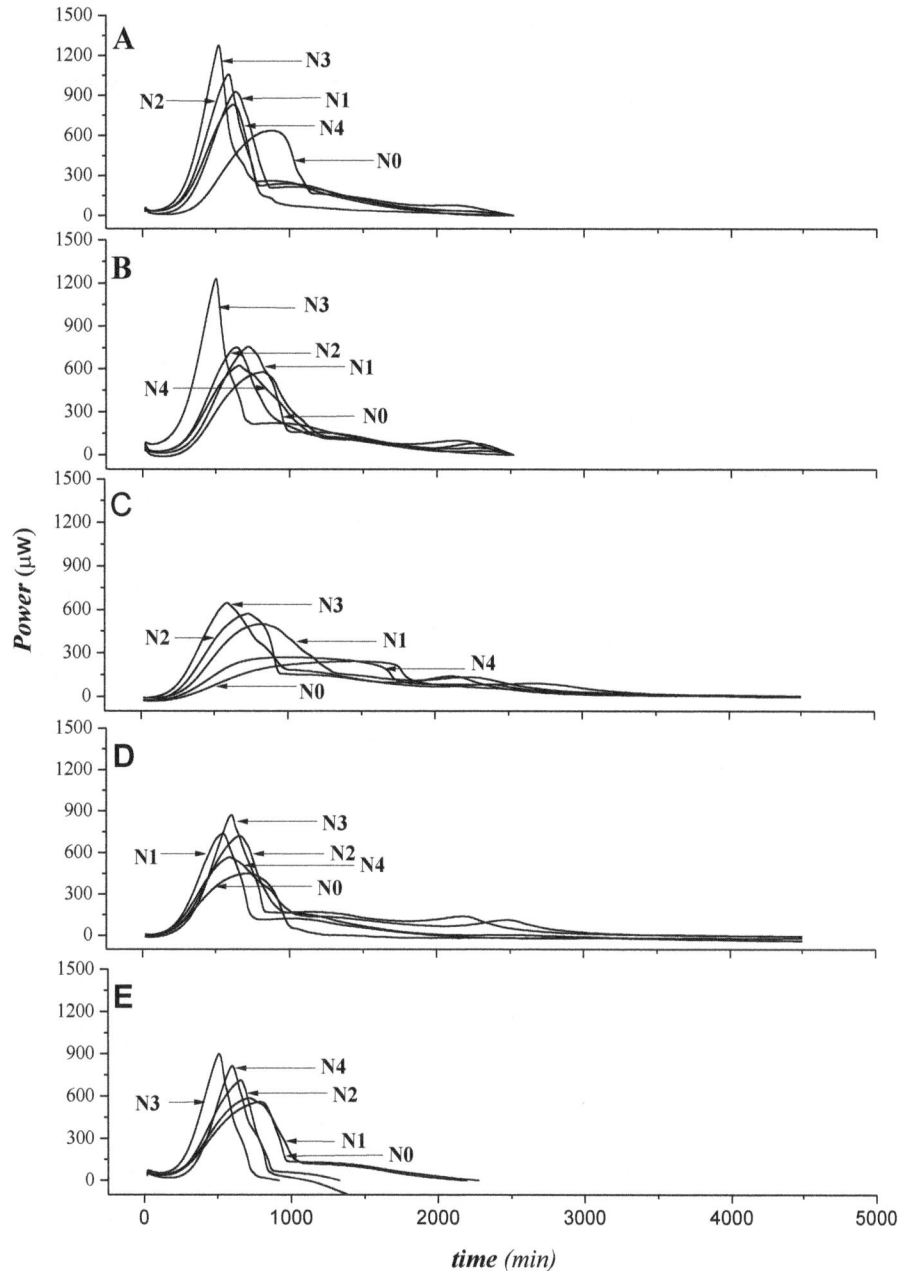

Fig. (1) Power-time of microorganism under different nitrogen fertilizer rate in the presence of rough winter stage (**A**), Reviving stage (**B**), Jointing stage (**C**), Blossoming stage (**D**) and Maturing stage (**E**).

growth. After it started to fall in blossoming stage, a little improve in maturing stage. These results clearly indicate that urease activity in N treatment increased compared to CK and N3 treatment of urease activity was higher than other N treatment. Thus, the different of soil urease activity related to the amount of nitrogen rate.

Continuous farming without the addition of N fertilizer over 3 years caused decrease in the SOM. The different content of SOM were due to the N fertilizer, but had no significant differences across N treatments and litter variability occurring at five stages (Fig. **3**). SOM of whole growth stage content ranged from 15.1 to 12.6 g kg^{-1}, 16.3 to 15.2 g kg^{-1}, 16.6 to 16.0 g kg^{-1}, 16.9 to 16.6 g kg^{-1}, 17.0 to

14.9 g kg^{-1} in the N0, N1, N2, N3 and N4 treatments. Five treatments from sowing to maturing stage the SOM decreased by 16.7%, 6.9%, 3.1%, 2.0%, 12.2%, respectively. SOM were lower by the maturing stage and N3 treatment had the lowest variation.

3.3. The Relationship between Soil Urease and Organic Matter

Variance analysis (Table **2**) showed that the effect of N fertilizer rate and growth stage on urease and organic matter is different. The effect of N fertilizer rate and growth stage on urease was greater than organic matter. For urease, the

Table 1.   **Thermokinetic parameters for microbial growth at 28℃ under different nitrogen fertilizer at different stage.**

	Treatment	Trough winter stage	Reviving stage	Jointing stage	Blossoming stage	Maturing stage
$J\ g^{-1}$	N0	27.95	25.75	20.6	7.43	22.8
	N1	29.2	27.55	30.57	4.4	23.6
	N2	30.72	25.4	27.44	28.3	16.7
	N3	30.24	30.71	29.66	10.21	14.8
	N4	21.08	26.77	22.64	24.68	14.7
P_{max} μW	N0	635	578	241	448	560
	N1	927	758	501	738	586
	N2	1057	753	572	719	713
	N3	1275	1230	647	870	899
	N4	831	623	269	566	815
T_{max} min	N0	873	816	1388	699	780
	N1	636	721	810	547	706
	N2	587	640	723	655	653
	N3	520	502	584	605	509
	N4	621	658	927	590	598
t min	N0	2526	2527	5228	1387	2275
	N1	2524	2525	4727	2012	2194
	N2	2527	2528	4983	3833	1330
	N3	2520	2524	6332	2664	920
	N4	2525	2530	4086	3662	1112

Table 2.   **Effect of nitrogen fertilizer rate and growth stage on soil urease and organic matter.**

Item	Urease	Organic Matter
N fertilizer rate	25.32	10.72
Growth stage	61.24	5.37

effect of growth stage was stronger than N fertilizer rate, for organic matter, N fertilizer rate is more influence on growth stage.

4. DISCUSSION

4.1. Effect of Nitrogen Rate on Soil Microbial Activity

Soil microbial activity is influenced by many factors (fertilizer, field management, soil physicochemical properties and so on) and microorganisms are important to most soil processes which depend on nutrient, carbon and energy of soil. From the result in the present study that change of microbial activity due to N fertilizer change and in different stage applied to soil (Fig. 1). This is in agreement [20]. N3 in five stages had the highest microbial activity, N0

of the power-time curves had the low activity compared to other N fertilizer treatments. In general, the rhizosphere microorganisms is sensitive to inorganic fertilizer, when crop root exudates contain a large number of C/N, would cause the rhizosphere of N in a deficiency environment. In this study, above or below N3' treatments all had lower microbial activity than N3 (Fig. 1). Because N fertilizer within a certain range that microbial quantity showed an increase with the increasing of N fertilizer rate, but excessiveness of N fertilizer was disadvantageous to carbon source, and the increasing of N fertilizer rate increased the growth of crop and water consumption, decreased soil water content and the number of microorganism.

In this paper, N fertilizer staging added to the consumption of N of soil, promoted the soil microbial breeding, thus improved the soil enzyme activity and soil

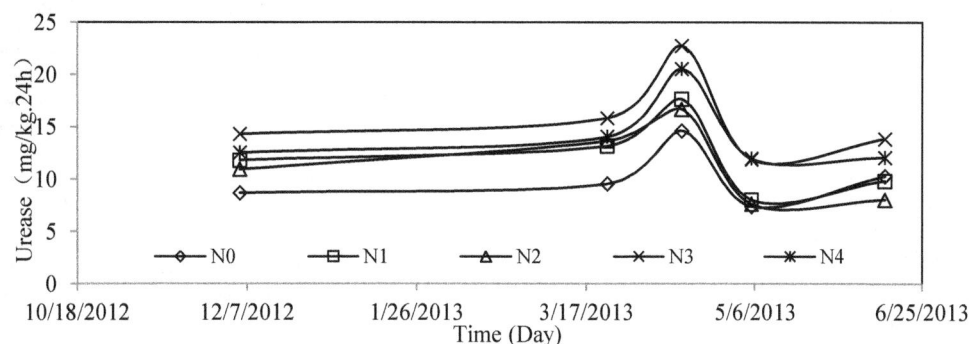

Fig. (2). The effect of nitrogen fertilizer on the urease at rough winter stage, reviving stage, jointing stage, blossoming stage and maturing stage.

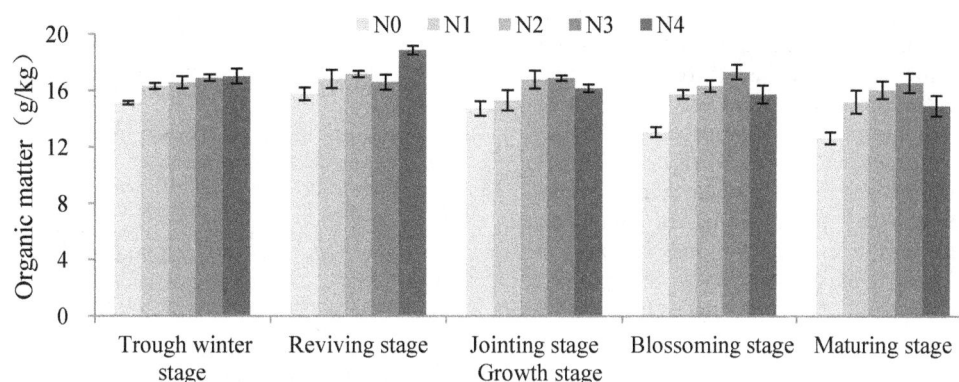

Fig. (3). Effect of nitrogen fertilizer on organic matter at different stage.

fertility. In the case of low N fertilizer, low nutrition stress increase wheat root exudates, strengthen the soil nutrition and balance of nutrition metabolism of biological, promoted the soil microbial community, quantity and activity.

4.2. Effect of Nitrogen Rate on Soil Urease and Organic Matter

Organic management increases overall enzyme activity [21, 22]. During the whole winter wheat growing season, from sowing stage to reviving stage is the vegetative growth phase of winter wheat which had less dry matter accumulation and N fertilizer rate, at this point urease activity slight low, so it need to reduce the N fertilizer rate [23]. From jointing to blossoming stage, with the dry matter accumulation increased, the demand for N fertilizer increased, and in jointing stage, the root activity and root weight density of winter wheat enhanced [24], root exudates is the important source of soil enzyme [25], result in urease activity rapidly increased in jointing stage and reached the peak of the whole growing season, therefore, increased the N fertilizer rate is necessary. In this study, urease activity in trough winter stage was higher than the mid-to late stage due to the irrigation, meanwhile, in the mid-to late stage soil permeability variation for irrigation and urease activity decreased. Urea applied into the soil by the urea enzyme hydrolysis into ammonia then absorbed by the winter wheat. Because ammonia is easily to volatilize and under the condition of high temperature and enough water that complete the conversion process in a short time. Than it would result in winter wheat can not absorb ammonia and N fertilizer loss. Besides, N fertilizer can inhibit urease

production. Therefore, a long term and high concentration of N fertilizer cause agricultural system that the effective nutrient mineralization and cycle ability reduced, leading to the decrease of available nitrogen.

The present dates demonstrate that under the addition of different N fertilizer rate over 3 years was slightly changed SOM content. [26] reported that fertilizer can improve crop growth, yield and organic inputs into soil, but a little positive impact on SOM [27]. The present study showed that SOM quality was improved by addition of proper N fertilizer. In trough winter stage, SOM content of five treatments had little difference. But N4 had the highest value of SOM in the reviving stage due to the high N fertilizer rate. To jointing stage, N4 of SOM decreased and similar trend continued to maturing stage, the same to N1. Moreover, N2 and N3 both had very little change in whole stages, this means that proper fertilizer is conducive to the preservation of SOM.

4.3. Nitrogen Fertilizer Management is Very Important in Agriculture System

Fertilizer plays an important role in agricultural system. M.J. Hawkesford reported that crop wanted high yield need nitrogen [28] which for storage proteins in the grain and photosynthesis, efficient and proper management of nitrogen not only was significant for sustainability of agricultural system, but also it can avoid adverse environmental impacts of pollution. Higher than recommended rate of N fertilizer would reduce microbial biomass C [29]. In this study, after applying N fertilizer to soil, the relationship with soil activity in the growth season dynamic changes and soil nutrient status. Under different N treatments the microbial activity was higher than control observed (Fig. **1**). N1, N2 and N4 had lower microbial activity than N3.

According to [30, 31], the quantity and quality of N and carbon (C) and SOM inputs are the overriding controls on soil microbial activity. Soil microorganisms play an important role in increasing soil fertility, improving soil structure and promoting the material circulation in nature. Some heterotrophic microorganisms can decompose animal and plant residues and organic fertilizer, then re-synthetic form humus. When soil temperature is low, poor ventilation, anaerobic microbial activity increased, humus synthetic speed is accelerated, and accumulated of humus, humus was influence on soil fertility. Under the condition of high temperature, adequate soil moisture, good ventilation, aerobic microbial activity increased, humus decomposition, releasing nutrients for plant uptake. Azotobacter fixed nitrogen in the air, as their protein, when these bacteria death and decomposition, the nitrogen can be absorbed by the plant and accumulated nitrogen in soil. Nitrifying bacteria decomposed organic fertilizer into ammonia then converted to effective nitrate for the plant. Some of fungi can decompose cellulose, lignin and pectin and so on. The accumulation of fungal hyphae can make the physical structure of the soil improvement. Actinomyces can produce antibiotics.

CONCLUSION

The present study showed that high soil microbial activity due to N fertilizer rate, while soil organic matter was not affected obviously, in contrast, soil urease activity had a difference in each stage and at the jointing stage reached the peak. The variance analysis results indicated N fertilizer rate and growth stage both can influence the soil urease activity and soil organic matter. It is concluded that proper N fertilizer rate is important for optimal microbial growth while topdressing N fertilizer at jointing stage it is beneficial to the soil urease activity.

ACKNOWLEDGEMENTS

This work is supported in part by grants from the International Joint Key Project from Chinese Ministry of Science and Technology (2010DFB23160), National Natural Science Foundation of China (41273092), Public welfare project of Chinese Ministry of Environmental Protection (201409042), and Overseas, Hong Kong and Macau Young Scholars Collaborative Research Fund (41328005).

REFERENCES

[1]   Shen JL, Tang AH, Liu XJ, Fangmeier A, Goulding KT, Zhang FS. High concentrations and dry deposition of reactive nitrogen species at two sites in the North China Plain. Environ Pollut 2009; 157: 3106-13.

[2]   Basso B, Cammarano D, Fiorentino C, Ritchie JT. Wheat yield response to spatially variable nitrogen fertilizer in Mediterranean environment. Eur J Agron 2013; 51: 65-70.

[3]   Li W, Li L, Sun J, et al. Effects of intercropping and nitrogen application on nitrate present in the profile of an Orthic Anthrosol in Northwest China. Agric Ecosyst Environ 2005; 105: 483-91.

[4]   Chen D, Suter H, Islam A, Edis R, Freney JR, Walker CN. Prospects of improving efficiency of fertiliser nitrogen in Australian agriculture: a review of enhanced efficiency fertilisers. Aus J Soil Res 2008; 46: 289-301.

[5]   Lu C, Fan L. Winter wheat yield potentials and yield gaps in the North China Plain. Field Crops Res 2013; 143: 98-105.

[6]   Costanza R, d'Arge R, Groot RD, et al. The value of the world's ecosystem services and natural capital. Nature 1997; 387: 253-60.

[7]   Akhtar A, Hisamuddin, Robab MI, Abbasi, Sharf R. Plant growth promoting Rhizobacteria An overview. J Nat Prod Plant Resour 2012; 2: 19-31.

[8]   Singh JS, Pandey VC, Singh DP. Efficient soil microorganisms: A new dimension for sustainable agriculture and environmental development. Agric Ecosyst Environ 2011; 140(3-4): 339-53.

[9]   de Vries FT, Hoffland E, van Eekeren N, Brussaard L, Bloem J. Fungal/bacterial ratios in grasslands with contrasting nitrogen management. Soil Biol Biochem 2006; 38(8): 2092-103.

[10]   Tejada M, Dobao MM, Benitez C, Gonzalez JL. Study of composting of cotton residues. Bioresour Technol 2001; 79: 199-202.

[11]   Ryals R, Kaiser M, Torn MS, Berhe AA, Silver WL. Impacts of organic matter amendments on carbon and nitrogen dynamics in grassland soils. Soil Biol Biochem 2014; 68: 52-61.

[12]   Lucchini P, Quilliam RS, De Luca TH, Vamerali T, Jones DL. Does biochar application alter heavy metal dynamics in agricultural soil? Agric Ecosyst Environ 2014; 184: 149-57.

[13]   Kerek M, Drijber RA, Gaussoin RE. Labile soil organic matter as a potential nitrogen source in golf greens. Soil Biol Biochem 2003; 35(12): 1643-9.

[14]   Di Bene C, Tavarini S, Mazzoncini M, Angelini LG. Changes in soil chemical parameters and organic matter balance after 13 years of ramie [Boehmeria nivea (L.) Gaud.] cultivation in the Mediterranean region. Eur J Agron 2011; 35(3): 154-63.

[15]   Li M, Zhou X, Zhang Q, Cheng X. Consequences of afforestation for soil nitrogen dynamics in central China. Agric Ecosyst Environ 2014; 183: 40-6.

[16]   Pandey D, Agrawal M, Bohra JS. Effects of conventional tillage and no tillage permutations on extracellular soil enzyme activities and microbial biomass under rice cultivation. Soil Tillage Res 2014; 136: 51-60.

[17]   Liang Q, Chen H, Gong Y, Yang H, Fan M, Kuzyakov Y. Effects of 15 years of manure and mineral fertilizers on enzyme activities in particle-size fractions in a North China Plain soil. Eur J Soil Biol 2014; 60: 112-9.

[18]   Guo H, Yao J, Cai M, et al. Effects of petroleum contamination on soil microbial numbers, metabolic activity and urease activity. Chemosphere 2012; 87(11): 1273-80.

[19]   Walkley A, Black IA. An examination of degtjareff method for determining soil organic matter and a proposed modification of the chromic acid titration method. Soil Sci 1934; 37(1): 29-39.

[20]   Geisseler D, Scow KM. Long-term effects of mineral fertilizers on soil microorganisms – A review. Soil Biol Biochem 2014; 75: 54-63.

[21]   Maeder P, Fliessbach A, Dubois D, Gunst L, Fried P, Niggli U. Soil fertility and biodiversity in organic farming. science 2002; 296: 1694-7.

[22]   B Moeskops, Sukristiyonubowo, D. Buchan, et al. Soil microbial communities and activities under intensive organic and conventional vegetable farming in West Java, Indonesia. Appl Soil Ecol 2010; 45: 112-20.

[23]   Dordas C. Dry matter, nitrogen and phosphorus accumulation, partitioning and remobilization as affected by N and P fertilization and source–sink relations. Eur J Agron 2009; 30: 129-39.

[24]   Shi Z, Jing Q, Cai J, Jiang D, Cao W, Dai T. The fates of 15N fertilizer in relation to root distributions of winter wheat under different N splits. Eur J Agron 2012; 40: 86-93.

[25]   Kotroczó Z, Veres Z, Fekete I, et al. Soil enzyme activity in response to long-term organic matter manipulation. Soil Biol Biochem 2014; 70: 237-43.

[26]   Hai L, Li XG, Li FM, Suo DR, Guggenberger G. Long-term fertilization and manuring effects on physically-separated soil organic matter pools under a wheat–wheat–maize cropping system in an arid region of China. Soil Biol Biochem 2010; 42: 253-9.

[27]   Lal R. Carbon sequestration in dryland ecosystems. Environ Manage 2004; 33: 528-44.

[28]   Hawkesford MJ. Reducing the reliance on nitrogen fertilizer for wheat production. J Cereal Sci 2014; 59: 276-83.

[29]   Lupwayi NZ, Lafond GP, Ziadi N, Grant CA. Soil microbial response to nitrogen fertilizer and tillage in barley and corn. Soil Tillage Res 2012; 118: 139-46.

[30]   Kallenbach C, Grandy AS. Controls over soil microbial biomass responses to carbon amendments in agricultural systems: A meta-analysis. Agric Ecosyst Environ 2011; 144, pp. 241-52.

[31]   Grandy AS, Strickland MS, Lauber CL, Bradford MA, Fierer N. The influence of microbial communities, management, and soil texture on soil organic matter chemistry. Geoderma 2009; 150: 278-86.

5

Long-term Effects of Elevated CO_2 on the Proliferation of Cyanophage PP

Cheng Kai[1,2], Shang Shi Yu[1], Gao Ying[2], Zhao Yi Jun[1] and Huang Z. Guang[3,*]

[1]*Key Laboratory of Ecological Remediation of Lakes and Rivers and Algal Utilization of Hubei Province, College of Resources and Environmental Engineering, Hubei University of Technology, Wuhan 430068, China*

[2]*Department of Life Sciences, Huazhong Normal University, Wuhan 430079, China*

[3]*South China Institute of Environmental Sciences, Guangzhou 510655, China*

Abstract: Much of the research effort focused on the impacts of elevated CO_2 on marine algae but very little work was done on freshwater algae, or on freshwater algal viruses. In this paper, we studied the impacts of elevated CO_2 on the infection of a freshwater cyanobacterium (wild *Leptolyngbya sp.*) by cyanophage PP that have a wide distribution in China. In a 12-month experiment, logarithmic-phase host cells were infected with cyanophage PP at 370 or 740 µatm pCO_2 concentrations; the burst size, lysing cycle and proportion of adsorption were measured. The results showed that the proportion of adsorption, and burst sizes of cyanophage PP increased significantly with elevated CO_2 concentrations, and the proportion of adsorption increased gradually within the 12 months with the gradual increment of cell width. The result indicated that elevated CO_2 concentration may have significantinfluences on the proliferation dynamics of cyanophage–host systems, and some of the influences may increase gradually in a long-term.

Keywords: Adsorption, burst size, global change, phage–host system.

1. INTRODUCTION

Atmospheric CO_2 has increased by 25% in the past 200 years, and projections indicate that it may reach 750 µatm pCO_2 by 2100 [1]. Aquatic ecosystems are an important component of the biosphere, where the amount of CO_2 fixed by planktonic algal photosynthesis accounts for approximately 50% of the total fixed by the entire biosphere. This proportion is expected to increase further with rising temperatures [2].

Algal viruses control the community structure, contribute toward the abundance of planktonic algae, regulate the nutrient cycling of certain elements [3], and mediate the evolution of their hosts *via* horizontal gene transfer [4].

The ecological functions of algal viruses depend on their abundance, andtheir abundance is closely related to the levels of virus proliferation [5]. These processes occur within host cells and are affected by the host physiological status. Much research has focused on the impacts of elevated CO_2 on marine and freshwater algae [6-8], but little is known about cyanophage–host systems, especially in freshwater. Thus, the present study aimed to improve our understanding of the effects of high atmospheric CO_2 on the proliferation of freshwater cyanophage PP.

2. MATERIAL AND METHODS

Phage-host system and culture conditions: Cyanophage PP is a podovirus with linear, double-stranded DNA, which has been frequently detected at high levels in many eutrophic lakes in China [9]. The following experiments used a cyanophage PP lysate with $>10^8$ PFU· mL-1(PFU, plaque-forming units), which was stored at 4 °C. The host cyanobacterial strain used was isolated from Donghu Lake (the East Lake) (30°31′N, 114°22′E) in Wuhan, China and identified by 16S rDNA analysis as *Leptolyngbya sp.* (the identity score is up to 99% in Genebank). The host was cultured in AA medium at 28 °C, with 3000 lx, and 14:10 h day/night conditions in an ambient CO_2 chamber (to supply a CO_2 concentration of approximately 370 µatm pCO_2, which was measured using a Telaire 7001 CO_2 sensor (USA), or in a 740µatm pCO_2 chamber. Within 1 year, the culture was re-inoculated by adding 35 mL of the cyanobacterium to 115 mL AA medium every 14 days, and the following one-step growth curve, adsorption, and photoreactivation experiments were repeated at 1, 2, 3, 6, and 12 months with in 1 year. Doubling the CO_2 concentration significantly accelerated host growth (P < 0.01, paired samples t-test), thus the cell densities of the cultures maintained in two different CO_2 concentrations were normalized to 5.6×10^6 cells· mL^{-1} at pH 7.6 by adding AA medium before the following experiments.

One-step growth experiment: One-step growth experiments were used to measure the lytic cycle and the average burst size, where 50 mL of prepared host cell suspensions were mixed with the cyanophage lysate at MOI of 10^{-5}(MOI,

---

*Address correspondence to this author at the South China Institute of Environmental Sciences, Guangzhou 510655, P.R. China;
E-mail: huangzhengguang@scies.org

multiplicity of infection). The MOI is low enough to avoid numerous phages attach to a single trichome even after the first generation offspring cyanophages were released. After 30min incubation without shaking to allow adsorption, the mixtures were centrifuged at 10000 ×g for 10 min at 28 °C. The pellets were then collected and washed twice in AA medium, before the washed precipitates were resuspended in 50 mL AA medium and incubated in the chambers, as described above. For the plaque assays, 1 mL samples from the culture were plated at 0, 30, 60, 90, 120, 180, and 240 min. After constructing the one-step growth curves, the average burst size was determined as the ratio of the titer after the burst relative to the initial titer (at 0 min), and the corresponding time duration was recorded as the lytic cycle.

Adsorption experiment: In the adsorption experiments, an adsorption mixture was prepared by adding the cyanophage lysate to the prepared host cell suspensions at MOI of 10^{-4}(the MOI is low enough to avoid numerous phages attach to a single trichome). The initial titers of the mixtures were measured using a plaque assay and recorded as P0. The mixtures were then incubated without shaking in the chambers described above. To measure the extent of adsorption, 3 mL aliquots of the adsorption mixture were sampled at 15, 30, and 60 min. The samples were centrifuged at 12000 ×g for 10 min at 4 °C. Then, the supernatant containing free cyanophages was removed, and the titers of the pellets containing adsorbed cyanophages were determined

and recorded as Pt. The proportion adsorbed at time t (At) (%) was calculated using the following formula: $At = P_t/P_0$.

Statistical analysis: 3 biological replicates were carried out in the above experiment,and SPSS 20.0 (IBM, USA) software was used for following tests: Paired sample t-test was performed to measure the differences between tested group and control group, ANOVA (with LSD post-hoc test) was performed to measure the differences among months, and Spearman's correlation test was performed to establish a relationship between proportion of adsorption and cell width.

3. RESULTS

Burst size and lytic cycle: According to the one-step growth curves (Fig. 1), the cyanophage titer was detected 120 min after adsorption and it reached its maximum at approximately 180 min, regardless of the CO_2 concentration. In contrast to the lytic cycle, the burst size (Table 1) was significantly higher when the CO_2 concentration was doubled than that in the controls (P < 0.05). From Table 1, doubled CO_2 seemed to cause a gradual increase in the burst size and cell width from the 1st month to the 12th month. However, ANOVA indicated that the difference in the burst size was not significant among months (P > 0.05), and the correlation between the burst size and cell width at the doubled CO_2 conditions was not significant too (P > 0.05).

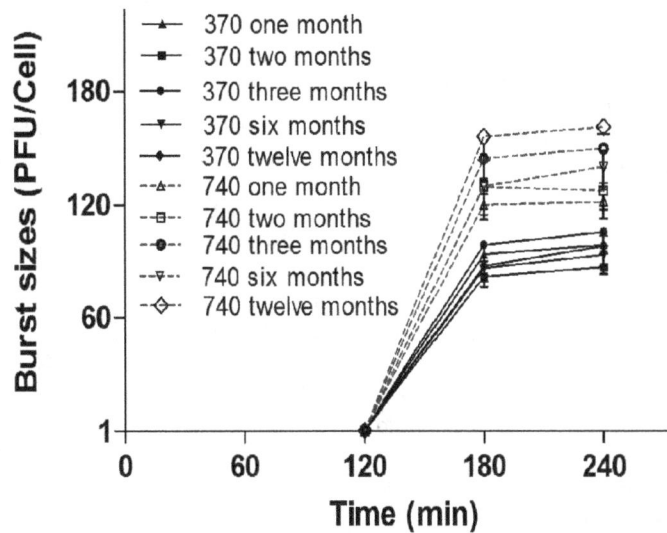

Fig. (1). One-step growth curves of cyanophage PP at different CO_2 concentrations(mean ± SD).

Table 1.   Effects of doubled CO_2 onthe burst size of cyanophage PP and the cell width of the host cell (mean ± SD).

Culture Time	370 µatm		740 µatm	
	Burst Size (PFU cell⁻¹)	Cell Width (µm)	Burst Size (PFU cell⁻¹)	Cell Width (µm)
1 month	93 ± 4	2.51±0.02	120 ± 8	2.52±0.05
2 months	84 ± 5	2.48±0.02	129 ± 26	2.49±0.01
3 months	99 ± 11	2.44±0.08	144 ± 11	2.71±0.07
6 months	93 ± 18	2.50±0.02	140 ± 15	3.02±0.07
12 months	87 ± 8	2.49±0.03	156 ± 15	3.31±0.11

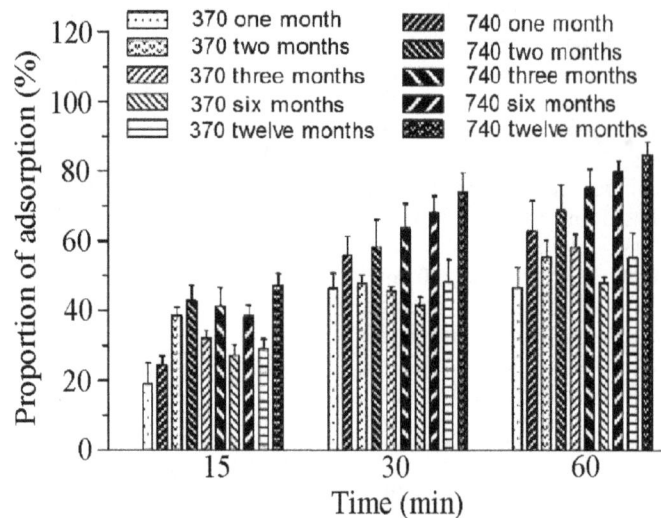

Fig. (2). Proportion of adsorption at different CO_2 concentrations(mean ± SD).

Proportion of adsorption: The results of the adsorption experiments are shown in Fig. (2). The proportion of adsorption that occurred when the CO_2 level was doubled was significantly higher than that in the controls ($P < 0.05$). Moreover, based on the proportion of adsorption at 60 min, it is clear that doubled CO_2 caused a gradual increase in the proportion of adsorption from the 1st month until the 12th month. ANOVA also indicated that the difference in the proportion of adsorption at 60 min was significant among months ($P < 0.05$). In addition to the continuous increase in the adsorption capacity throughout the year, a continuous increase in the cell width was observed during the year (Table 1, $P < 0.05$, ANOVA), and there exists a significant positive correlation between the proportion of adsorption at 60 min and cell width at the doubled CO_2 conditions ($P < 0.05$).

4. DISCUSSION

Adsorption and the burst size, which are important indicators of cyanophage infectivity, can be affected by the host's physiological status [10] in numerous ways. Firstly, viruses had higher adsorption rates and burst sizes when the host was supplied with sufficient nutrients or was in the rapid growth phase [11], and we observed a doubling the CO_2 concentration significantly accelerated host growth in the present study.Secondly, enlarged host cells may provide a greater surface area for adsorption [12], and we observed a significant positive correlation between the proportion of adsorption and cell width at the doubled CO_2 conditions in the present study.

Many studies have reported the effects of elevated CO_2 concentrations on marine and freshwater algae, but only three previous studies have considered the effects of different CO_2 concentrations on algal viruses. In a 30-day mesocosm experiment, Larsen et al. [13] showed that the abundance of small viruses (mostly phages) did not respond to changes in the CO_2 levels, whereas the abundance of the Emilianiahuxleyi virus and an unidentified large dsDNA virus decreased with increasing CO_2 levels. In another short-term (only a few weeks) laboratory batch monoculture experiment [14], no impact was found on viral lysis of Phaeocystispouchetii while

increased burst size and slightly delayed lysis was observed for E. huxleyi with increased CO_2. In a 39-day laboratory batch culture experiment [15], CO_2 and NaOH were used to adjust the pH concentration, where the eclipse period of cyanophage S-PM2 increased as the pH decreased, whereas the latent period and burst size decreased. The present study was conducted over the course of a year, and a continuous increase in the adsorption capacity was observed throughout the year, which suggests that long-term experiments are more representative and useful than short-term experiments for understanding the effects of long-term global change [14].

Changes in the species composition of freshwater phytoplankton communities have already been observed in areas with elevated CO_2 levels [16]. However, the roles of phytoplanktonic viruses in these changes are poorly understood. The contributions of phytoplanktonic viruses to changes in aquatic ecosystems during global change needs to be studied in terms of significant shifts in viral infectivity (as observed in the present study) and the species-specific viral response to increasing CO_2 concentrations [14]. It is also necessary to establish more accurate climate and ecosystem models [15], which have previously excluded the abundant phytoplanktonic viruses and their major ecological functions [5].

CONCLUSION

Elevated CO_2 concentration influenced the infectivity of cyanophage PP significantly, and some of the influences may increase gradually in a long-term.

ACKNOWLEDGEMENTS

This work was supported by National Science Foundation of China [grant numbers 31200385, 31370148] and Science Foundation of Hubei Province [2013CFA108].

REFERENCES

[1]   Stocker T, Qin DH, Plattner GK, *et al*. Climate change 2013: the physical science basis: Summary for Policymakers, IPCC, Switzerland, 2013.

[2]   Behrenfeld MJ, O'Malley RT, Siegel DA, *et al*. Climate-driven trends in contemporary ocean productivity. Nature 2006; 444: 752-5.

[3]   Fuhrman JA. Marine viruses and their biogeochemical and ecological effects. Nature 1999; 399: 541-8.

[4]   Thompson LR, Zeng Q, Kelly L, *et al*. Phage auxiliary metabolic genes and the redirection of cyanobacterial host carbon metabolism. Proc Nat Acad Sci USA 2011;108: E757-64.

[5]   Suttle CA, Marine viruses - major players in the global ecosystem. Nat Rev Microbiol 2007; 5: 801-12.

[6]   Riebesell U. Effects of CO_2 enrichment on marine phytoplankton. J Oceanogr 2004; 60: 719-29.

[7]   Fu F, Warner ME, Zhang Y, Feng Y, Hutchins DA. Effects of increased temperature and CO_2 on photosynthesis, growth, and elemental ratios in marine Synechococcus and Prochlorococcus (Cyanobacteria). J Phycol 2007; 43: 485-96.

[8]   McCarthy A, Rogers SP, Duffy SJ, Campbell DA. Elevated carbon dioxide differentially alters the photophysiology of Thalassiosira Pseudonana (Bacillariophyceae) and Emiliania Huxleyi (Haptophyta). J Phycol 2012; 48: 635-46.

[9]   Cheng K, Zhao Y, Du X, Zhang Y, Lan S, Shi Z. Solar radiation-driven decay of cyanophage infectivity, and photoreactivation of the cyanophage by host cyanobacteria. Aqua Microbial Ecol 2007; 48: 13-8.

[10]   Short SM. The ecology of viruses that infect eukaryotic algae. Environ Microbiol 2012; 14: 2253-71.

[11]   Gnezda-Meijer K, Mahne I, Poljsak-Prijatelj M, Stopar D. Host physiological status determines phage-like particle distribution in the lysate. Fems Microbiol Ecol 2006; 55: 136-45.

[12]   Hadas H, Einav M, Fishov I, Zaritsky A. Bacteriophage T4 development depends on the physiology of its host Escherichia coli. Microbiology 1997; 143: 179-85.

[13]   Larsen JB, Larsen A, Thyrhaug R, Bratbak G, Sandaa RA. Response of marine viral populations to a nutrient induced phytoplankton bloom at different pCO2 levels. Biogeosciences 2008; 5: 523-33.

[14]   Carreira C, Heldal M, Bratbak G. Effect of increased pCO2 on phytoplankton–virus interactions. Biogeochemistry 2013; 114: 391-7.

[15]   Traving SJ, Clokie MRJ, Middelboe M. Increased acidification has a profound effect on the interactions between the cyanobacterium Synechococcus sp WH7803 and its viruses. Fems Microbiol Ecol 2014; 87: 133-41.

[16]   Verschoor AM, van Dijk MA, Huisman J, van Donk E. Elevated CO_2 concentrations affect the elemental stoichiometry and species composition of an experimental phytoplankton community. Freshwater Biol 2013; ED-58: 597-611.

6

In-situ Product Recovery as a Strategy to Increase Product Yield and Mitigate Product Toxicity

Yuen Ling Ng* and Yi Yang Kuek

Department of Chemical and Environmental Engineering, University of Nottingham, University Park, Nottingham NG7 2RD, United Kingdom

Abstract: Product inhibition is often the cause limiting the maximum product concentration attainable in fermentation. This study showed the product yield of *p*-cresol could be improved by *in-situ* product recovery (ISPR). *Escherichia coli* transformed with the *hpd BCA* operon from *Clostridium difficile* was shown in this study to express *p*-hydroxyphenylacetate decarboxylase which converted *p*-hydroxyphenylacetate into *p*-cresol under anaerobic fermentation. Toxicity of *p*-cresol found at a concentration as low as 5 mM in a broth spiked with *p*-cresol was shown to have limited the maximum product concentration at 1 ± 0.1 mM after 30 hours of batch fermentation. Product yield was however shown to increase by 51% when activated carbon was used to remove *p*-cresol *in-situ* production. The activated carbon concentrated *p*-cresol on the solid adsorbent which was subsequently separated by sedimentation and *p*-cresol recovered by ultrasonic-assisted solvent extraction. Desorption of *p*-cresol from the spent activated carbon allowed the adsorbent to be regenerated for further product recovery. The ISPR strategy reported here was shown to improve the yield of a toxic product, was sustainable, and when adapted to a continuous process would increase productivity.

Keywords: adsorption, *Escherichia coli*, glycyl radical enzyme, *in-situ* product recovery, *p*-cresol, product toxicity.

1. INTRODUCTION

Para-cresol (*p*-cresol) is an important intermediate for the production of antioxidants in food, personal care products, pharmaceuticals, materials, fuels, oils, herbicides, flavours, fragrances, dyes, resins, disinfectants and preservatives. Production of *p*-cresol by *Clostridium difficile* (*C. difficile*) was via the expression of the *hpd* BCA operon encoding the highly oxygen-sensitive Glycyl Radical Enzyme (GRE) *p*-hydroxyphenylacetate decarboxylase [1]. Although *C. difficile* was able to produce *p*-cresol, it cannot be used in large-scale production due to its virulence in causing life-threatening gastrointestinal infection in humans [2, 3]. This study therefore explored the possibilities of producing *p*-cresol by genetically modifying the Generally Recognised As Safe (GRAS) host *Escherichia coli* (*E. coli*) to express *p*-hydroxyphenylacetate decarboxylase, and to convert *p*-hydroxyphenylacetate (HPA) into *p*-cresol. The *hpd* BCA operon from *C. difficile* was cloned into plasmids which were then used to transform *E. coli* to produce *p*-cresol in whole-cell fermentation. Since *E. coli* was a facultative anaerobe and was most commonly used in recombinant DNA technology [4], it was postulated that the expression as well as the enzymatic activity of the GRE would be possible during anaerobic fermentation.

P-cresol was reported to be toxic to cells whereby the maximum concentration tolerable by *C. difficile* was reported at 35 mM [5]. It was this toxicity that attributed *C. difficile* its virulent characteristics [6], and rendered the bacteria a competitive advantage over other microorganisms in its natural habitat [7]. This study explored the possibilities of mitigating the toxicity imposed by *p*-cresol by implementing an ISPR strategy in order to improve product yield and productivity. The model with *p*-cresol would allow adaptation to other bioproducts, thus providing solutions to product inhibition and improve product yield [8-12].

Solvent extraction using a biphasic extraction system was reported in [13] for product recovery. [14] and [15] suggested the requirement for extraction solvents to have log P values of at least 4 in order to ensure biocompatibility and non-toxicity to the cells during product separation. Alkanes containing 7 carbons and above [16, 17], oleyl alcohol [18], dibutyl phthalate [19, 20] and ethyl laurate [21] were some common organic solvents used for product recovery. Although solvent extraction could be used for product separation, possible formation of emulsion during the product recovery step could result in subsequent difficulty in a clean phase separation. High energy consumption required in the final product purification step due to the high boiling points of extraction solvents, and the reported decrease in the growth rates of cells exposed to toxic solvents [17] rendered solvent extraction to be a non-ideal product separation and purification method.

[22] suggested adsorption to be a more viable separation method for biological products where polymeric resins were reported by [23] to be used for the ISPR of second-generation biofuels. Adsorption of *p*-cresol by different types of adsorbents in the applications of hemodialysis, wastewater

*Address correspondence to this author at the Department of Chemical and Environmental Engineering, University of Nottingham, University Park, Nottingham NG7 2RD, United Kingdom;
E-mail: yuen.ng@nottingham.ac.uk

treatment and as first-principle adsorption study was reported by various authors in Table 1.

This study explored the feasibility of using activated carbon (AC) adsorption for the separation and recovery of p-cresol from fermentation broth which was subsequently recovered by solvent extraction. The decoupling of the solvent extraction process from fermentation via adsorption allowed the flexibility of using an extraction solvent with a lower

Table 1. Adsorption of p-cresol by Various Types of Adsorbents

Adsorbent	Adsorption characteristics	Application	Reference
High silica MFI zeolite	The adsorbent was able to remove 80-85% of p-cresol from solution, although the removal performance was reduced in the presence of serum.	Hemodialysis	[24]
Hypercross-linked polymer, bituminous coal activated carbon, phartenium-based activated carbon, fly ash, clay and silicalite	Maximum adsorption capacity was achieved with the hypercross-linked polymer at 1.8 mmol/g of polymer.	Adsorption of p-cresol from water.	[25]
Polyaniline-modified mesoporous carbon (CMK-1/PANI)	Maximum adsorption capacity for p-cresol was 1.5 mmol/g of adsorbent. Acidic-alkaline interaction between the amine groups on the surface of the adsorbent and the phenolic compounds was hypothesised to have improved the performance of the adsorbent.	Adsorption of phenolic compounds including p-cresol from aqueous solutions.	[26]
Crosslinked zinc chloride, chloromethylated poly(styrene-co-divinylbenzene) resin (HJ-1), and Amberlite XAD-4 resin	Adsorption capacities of p-cresol were reported at 1.3 mmol/g of HJ-1, and 0.6 mmol/g of Amberlite XAD-4.	Adsorption of p-cresol from aqueous solutions at 25 to 40 °C.	[27]
Amberlite XAD-4 and NDA polymeric adsorbents	Maximum adsorbate concentration was at 1.8 mmol/g of adsorbent. The basic functional groups and oxygen groups on the surface of NDA-99 were reported to enhance adsorption.	Adsorption study with different types of adsorbents.	[28]
Fly ash made from wood	Adsorption capacity for p-cresol was at 0.49 mmol/g of adsorbent.	The application of low-cost adsorbents for the removal of phenolic compounds from wastewater.	[29]
Sulphuric acid-treated parthenium-based activated carbon	Optimum adsorption capacity of p-cresol was reported at 0.56 mmol/g of parthenium-based activated carbon, compared to 0.78 mmol/g of commercial activated carbon.	Adsorption of p-cresol from aqueous solutions at pH 6. Maximum concentration of p-cresol used in solution was 1000 mg/L.	[30]
Cellulose diacetate and triacetate membranes, synthetic polyamide, polysulfone, polyacrylonitrile, polymethylmethacrylate, zeolite and silicalite	Adsorption performance of silicalite adsorbent was the best, where a 2-min equilibration time was adequate for the adsorbate uptake. The maximum adsorption capacity was 0.98 mmol/g of silicalite.	Hemodialysis	[31]
Spectracarb activated carbon cloth	Adsorption capacities of p-cresol were 1.9 mmol/g of activated carbon in water, 1.7 mmol/g of activated carbon in sulphuric acid solution, and 0.65 mmol/g of activated carbon in sodium hydroxide solution.	Adsorption of p-cresol at 30 °C in water, in 1 M sulphuric acid and in 0.1 M sodium hydroxide solution.	[32]
Adsorbents prepared from fertiliser waste	Adsorption capacity of p-cresol was 0.38 mmol/g of fertiliser waste adsorbent, compared to 0.85 mmol/g of commercial activated charcoal.	Adsorption of p-cresol in water.	[33]
Amberlite XAD-4 and chloromethylated styrene-divinylbenzene copolymer beads (NJ-8).	Adsorption capacities of p-cresol were 2.5 mmol/g of NJ-8 and 1.2 mmol/g of XAD-4.	Adsorption of p-cresol in water.	[34]

boiling point to separate and recover the product, thus enabling energy savings in subsequent product purification processes. Desorption of *p*-cresol from the spent adsorbent allowed the recovery and reuse of the adsorbent for further product separation and recovery. This study was divided into various phases in order to establish the toxicity levels of *p*-cresol on *E. coli*, to test and establish whether the productivity and yield of *p*-cresol could be improved by ISPR, and to determine the suitability and capacity of the integrated adsorption-cum-solvent extraction ISPR strategy in the recovery of *p*-cresol.

2. MATERIALS AND METHODS

2.1. Cloning

Genomic DNA was extracted from *C. difficile* 630Δerm as described in [35, 36]. The DNA was PCR-amplified, cloned into the modular vectors pMTL84151 and pMTL84251, and were then used to transform Top10 *E. coli*. The transformed cells were cultured aerobically from where the positive clones with chloramphenicol (CM) and erythromycin (EM) resistance were screened and isolated. The clones were maintained aerobically and were switched to anaerobic fermentation for the production of *p*-cresol.

2.2. Production of *p*-cresol Using Transformed *E. coli* Cells

Both CM and EM clones were maintained on Terrific Broth (TB) media containing 12 g/L Bacto-Tryptone, 24 g/L yeast extract, 4 ml/L glycerol, 2 ml/L Vishniac trace element solution, 20 ml/L $FeSO_4$, 4 g/L K_2HPO_4 and 2 g/L KH_2PO_4. The selective media was supplemented with either chloramphenicol at a concentration of 12.5 µg/ml, or erythromycin at a concentration of 500µg/ml, for the CM and EM clones respectively. The Vishniac trace element solution was made up of 0.1 g/L EDTA-disodium salt, 2.2 g/L $ZnSO_4$, 5.54 g/L $CaCl_2$, 5.06 g/L $MnCl_2.4H_2O$, 5 g/L $FeSO_4.7H_2O$, 1.1 g/L $(NH_4).6Mo_7O_{24}.4H_2O$, 1.57 g/L $CuSO_4.5H_2O$, and 1.61 g/L $CoCl_2.6H_2O$. The trace element solution was adjusted to pH 6 with KOH and stored at 4 °C until the solution was used. The cell stocks were frozen down in the same culture media formulation with the addition of 100 ml/L glycerol for cryopreservation.

The cells were maintained at 37 °C on agar plates and were cultured in liquid broth for the production of *p*-cresol. 100 mM of *p*-hydroxyphenylacetic acid stock solution was neutralised to pH 7 with 10 M of sodium hydroxide solution before it was used as a substrate for the production of *p*-cresol. The neutralised sodium *p*-hydroxyphenylacetate (HPA) stock solution was added to 20 ml of liquid culture media to the required final concentrations for the respective experiments. The culture medium and substrates were autoclaved and placed in the Don Whitley anaerobic workstation while hot. The solutions were left in the anaerobic workstation overnight for equilibration prior to inoculation the next day. Cells that were cultured to the mid-exponential growth phase in liquid culture media were used to inoculate the fermentation broth at 10% (volume by volume) (v/v). The culture universal bottles were capped and taped down with parafilm for anaerobic fermentation at 37 °C and 200 rpm. Samples were taken at regular intervals to determine culture growth, product formation and substrate consumption.

2.3. Analytical Methods

Culture growth was quantified by optical density (OD) measurement at 600 nm using the UV Mini 1240 Shimadzu UV-Vis Spectrometer. Colony forming units (CFU) were used to determine culture growth when AC was used in the experiments in order to avoid interference in the OD measurement caused by fines generated from the AC during agitation. CFU was determined by diluting the cell sample and spreading on agar plates. The agar plates were incubated at 37 °C overnight and the number of colonies was counted. The values of CFU/ml were then calculated, taking into account the dilution factors used in the experiments.

The concentrations of *p*-cresol and HPA were quantified using the High Performance Liquid Chromatography (HPLC) (Agilent 1200 HPLC) with an Agilent Eclipse XDB-C18 column, a detector at 270 nm, and a temperature controller at 25 °C. The mobile phase containing 70% (v/v) acetonitrile, 30% (v/v) water and 0.5% (v/v) acetic acid was used at 0.4 ml/min. The samples were taken from the fermentation broth and were centrifuged repeatedly at 6000 rpm for 10 min at 4 °C until no AC fines were left in the sample. 1 µl of the supernatant was injected into the HPLC for analysis. Samples that were not analysed immediately were frozen down to -20 °C and were thawed prior to analysis. Internal standards containing 20 µg/ml of nitrobenzene were used in the sample analysis.

The production of *p*-cresol in the fermentation broth was confirmed by qualitative analysis using the Agilent 7890A Gas Chromatography (GC) coupled to the Agilent 5975C Mass Spectrometer Detector (MSD). A HP5MS column with a N_2 mobile phase at 0.5 ml/min was used for the analyses. The GC program was set at 70 °C for 2 min, followed by a 10 °C/min ramp to 325 °C, and was held at 325 °C for 10 min. The samples were centrifuged at 6000 rpm for 10 min at 4 °C from which 3 mL of the supernatant was acidified with 0.2 mL of 0.1 M HCl. The acid-treated samples were extracted with an equal volume of dichloromethane and the organic phase was injected into the GC-MSD for analysis.

2.4. Recovery of *p*-cresol from the Fermentation Broth Using Activated Carbon

The AC used for the experiments were obtained from Sigma Aldrich. C3014 was wetted and equilibrated with deionised water at 1 g per 10 ml of water in the Certomat BS-1 Sartorius orbital shaker at 200 rpm and 37 °C for 24 h before autoclave. The sterilised AC was then neutralised to pH 7 with sterile 0.1 M HCl before it was washed to remove fines. The AC was then coated with Dextran, based on a modification of the method described by [37]. The coating solution consisted of 0.25 M sucrose, 1.5 mM $MgCl_2$, 10 mM HEPES, and 0.05% (weight by volume) (w/v) Dextran T-70 obtained from Sigma Aldrich. The coating was performed at 180 rpm and 30 °C for 3 days before the AC was washed and used for experiments.

2.5. Desorption of *p*-cresol from the Activated Carbon Using Ultrasonic-assisted Solvent Extraction

Solvent extraction was used to recover the *p*-cresol concentrated on the AC. Ethanol was selected as the extraction solvent based on its miscibility and good solubility with many chemicals including those containing the hydroxyl functional group. The extraction was performed batch wise with incremental volumes of 5 ml of solvent for the first 4 cycles of extraction, followed by 20 ml each for the subsequent 6 more cycles of extraction. The saturated AC was immersed with the extraction solvent in glass tubes, and was heated to 50 °C in the U300H Ultra Wave Heated Ultrasonic Bath before sonication at 44 kHz for 15 min. 1.2 ml of sample was taken from the extraction solvent and was analysed for *p*-cresol using the HPLC. The % cumulative recovery was calculated based on the amount of *p*-cresol extracted into the solvent with respect to the original amount of *p*-cresol pre-adsorbed on the AC.

3. RESULTS AND DISCUSSION

The *hpd BCA* operon was cloned from *C. difficile* and expressed in *E. coli*, where the expressed *p*-hydroxyphenylacetate decarboxylase was shown to convert HPA into *p*-cresol under anaerobic conditions (Fig. **1**). The production of *p*-cresol was correlated with the depletion of HPA in the fermentation broth, and reached a maximum concentration of 1 ± 0.1 mM with the EM clone. The CM clone showed a lower product concentration compared to the EM clone, and hence it was not used in subsequent experiments. Toxicity of *p*-cresol was observed to take effect on the cells at 5 mM of spiked *p*-cresol concentration as shown in Fig. (**2**). This toxicity was hypothesised to have caused the low maximum

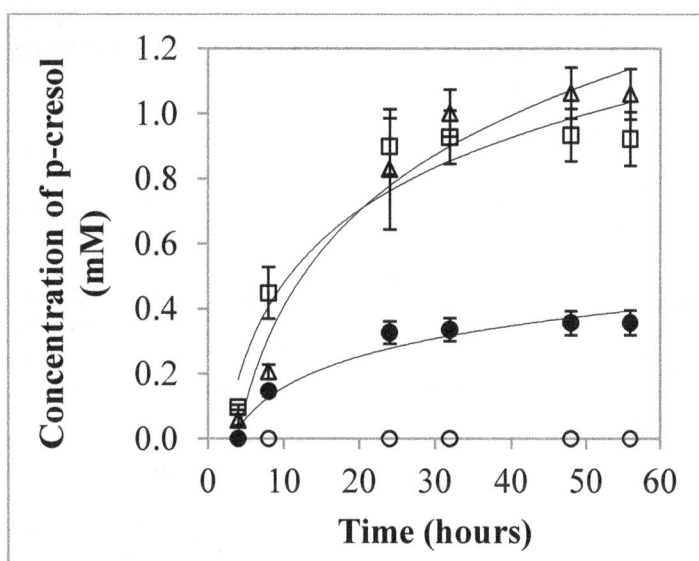

Fig. (1). Production of *p*-cresol by EM clone using various concentrations of HPA without AC. HPA was used at concentrations of: 0 mM (O), 10 mM (□), 50 mM (△) and 90 mM (•).

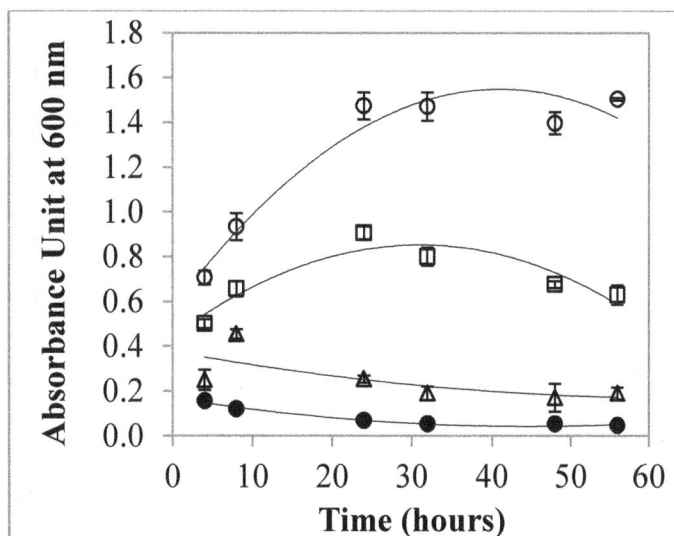

Fig. (2). Toxicity test of *p*-cresol on EM clone at various concentrations of *p*-cresol without HPA and without AC. *P*-cresol was used at concentrations of: 0 mM (O), 2 mM (□), 5 mM (△) and 10 mM (•).

concentration of *p*-cresol attainable in the fermentation as shown in Fig. (**1**).

Granular AC was used in the ISPR strategy to explore the possibility of increasing the productivity of *p*-cresol by removing the accumulation of the toxic product from the fermentation broth. The AC used in these experiments was commercially supplied to treat aqueous solutions, with applications in serum purification to remove steroid hormones. Since the AC was supplied to treat serum, it was postulated to be suitable for microbial cell culture without causing poisoning to the cells. Dextran T-70 was used to coat the AC in order to prevent any possible competitive adsorption imposed by nutrients in the fermentation broth. Experiment was conducted to determine whether there was any effect attributed by the coating on the adsorption of *p*-cresol.

Batch wise adsorption was performed at various concentrations of *p*-cresol in order to determine the performance characteristics and the adsorption capacities. A 500-mM concentration of *p*-cresol was used in the adsorption study, notwithstanding this concentration was above its solubility limit. This was to ensure that the AC has reached saturation during the determination of the adsorption capacity. A 3-phase mixture including a solid phase with 2 aqueous and organic phases was formed at the onset of the experiment with 500 mM of *p*-cresol. The 3-phase mixture became a 2-phase mixture as the adsorption progressed and as the AC adsorbed the *p*-cresol from the liquid phase. The adsorption of *p*-cresol by the AC was observed to be very rapid when 5 and 148 mM of *p*-cresol were used respectively. All of the *p*-cresol in the initial 5 mM concentration was taken up by the AC within the first 5 h of experiment. The adsorption capacities of both dextran-coated and non-coated AC were 2.6 ± 0.088 mmol/g of AC, where the Dextran coating showed negligible effect on the adsorption of *p*-cresol. The AC used in this study was shown to have comparable, if not higher, adsorption capacity compared to those of other adsorbents shown in Table **1**.

Solvent extraction assisted with ultrasonication at low frequency was used to recover *p*-cresol from the AC. The strategy was to exploit the cleaning function of ultrasonica-tion to desorb *p*-cresol from the spent AC. Fig. (**3**) shows ethanol extraction of *p*-cresol with 10 batches of fresh extraction solvent. Extraction cycles 1 to 4 were done with 5 ml of ethanol each, while cycles 6 to 10 were done with 20 ml of the solvent each. 85% of the adsorbed *p*-cresol was shown to be recoverable into the extraction solvent at cycle 6, with a total of 60 ml of solvent used. It was hypothesised that strong interaction of *p*-cresol in the micro pores of the AC could have caused the remaining 15% of *p*-cresol to be non-desorbable.

Figs. (**4** and **5**) show the culture growth curve and the *p*-cresol production curve respectively for the EM clone. The cultures were inoculated with 10% (v/v) inoculum and were fed with 30 and 50 mM HPA respectively. Fermentation broth spiked with 5 mM of *p*-cresol (predetermined to be the lowest concentration of *p*-cresol causing observable cellular toxicity as shown in Fig. **2**), was used with AC to determine whether the presence of AC would mitigate toxicity and enable culture growth. Fermentation broth with AC but without HPA and without *p*-cresol was also used as experimental control to determine whether there was possible toxicity caused by the AC itself.

From the experimental results, the presence of AC in the fermentation broth did not cause toxicity to the cells. On the contrary, the expansion ratio of cells with AC but without HPA and without *p*-cresol was higher at 30.7 times as shown in Fig. (**4**), compared to 10 times when cells were cultured without AC, without HPA and without *p*-cresol as shown in Fig. (**2**). The expansion ratio of 30.7 times with AC in Fig. (**4**) was calculated by dividing the maximum CFU at 1.87×10^9 with the CFU at inoculation at 6.10×10^7, both normalised on per ml basis. The expansion ratio without AC in Fig. (**2**) was calculated by dividing the maximum OD at 1.50 with the OD at inoculation at 0.15, giving an expansion ratio of 10 times. In addition, the specific growth rate of cells with AC but without HPA and without *p*-cresol was observed to be highest at 0.12 /h; compared to the growth rate of cells cultured without AC, without HPA and without *p*-cresol at 0.034 /h; and the growth rate of cells cultured without AC,

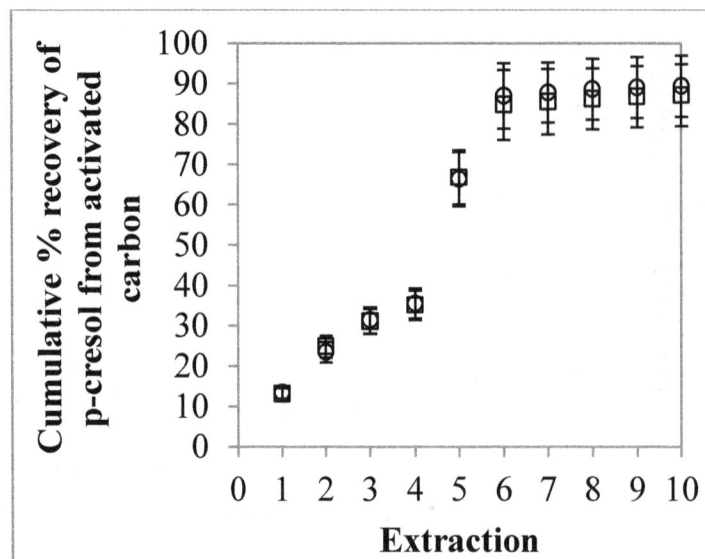

Fig. (3). Ultrasonic-assisted ethanol extraction of *p*-cresol from non-coated AC (O) and Dextran-coated AC (□).

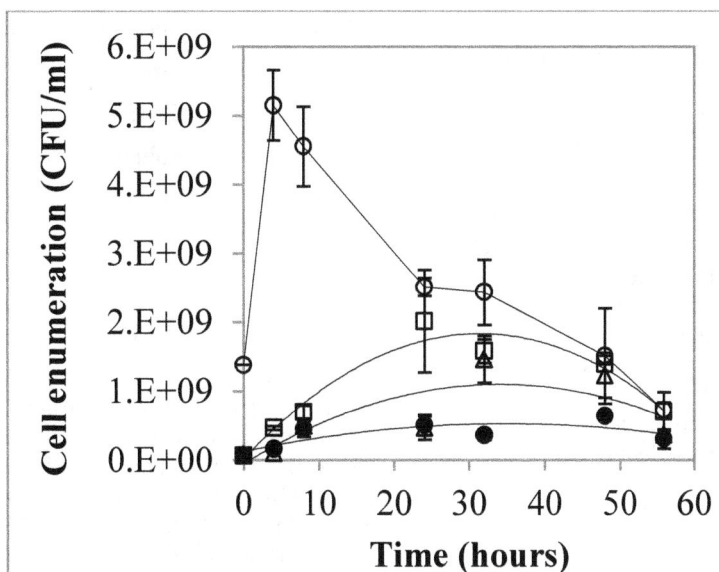

Fig. (4). Growth curves of EM clone at various concentrations of HPA with Dextran-coated AC. 5 mM of *p*-cresol was spiked into the fermentation broth as experimental control without HPA (O). Cells were cultured with various concentrations of HPA at: 0 mM as experimental control (□), 30 mM (△) and 50 mM (•).

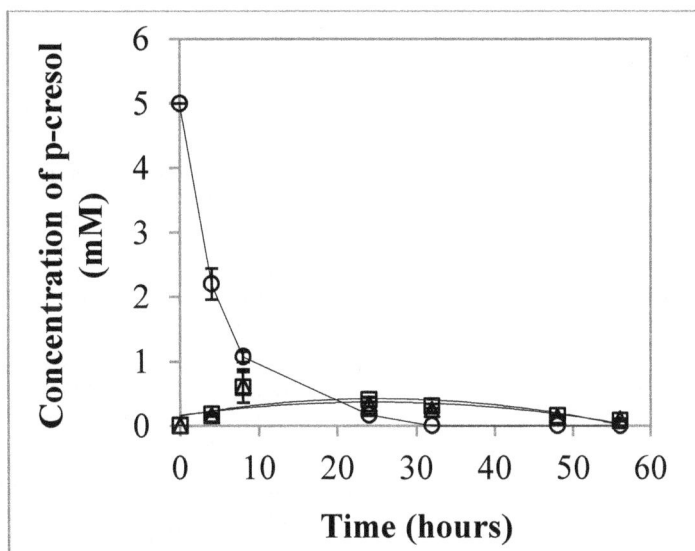

Fig. (5). Production of *p*-cresol by EM clone induced with various concentrations of HPA, and with Dextran-coated AC added to the fermentation broth. 5 mM of *p*-cresol was spiked into the fermentation broth as experimental control without HPA (O). Cells were cultured with various concentrations of HPA at: 30 mM (□) and 50 mM (△).

without HPA and with 2 mM *p*-cresol at 0.027 /h. Toxicity of *p*-cresol was observed to have reduced the specific growth rate of the cells, in agreement with the culture growth curves as shown in Fig. (**2**).

The presence of AC enabled the cells in the fermentation broth spiked with 5 mM of *p*-cresol to multiply, giving a higher CFU count of 4.79×10^9 CFU/ml compared to the initial 1.38×10^9 CFU/ml at inoculation as shown in Fig. (**4**). The concentration of *p*-cresol in the fermentation broth spiked with 5 mM of *p*-cresol was observed to reduce from the initial 5 mM to 0.58 mM by the 4th h, and further decreased to zero by the 32nd h as shown in Fig. (**5**). The removal of *p*-cresol from the fermentation broth by the AC was hypothesised to have reduced toxicity, thus enabled cul-

ture growth to reach a maximum CFU count of 4.79×10^9 CFU/ml at the 4th h (Fig. **4**).

Product removal *in-situ* fermentation by AC was observed in this study to have mitigated product toxicity and allowed culture growth and continuous production of *p*-cresol as shown in Figs. (**4 and 5**). *P*-cresol produced by the EM clone was found to be 3605 µg from 50 mM of HPA in 20 ml of fermentation broth added with AC. 3396 µg of the total of 3605 µg of *p*-cresol was recovered from the adsorbent during solvent extraction, while the remaining 209 µg of *p*-cresol was detected in the fermentation broth at the time of sampling. On the other hand, 2386 µg of *p*-cresol was quantified in the fermentation broth without AC. This showed a

51% increase in *p*-cresol production was possible when AC was used to remove *p*-cresol from the fermentation broth. The prevention of the accumulation of the toxic product in the broth by ISPR allowed the cells to survive and continue production, thus enabling a higher yield to be achieved.

The ISPR strategy demonstrated here in batch fermentation constituted a methodology that could potentially be adapted into a continuous process. As the toxic product was removed from the broth, production could be changed from a product-inhibited to a substrate-limited process, thus enabling higher productivity to be feasible for a product that could otherwise be toxic to the cells. As the AC was produced from renewable materials, and was shown to be recyclable by solvent extraction, the ISPR methodology shown here constituted a production process that was low-cost and sustainable.

4. CONCLUSIONS

E. coli cloned with the *hpd BCA* operon from *C. difficile* was shown in this study to be able to express *p*-hydroxyphenylacetate decarboxylase and converted the HPA substrate into *p*-cresol under strict anaerobic fermentation. Product toxicity of *p*-cresol at a concentration as low as 5 mM determined in a separate experiment spiked with *p*-cresol was found to have limited the maximum product concentration at 1 ± 0.1 mM. Product yield was however observed to increase by 51% when AC was used to remove *p*-cresol *in-situ* fermentation. The adsorption capacity for *p*-cresol was determined at 2.6 ± 0.088 mmol/g AC. The accumulated *p*-cresol on the AC was shown to be recoverable by subsequent ultrasonic-assisted ethanol extraction. The decoupling of the solvent extraction process from fermentation via adsorption allowed a solvent with a lower boiling point to be used for product recovery. The ISPR strategy demonstrated here constituted a sustainable production process that could mitigate product inhibition and thus improve product yield.

ABBREVIATIONS

AC	=	activated carbon
C. difficile	=	*Clostridium difficile*
CFU	=	colony forming unit
CM	=	positive clone with chloramphenicol resistance
E. coli	=	*Escherichia coli*
EM	=	positive clone with erythromycin resistance
GC	=	gas chromatography
GRAS	=	Generally Recognised As Safe
GRE	=	glycyl radical enzyme
HPA	=	*p*-hydroxyphenylacetate
Hpd	=	*p*-hydroxyphenylacetate decarboxylase
HPLC	=	High Performance Liquid Chromatography
ISPR	=	*in-situ* product recovery

MSD	=	mass spectrometer detector
OD	=	optical density
TB	=	Terrific Broth
v/v	=	volume by volume
w/v	=	weight by volume

ACKNOWLEDGEMENTS

The authors are grateful to Professors N. Minton and K. Winzer of the Clostridium Group, University of Nottingham, for the bacterial cells, the consumables for the cloning work, Dr K. Schwarz for the discussions on the cloning work, Professor Gill Stephens for the discussion on the fermentation work, and the Department of Chemical and Environmental Engineering for the support and consumables for the fermentation, *in-situ* product recovery, product analysis and quantification work.

REFERENCES

[1]   Yu L, Blaser M, Andrei PI, Pierik AJ, Selmer T. 4-hydroxyphenylacetate decarboxylases: properties of a novel subclass of glycyl radical enzyme systems. Biochemistry 2006; 45: 9584-92.

[2]   Selmer T, Andrei PI. *p*-hydroxyphenylacetate decarboxylase from *Clostridium difficile*. Eur J Biochem 2001; 268: 1363-72.

[3]   Deneve C, Janoir C, Poilane I, Fantinato C, Collignon A. New trends in *Clostridium difficile* virulence and pathogenesis. Int J Antimicrob Agents 2009; 33(S1): S24-8.

[4]   Ozkan P, Sariyar B, Utkur FO, Akman U, Hortacsu A. Metabolic flux analysis of recombinant protein overproduction in *Escherichia coli*. Biochem Eng J 2005; 22: 167-95.

[5]   Hafiz S, Oakley CL. *Clostridium difficile*: isolation and characteristics. J Med Microbiol 1976; 9: 129-37.

[6]   Martins BM, Blaser M, Feliks M, Ullmann GM, Buckel W, Selmer T. Structural basis for a Kolbe-type decarboxylation catalysed by a glycyl radical enzyme. J Am Chem Soc 2011; 133: 14666-74.

[7]   Dawson LF, Donahue EH, Cartman ST, *et al*. The analysis of para-cresol production and tolerance in *Clostridium difficile* 027 and 012 strains. BMC Microbiol 2011; 11: 86.

[8]   Khan NS, Mishra IM, Singh RP, Prasad B. Modeling the growth of *Corynebacterium glutamicum* under product inhibition in L-glutamic acid fermentation. Biochem Eng J 2005; 25: 173-8.

[9]   Tian Y, Kasperski A, Sun K, Chen L. Theoretical approach to modelling and analysis of the bioprocess with product inhibition and impulse effect. Biosystems 2011; 104: 77-86.

[10]   Malinowski JJ. Two-phase partitioning bioreactors in fermentation technology. Biotechnol Adv 2001; 19(7): 525-38.

[11]   Wang Z, Dai Z. Extractive microbial fermentation in cloud point system. Enzyme Microb Technol 2010; 46: 407-18.

[12]   Heerema L, Wierckx N, Roelands M, *et al*. *In situ* phenol removal from fed-batch fermentations of solvent tolerant *Pseudomonas putida* S12 by pertraction. Biochem Eng J 2011; 53(3): 245-52.

[13]   Trivunac K, Stevanovic S, Mitrovic M. Pertraction of phenol in hollow-fiber membrane contactors. Desalination 2004; 162: 93-101.

[14]   Sardessai YN, Bhosle S. Industrial potential of organic solvent tolerant bacteria. Biotechnol Prog 2004; 20(3): 655-60.

[15]   Yang ST. Bioprocessing-from biotechnology to biorefinery. In: Bioprocessing for Value-Added Products from Renewable Resources Amsterdam: Elsevier 2007; pp. 1-24.

[16]   Leon R, Martin M, Vigara J, Vilchez C, Vega JM. Microalgae mediated photoproduction of β-carotene in aqueous-organic two phase systems. Biomol Eng 2003; 20: 177-82.

[17]   Rodriguez-Martinez MF, Kelessidou N, Law Z, Gardiner J, Stephens G. Effect of solvents on obligately anaerobic bacteria. Anaerobe 2008; 14(1): 55-60.

[18]   Yabannavar VM, Wang DIC. Strategies for reducing solvent toxicity in extractive fermentations. Biotechnol Bioeng 1991; 37: 716-22.

[19]   Guillot S, Kelly MT, Fenet H, Larroque M. Evaluation of solid-phase microextraction as an alternative to the official method for the analysis of organic micro-pollutants in drinking water. J Chromatogr A 2006; 1101: 46-52.

[20]   Zhang W, Ni Y, Sun Z, et al. Biocatalytic synthesis of ethyl (R)-2-hydroxy-4-phenylbutyrate with Candida krusei SW2026: a practical process for high enantiopurity and product titer. Process Biochem 2009; 44(11): 1270-5.

[21]   Cardoso VM, Solano AGR, Prado MAF, Nunan EdA. Investigation of fatty acid esters to replace isopropyl myristate in the sterility test for ophthalmic ointments. J Pharm Biomed Anal 2006; 42: 630-4.

[22]   Embree HD, Chen T, Payne GF. Oxygenated aromatic compounds from renewable resources: motivation, opportunities, and adsorptive separations. Chem Eng J 2001; 84: 133-47.

[23]   Nielsen DR, Amarasiriwardena GS, Prather KLJ. Predicting the adsorption of second generation biofuels by polymeric resins with applications for in situ product recovery (ISPR). Bioresour Technol 2010; 101: 2762-9.

[24]   Berge-Lefranc D, Vagner C, Calaf R, et al. In vitro elimination of protein bound uremic toxin p-cresol by MFI-type zeolites. Microporous Mesoporous Mater 2012; 153: 288-93.

[25]   Hadjar H, Hamdi B, Ania CO. Adsorption of p-cresol on novel diatomite/carbon composites. J Hazard Mater 2011; 188: 304-10.

[26]   Anbia M, Ghaffari A. Adsorption of phenolic compounds from aqueous solutions using carbon nanoporous adsorbent coated with polymer. Appl Surf Sci 2009; 255: 9487-92.

[27]   Huang J. Treatment of phenol and p-cresol in aqueous solution by adsorption using a carbonylated hypercrosslinked polymeric adsorbent. J Hazard Mater 2009; 168: 1028-34.

[28]   Liu FQ, Xia MF, Yao SL, Li AM, Wu HS, Chen JL. Adsorption equilibria and kinetics for phenol and cresol onto polymeric adsorbents: effects of adsorbents/adsorbates structure and interface. J Hazard Mater 2008; 152: 715-20.

[29]   Ahmaruzzaman M. Adsorption of phenolic compounds on low-cost adsorbents: a review. Adv Colloid Interface Sci 2008; 143: 48-67.

[30]   Singh RK, Kumar S, Kumar S, Kumar A. Development of parthenium based activated carbon and its utilisation for adsorptive removal of p-cresol from aqueous solution. J Hazard Mater 2008; 155: 523-35.

[31]   Wernert V, Schaf O, Faure V, et al. Adsorption of the uremic toxin p-cresol onto hemodialysis membranes and microporous adsorbent zeolite silicalite. J Biotechnol 2006; 123: 164-73.

[32]   Ayranci E, Duman O. Adsorption behaviours of some phenolic compounds onto high specific area activated carbon cloth. J Hazard Mater 2005; B124: 125-32.

[33]   Jain AK, Bhatnagar S, Bhatnagar A. Methylphenols removal from water by low-cost adsorbents. J Colloid Interface Sci 2002; 251: 39-45.

[34]   Li A, Zhang Q, Zhang G, Chen J, Fei Z, Liu F. Adsorption of phenolic compounds from aqueous solutions by a water-compatible hypercrosslinked polymeric adsorbent. Chemosphere 2002; 47: 981-9.

[35]   Heap JT, Pennington OJ, Cartman ST, Minton NP. A modular system for Clostridium shuttle plasmids. J Microbiol Methods 2009; 78: 79-85.

[36]   Cartman ST, Kelly ML, Heeg D, Heap JT, Minton NP. Precise manipulation of the Clostridium difficile Chromosome reveals a lack of association between the tcdC genotype and toxin production. Appl Environ Microbiol 2012; 78(13): 4683-90.

[37]   Oakey RE, Ed. Steroid hormones: A practical approach. Oxford: IRL Press, 1988; p. 278.

7

Analysis Fatty Acids Profile in *Tabanus bivittatus* Mats with Gas Chromatography-Mass Spectrometry

Wang Yanhua[*], Wu Fuhua, Guo Zhaohan, Peng Mingxing, Xia Min, Pang Zhenling, Wang Xiaoli, Liang Zian and Zhang Naiqun

Life Science and Technology College, Nanyang Normal University, Henan Province, China

Abstract: *Tabanus bivittatus* Mats., a traditional Chinese medicine, is commonly used for cardiovascular disorders treatment including atherosclerosis. There have been only a few researches on its chemical components, and no detailed report has appeared on its fatty acids. To develop a simple and effective method for the extraction of total fatty acids from *Tabanus bivittatus* Mats., the Soxhlet extraction (SE) condition was optimized with response surface methodology. The fatty acid composition of the extract were determined by GC-MS with previous derivatization to fatty acid methyl esters (FAMEs). The major fatty acids in *Tabanus bivittatus* Mats. were oleic acid, palmitic acid, linoleic acid, palmitoleic acid, and stearic acid, and the unsaturated fatty acids occupy 63.9% of the total fatty acids.

Keywords: Fatty acid, gas chromatography, mass spectrometry, response surface methodology, *Tabanus bivittatus Mats.*

1. INTRODUCTION

As a traditional Chinese medicine , *Tabanus bivittatus* Mats. is known to have a regulatory property for blood circulation and inflammatory disease [1, 2]. This natural product is one of the major drugs used in the Chinese traditional medicine Da Huang Zhe Chong pill to treat hepatic cirrhosis [3], atherosclerosis, and menstrual disturbance [4]; these actions are probably produced by pharmacodynamic activity on vascular system. Up to now, there have been only a few researches on the chemical componets of *Tabanus bivittatus* Mat. [5], but no detailed report has appeared on its fatty acid components.

A variety of chromatographic techniques have been employed in the analysis of fatty acids, such as high performance liquid chromatography [6, 7], gas chromatography[8, 9]. A way to decrease the limit of quantification and provide a higher level of information is to use a gas chromatograph (GC) equipped with an MS detector, which allows quantification of each individual compound [10].

Soxhlet extraction (SE) is a classical method for decades in extraction of organic compounds from solid sample, and this apparatus has been developed to several types for special used. It is considered to be a "thorough" extraction method because th organic phase cooled from condensation tube continuously passed through the target solid sample for hours. Therefore it is a popular technique to analysis the minor composition file of the solid sample although high organic solvent volume, extensive extraction time, and intensive manpower are required [11].

In this paper, SE followed by GC-MS was developed for the rapid analysis of fatty acids in Tabanus bivittatus Mats.. The experimental parameters were optimized with RSM in order to obtain the greatest extraction yield of fatty acids, and the fatty acid compositions of extract were determined by GC–MS with previous derivatisation to FAMEs.

2. EXPERIMENT DESIGN OF THE OPTIMIZATION OF EXTRACTION CONDITION WITH RSM

Response surface methodology (RSM) is a collection of statistical and mathematical techniques useful for developing, improving, and optimizing processes in which a response of interest is influenced by several variables [12], and has been widely applied to the optimization of the extraction procedureparameters [13, 14].

A three-level-three-variable Box–Behnken design (BBD) was adopted to optimize the extraction procedure. The independent variables were extractant volume (V, mL), extraction time (T, min), and ratio of acetone and petroleum ether (R). Three levels of each variable were coded as -1, 0, and +1 (Table 1). The extraction yield, represented as the weight of oil extracted form *Tabanus bivittatus* Mats., was taken as response, Y. A regression analysis was carried out in order to fit the experimental data into an empirical second-order polynomial model.

The variables and levels for each variable in BBD were determined according to the results of preliminary experiments. Based on the experimental results of the BBD (Table 1), the extraction yield followed a second-order polynomial model.

Where Y represented the extraction yield, calculated by the weight of oil extracted form *Tabanus bivittatus* Mats.; V, R and T correspond to three independent variables,

---

*Address correspondence to this author at the School of Life Science and Technology, Nanyang Normal University, Nanyang, Henan,473000, P.R. China; E-mail: wanga_yanhua@163.com

Table 1. Box-Behnken design and experiment results for the optimization of SE

Run	Petroleum ether /acetone(R)	Extraction volume (V, ml)	Extraction time (t, h)	Response (extraction yield)	
				Observed	Predicted
1	1(-1)	50(0)	2(+1)	0.43	0.38
2	9(+1)	50(0)	0.3(-1)	0.48	0.44
3	5(0)	50(0)	1.15(0)	0.59	0.49
4	9(+1)	80(+1)	1.15(0)	0.52	0.51
5	5(0)	50(0)	1.15(0)	0.56	0.40
6	5(0)	80(+1)	2(+1)	0.63	0.44
7	1(-1)	20(-1)	1.15(0)	0.37	0.47
8	5(0)	20(-1)	0.3(-1)	0.41	0.51
9	1(-1)	80(+1)	1.15(0)	0.5	0.44
10	5(0)	50(0)	1.15(0)	0.61	0.50
11	5(0)	20(-1)	2(+1)	0.54	0.49
12	5(0)	50(0)	1.15(0)	0.58	0.60
13	5(0)	80(+1)	0.3(-1)	0.45	0.58
14	1(-1)	50(0)	0.3(-1)	0.44	0.58
15	9(+1)	20(-1)	1.15(0)	0.43	0.58
16	9(+1)	50(0)	2(+1)	0.47	0.58
17	5(0)	50(0)	1.15(0)	0.58	0.58

$$Y=0.126+0.066P+5.561V+0.139T-8.33E-005P*V+0.00P*T+4.901E-004V*T-5.671E-003P2-4.250E-005V2-0.529T2$$

extractant volume (mL, for 10.0 g sample), ratio of acetone and petroleum ether, and extraction time (min).

The statistical analysis of our model variance and experimental results was included in Table 2. The F-value of our model is 3.96 which is significant. The chance that a 'Model F-Value' this large could occure due to noise was only 4.15%. The linear terms of extractant volume, and quadratic terms of proportion of acetone and petroleum ether

had statistically significant effects on extraction yield. This is indicated by the B, A2 were significant model terms ($p < 0.05$). The coefficient determination in our model was 0.8360.

The best way to visualize the influence of independent variables on the dependent one is to draw a surface response plot of the model. With a fixed third independent variable at the central experimental level of zero, Fig. (1) showed the

Table 2. Analysis of variance (ANOVA) for the model

Source	Sum of squares	df[a]	Mean square	F Value	p-value
Model	0.081	9	9.031E-003	3.96	0.0415[b]
R	3.200E-003	1	3.200E-003	1.40	0.2746
V	0.015	1	0.015	6.72	0.0358[b]
t	0.011	1	0.011	4.62	0.0688
RV	4.000E-004	1	4.000E-004	0.18	0.6877
Rt	0.000	1	0.000	0.000	1.0000
Vt	6.250E-004	1	6.250E-004	0.27	0.6166
R^2	0.035	1	0.035	15.22	0.0059[b]
V^2	6.160E-003	1	6.160E-003	2.70	0.1441
T^2	6.160E-003	1	6.160E-003	2.70	0.1441
Residual	0.016	7	2.278E-003		
Lack of fit	0.015	3	4.875E-003	14.77	0.0125
Pure error	1.320E-003	4	3.300E-004		
Cor total	0.097	16			
R-Squared	0.8360			C.V.%	9.45

a df, degree of freedom

b Significance, $p < 0.05$

effect and interaction of two independents on the responding variable, extraction yield. The extraction yield was increased with the increase of extractant volume up to about 70.0 ml, and then decreased. In contrast, the effects of ratio of acetone and petroleum ether and extraction time were not as important as the extractant volume. These results revealed that the extraction yield of total fatty acids was depended more on extractant volume than on ratio of acetone and petroleum ether and extraction time.

3. IMPACT OF EXTRACTING TIMES

The influence of multiple extracting was explored on the total extraction rate of fatty acids at the optimized condition. Fig. (2). showed that the total extraction rate of fatty acids increased with the extracting times, and the fatty acids were extracted entirely after 5 times. Considering the extraction cost and the total extraction rate increasing slightly over 3 times, extraction 3 times was selected as the optimum condition, at which the rate was 98.9% of that of 5 times.

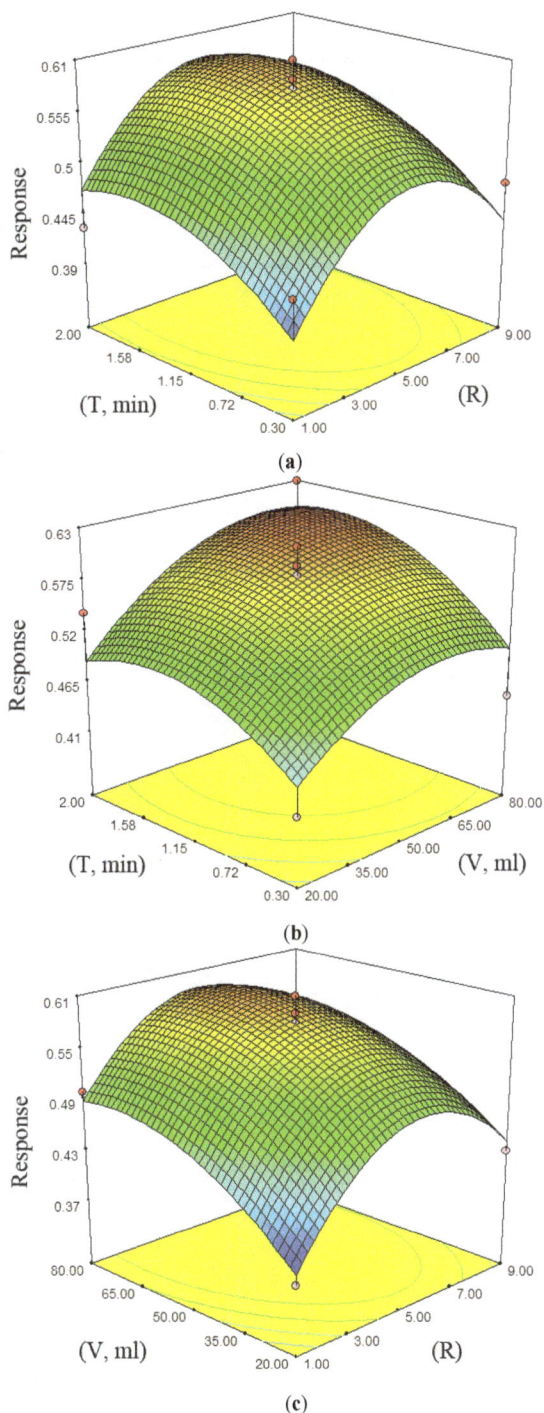

Fig. (1). (**a**) Response surface for the effects of extract time (T, min) and ratio of acetone and petroleum ether (R) at extractant volume of 50 ml on the extraction yield. (**b**) Response surface for the effects of extract time (T, min) and extractant volume (V, mL) at ratio of acetone and petroleum ether (R) of 5 on the extraction yield. (**c**) Response surface for the effects of extractant volume (V, mL) and ratio of acetone and petroleum ether (R) at extractant time 75 min on the extraction yield.

4. SYSTEM ANALYSIS OF CERAMIC DESIGN SYSTEM

After soxhlet extraction, the fatty acids in the extract were derivatized to FAMEs with sulphuric acid–methanol

complex, and then analyzed with GC-MS. Complex, and then analyzed with GC-MS. The analytical results were listed in Table **3**. The relative contents were calculated by using the area normalization method, without considering response factors. Twenty-one kinds of fatty acids in the SE extract were identified and the sum of identified fatty acids occupied 95.95% of the area normalized. It can be seen from Table **3**. that *Tabanus bivittatus* Mats. is remarkably rich in oleic acid (33.38%) and palmitic acid (20.96%) followed by linoleic acid (19.03%) , palmitoleic acid (7.59%), stearic acids (7.53%). the unsaturated fatty acids occupy 63.92% of the total fatty acids. This result is consistent with the regulatory property of *Tabanus bivittatus* Mats. for blood circulation. Researches have identified that unsaturated fatty acids are effective in preventing cardiovascular events, cardiac death and coronary events, especially in persons with high cardiovascular risk. The high content of unsaturated fatty acids in Tabanus bivittatus Mats maybe the reason for it was used in cardiovascular disease treatment in China.

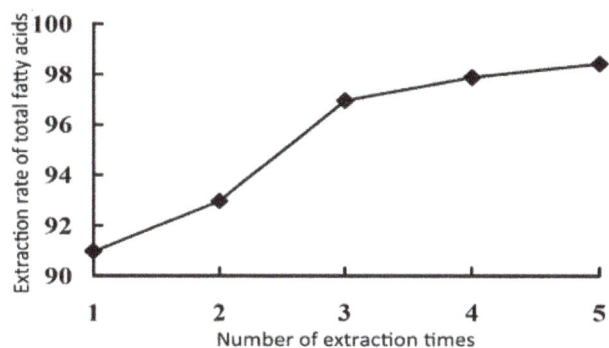

Fig. (2). Effect of extraction times on extraction rate of total fatty acids

5. Analysis of Fatty Acids in *Tabanus bivittatus* Mats.

The current ceramic enterprises in ceramic product design process and production process of detailed study, the design process of ceramic products for further analysis and decomposition, Exploring the project design of ceramic needs human interaction steps in the 3d CAD system, and needs to be done by the system automatically in order to better improve enterprise's key steps in the efficiency of product design and in the decomposition on the basis of the design process to classify modeling of ceramic products, and finished components decomposition of complex products. Ceramic products decomposition is different from the mechanical parts and components industry and other industry products, ceramic products decomposition lies mainly in the design of the components in the process of decomposition, and the final product in general is not an integral whole, for example, in the design process for more complex products such as "pot" will be broken down parts into the pot body, the pot, a spout, the lid and lid knob, a girder of the pot, an ear piece [10]. And set up all kinds of parts in the three-dimensional model of the material library.

CONCLUSION

To extract fatty acid of Tabanus bivittatus Mats., the best condition including extractant volume of 69.71 mL, ratio of acetone and petroleum ether of 5.3 and extraction time of

Table 3.Analytical results of fatty acids in *Tabanus bivittatus* Mats. By GC-MS(n=3)

Retention Time(min)	Fatty acid	Molecular formula	Relative Molecular weight	Relative content
3.49	undec-10-enoic acid	$C_{11}H_{20}O_2$	184	0.22
3.86	Lauric acid	$C_{12}H_{24}O_2$	198	0.07
4.62	Tridecylic acid	C14H28O2	228	0.06
4.73	Tetradecanoic acid	C14H28O2	228	0.55
5.14	i-Pentadecanoic acid	C15H30O2	242	0.20
5.32	Pentadecanoic acid,	C15H30O2	242	0.42
5.77	iso-hexadecenoic acid	C16H32O2	250	0.32
5.91	palmitoleic acid	C16H30O2	254	7.59
5.98	(Z)-9-Hexadecenoic acid	C15H30O2	254	1.13
6.07	palmitic acid	C16H32O2	256	20.96
6.94	margaric acid	C17H34O2	270	0.82
7.70	linoleic acid	C18H32O2	280	19.03
7.78	oleic acid	C18H34O2	282	33.38
7.92	Linolenic acid	C18H30O2	278	0.29
8.03	stearic acids	C18H36O2	284	7.53
8.53	11,14-Eicosadienoicacid	C20H36O2	308	0.06
8.99	(E)7- Nonadecenoic	C19H36O2	296	0.84
9.36	arachidic acid	C20H40O2	312	0.16

98.4 min was optimized using Response surface methodology. the unsaturated fatty acids occupy 63.92% of the total fatty acids.The analytical results of GC-MS showed that abundant unsaturated fatty acids such as oleic acid and linoleic acid accounted for 63.92% of total fatty acids.

ACKNOWLEDGEMENTS

The authors acknowledge the financial support provided by grants from the National Natural Science Foundation of China (No.11102092); Science and Technology Key Project of Education Department in Henan Province (No. 12B130003); and Biochemistry and Molecular Biology Key Discipline of Henan Province.

REFERENCES

[1]   Bailey-Shaw YA, Golden KD, Pearson AG, Porter RB. Characterization of Jamaican agro-industrial wastes. Part II, fatty acid profiling using HPLC: precolumn derivatization with phenacyl bromide. J Chromatogr Sci 2012; 50(8): 666-72.

[2]   D'Oca MG, Moron-Villarreyes JA, Lemoes JS, Costa CS. Fatty acids composition in seeds of the South American glasswort Sarcocornia ambigua. An Acad Bras Cienc 2012; 84(3): 865-70.

[3]   de Paiva SR, Lima LA, Figueiredo MR, Kaplan MA. Plumbagin quantification in roots of Plumbago scandens L. obtained by different extraction techniques. An Acad Bras Cienc 2004; 76(3): 499-504.

[4]   Li P, Lu S, Shan T, et al. Extraction optimization of water-extracted mycelial polysaccharide from endophytic fungus fusarium oxysporum dzf17 by response surface methodology. Int J Mol Sci 2012; 13(5): 5441-53.

[5]   Li X, Wang Z, Wang L, Walid E, Zhang H. Ultrasonic-assisted extraction of polysaccharides from Hohenbuehelia serotina by response surface methodology. Int J Biol Macromol 2012; 51(4): 523-30.

[6]   Liu N, Liu JT, Ji YY, Lu PP. Dahuang zhechong pill containing serum inhibited platelet-derived growth factor-stimulated vascular smooth muscle cells proliferation by inducing G1 arrest partly via suppressing protein kinase C alpha-extracellular regulated kinase 1/2 signaling pathway. Chin J Integr Med 2012; 18(5): 371-7.

[7]   Ma D, Li Y, Dong J, et al. Purification and characterization of two new allergens from the salivary glands of the horsefly, Tabanus yao. Allergy. 2011; 66(1): 101-9.

[8]   Ma D, Wang Y, Yang H, et al. Anti-thrombosis repertoire of blood-feeding horsefly salivary glands. Mol Cell Proteomics 2009; 8(9): 2071-9.

[9]   Ma D, Xu X, An S, et al. A novel family of RGD-containing disintegrins (Tablysin-15) from the salivary gland of the horsefly Tabanus yao targets alphaIIbbeta3 or alphaVbeta3 and inhibits platelet aggregation and angiogenesis. Thromb Haemost 2011; 105(6): 1032-45.

[10]   Maldonado RA, Kuniyoshi RK, Linss JG, Almeida IC. Trypanosoma cruzi oleate desaturase: molecular characterization and comparative analysis in other trypanosomatids. J Parasitol 2006; 92(5): 1064-74.

[11]   McNichol J, MacDougall KM, Melanson JE, McGinn PJ. Suitability of Soxhlet extraction to quantify microalgal Fatty acids as determined by comparison with in situ transesterification. Lipids 2012; 47(2): 195-207.

[12]   Naqvi AZ, Davis RB, Mukamal KJ. Dietary fatty acids and peripheral artery disease in adults. Atherosclerosis 2012; 222(2): 545-50.

[13]   Paiva SR, Lima LA, Figueiredo MR, Kaplan MA. Chemical composition fluctuations in roots of Plumbago scandens L. in relation to floral development. An Acad Bras Cienc 2011; 83(4): 1165-70.

[14]   Rao G. Optimization of ultrasound-assisted extraction of cyanidin 3-rutinoside from litchi (Lichi chinensis Sonn.) fruit pericarp. Anal Methods 2010; 2(8): 1166-70.

8

Analysis of the Ecological Sensitivity of Pengyang County Based on Key Factors

Shi Yun[*,1,2], Geng Sha1, Mi Wenbao[1,2] and Su Wei[1]

[1]School of Agriculture, Ningxia University, Yinchuan, Ningxia, 750021, P.R. China

[2]School of Resources and Environment, Ningxia University, Yinchuan, Ningxia, 750021, P.R. China

Abstract: Through selecting key factors such as gradient, exposure, erosive channel, soil erosion and vegetation coverage, the paper establishes the factor index system for the evaluation of the ecological system of Pengyang County, determines the weight of evaluation factors, makes analysis on key ecological factors, sets up the comprehensive sensitivity evaluation model of the ecological system, and obtains the ecological sensitivity grade, area, and spatial distribution of the study area through GIS spatial analysis function, factor superposition method, and analytic hierarchy process. The findings are as follows: the zone extremely sensitive to the ecology covers 82956.47 hm², accounting for 32.98%. Landforms in the area usually include steep slopes with rich vegetation, and high ecological value, erosive channels, and zones affected by erosive channels, all of which are under the key protection; the zone moderately sensitive to the ecology cover 104613.69 hm², accounting for 41.59%. Landforms in the area usually include forestlands of gentle relief and diversified plants. A good many factors should be taken into consideration in the development of the areas; the zone slightly sensitive to the ecology covers 63965.53 hm², accounting for 25.43%. Landforms in the area are usually of low elevation and flat, mainly ordinary greenbelt with singular vegetation, sections with poor vegetation, and farmlands. The area is resistant to human disturbance and able to go through development and construction of the specific intensity. The lands can be developed for a variety of purposes, but attention should be paid to the development intensity.

Keywords: Ecological sensitivity, ecological factors, Pengyang County.

1. INTRODUCTION

Ecological sensitivity analysis is to evaluate the possibility of a concrete ecological process giving rise to potential ecological and environmental problems under the natural status. Combining ecological sensitivity analysis methods such as ordinary analytic hierarchy process with GIS technology and utilizing GIS spatial data processing and computing capabilities can make the analysis process more concise and the analysis result more accurate, thus offering the reliable data support for related regions [1-3].

Taking Pengyang County located in the loess hilly and gully area as the example, and in the light of the region's research focus and objective conditions, the paper selects factors closely related to the ecological environment and able to produce data, especially natural factors, as key factors for evaluating the ecological sensitivity, including gradient, exposure, erosive channel, soil erosion, and vegetation coverage. The paper determines the weight of evaluation factors through analytic hierarchy process, constructs the comprehensive sensitivity evaluation model of the ecological system of Pengyang County according to GIS spatial analysis, and finally finds out the spatial distribution of different ecological sensitive areas, thus offering the basis for understanding the region's ecological environment status and making scientific ecological planning.

*Address correspondence to this author at the School of Agriculture, Ningxia University, Yinchuan, Ningxia, 750021, P.R. China;
E-mail: shiysky@163.com

2. THE STUDY AREA

Pengyang County is located at the southeast of Ningxia Hui Autonomous Region, with east longitude being 106° 32'-106° 58', north latitude 35° 41'-36° 17', and elevation 1,248-2,483m. The county is of the typical landform of the second sub-region of the loess hilly and gully area, mainly loess hills. The topography is shattered and disconnected, with steep hills and crisscross gullies [4]. The landform is composed of ridge, tableland, valley, and gully. Gully density reaches 2.4 km/km². Therefore, Pengyang County is one of the counties with the most serious water and soil erosion in Ningxia and also the key water and soil conservation zone in China.

3. STUDY FOUNDATION

3.1. Data Source and Processing

The paper takes 1:50,000 present land-use map (2010), 1:100,000 topographic map, soil erosion data, and statistical yearbook of Pengyang County as basic data for the study. First of all, it makes use of ArcGIS10.0 software to geometrically proofread the present land-use map, then rectifies the images to obtain the landscape data of the study area, and converts vector data into raster data. Raster data are defined as 25m × 25m. Via the analytic function of ArcGIS software, the paper produces the gradient and exposure raster figure with 1:100, 000 elevation model as data source. Raster data are defined as 30 m × 30 m.

3.2. Methodology

The research is to analyze the ecological sensitivity of the study area on the basis of GIS spatial analysis and weighted stacking method. In the light of original data collected and evaluation goal, the paper analyzes and selects key single factors for evaluating the ecological sensitivity of the study area, grades each single factor, draws out the single factor rating map [5-7], and determines the weight of each single factor through AHP. On the basis of DEM and via D8 algorithm, the paper automatically retrieves data of erosive channel, gradient, exposure and soil erosion in the northwestern loess plateau region, makes use of GIS spatial analysis methods such as raster data analysis, vector data analysis, and 3D spatial analysis to stack weights of key factors, and finally acquires the stacking result rating, namely the ecological sensitivity rating map, for the purposes of analyzing the spatial distribution of the ecological sensitivity of Pengyang County and further proposing related measures of ecological planning [8-11].

3.3. Selection of Key Factors for the Evaluation of Ecological Sensitivity

Ecological sensitivity analysis is to evaluate the possibility of a concrete ecological process giving rise to potential ecological and environmental problems under the natural status. The most serious ecological problem of Pengyang County finds expression in water and soil erosion and crisscross gullies. According to the implication of ecological sensitivity analysis, the paper selects related factors in terms of natural ecology, such as geology, landform, gradient, soil, hydrology, vegetation, biological diversity, climate, etc., and then distinguishes those resulting in water and soil erosion from the factors. Key factors for the evaluation of the ecological sensitivity of Pengyang County include gradient, exposure, erosive channel, soil erosion, vegetation coverage, etc.

In the light of the rating of the ecological sensitivity, the study area is divided into three rating zones in the evaluation, namely strong ecological sensitive zone, medium ecological sensitive zone, and slight ecological sensitive zone. In the light of the ecological sensitivity analysis, related ecological factors fall into different sensitivity value from 0-10..Rating scale of related ecological sensitive factors is shown as Table 1.

4. ANALYSIS ON GRADIENT FACTOR

Pengyang County is of varied landforms such as valley, ridge, and flat ground. Gradient factor is determined according to the provision of the Law of the People's Republic of China on Water and Soil Conservation: Reclamation of hillsides with a slope of over 25° for cultivation of crops shall be prohibited [13]. In addition, as a demonstrative county of returning farmland to forest nationwide, Pengyang County focuses on returning farmland in mountainous areas of gradient being above 15° and dry farmland in serious windy and sandy areas to forest and grassland with small watershed as the unit for scale treatment. The county combines basic construction of farmlands with structure adjustment of agriculture to return

and treat the farmland one by one. Therefore, farmlands in mountainous areas of gradient being above 15° are also covered in the scope of grain for green. Distribution of different gradients of Pengyang County is shown as the figure, and areas shown as Table 2.

4.1. Exposure Factor Analysis

As an important part of the microclimate, exposure has a bearing on the variety and distribution of vegetation and animals, lighting of buildings, and utilization efficiency of energies. As far as construction is concerned, south slope of a hill is superior to the north slope. The most favorable exposure is the south-by-east slope, where plants can receive sufficient lighting and ventilation, thus being favorable for their growth. Hence, exposure factor can be determined. Pengyang County's flat ground covers an area of 27,538.26 hm², mainly valley areas and southwest water areas as well as the surrounding areas, accounting for 10.95% of the total; true north region covers an area of 24,309.68 hm², accounting for 9.66% of the total; northeast and northwest regions cover an area of 54,393.65 hm², accounting for 21.62% of the total; true west and true east regions cover an area of 49,985.72 hm², accounting for 19.87% of the total; southeast and southwest regions cover an area of 59,187.60 hm², accounting for 23.53% of the total; and true south region covers an area of 36,120.78 hm², accounting for 14.36% of the total.

4.2. Channel Factor Analysis

Channel erosion is the major reason for water and soil loss in loess plateau. Since loess is vulnerable to the erosion of flowing water, there form a mass of widely distributed erosive channels [12]. Channel properties are also major factors affecting channel erosion, such as channel length, elevation difference, and vertical gradient, of which elevation difference can generate the biggest influence. In general conditions, the bigger the elevation difference is, the bigger the surface flow speed is, and the more serious channel erosion is. Therefore, the study decides on the ecological sensitivity on the basis of related channel properties, extracting catchment area of erosive channels with DEM as data source and processing properties to calculate elevation difference and vertical gradient. Channel vertical gradient is the key factor considered in the treatment of erosive channels. Generally speaking, the bigger the vertical gradient is, the more serious the erosion is.

According to the above findings, the paper carries out statistical analysis, and obtains related data such as channel length, elevation difference, and vertical gradient, and also rating distribution of different drainage areas in the light of catchment area of channels. Channel vertical gradient is the key factor considered in the treatment of erosive channels. Generally speaking, the bigger the vertical gradient is, the more serious the erosion is. Vertical gradients of different channels and the rating distribution of different drainage areas of Pengyang County are shown as Table 3.

4.3. Soil Erosive Factor Analysis

Soil erosion data are divided into five grades of minor, slight, medium, intense, extremely intense-severe in terms of

Table 1.   Rating scales of ecological sensitive factors.

Valuation Factor	Rating Standard	Ecological Factor Rating Scale Evaluation
Gradient/ (°)	0-8	0
	8-15	2
	15-25	4
	25-35	6
	>35	8
Exposure	Flat	0
	True north	2
	Northeast, northwest	4
	True west, true east	6
	Southeast, southwest	8
	True south	10
Channel factor (Channel vertical gradient) (%)	<1	0
	1-5	2
	5-10	4
	10-20	6
	>20	8
Soil erosion	Minor	0
	Slight	2
	Moderate	4
	Intense	6
	Extremely intense-severe	8
Vegetation coverage	High coverage	0
	Medium-high coverage	2
	Medium coverage	4
	Medium-low coverage	6
	Low coverage	8
	No coverage	10

Table 2.   Distribution of gradients of pengyang county (hm^2).

Gradient rating	0-8°	8-15°	15-25°	25-35°	>35°
Area	81953.25	78274.32	70181.90	18789.15	2337.07

erosion intensity, and area distribution of different soil erosion intensities is shown as Table **4**.

4.4. Vegetation Coverage Factor Analysis

Vegetation is the most important part of biological resources and one of the most important factors affecting the ecological sensitivity. Vegetation resources can be upgraded continuously in natural conditions or artificial maintenance; on the contrary, they may degrade or disappear in harsh environmental conditions or under artificial destruction or unreasonable utilization, and the process is sometimes irreversible. Vegetation resource distribution varies from region to region, so do their species composition and structural features. Areas of vegetation coverage are shown as Table **5**.

5. RESULTS AND ANALYSIS

5.1. Weight Determination

Using AHP, based on 5 experts to build judgment matrix of soil and water conservation, calculate the Exposure, Gradient and other factors Weight value , which are shown as Table **6**.

λ max=6.139 566;

IC. I=0. 027 913;

RC.R=0.031.

According to the calculation results above, if RC.R<0.1, the matrix is deemed to have the satisfactory consistency. Otherwise, it should be adjusted. If RC.R<0. 1, it is unnecessary to adjust the matrix.

5.2. The Evaluation Model

Transform the ecological sensitivity single factor layer into raster data format, then apply multiple factor model of weighted summation to do superposition calculation, generate the ecological sensitivity evaluation map of Pengyang county. The specific calculation formula is:

$$S=\sum_{i=1}^{n} w_i p_i$$

where,

S --- the ecological sensitivity level of the land use;

w ---the weight of the i evaluation factor;

p_i ---the numerical value of the i ecological sensitivity evaluation factor;

n ---the number of evaluation factors, n=1,2,...,5.

Table 3. Relations between channel vertical gradient and drainage area (hm²).

Rating	Channel vertical gradient(%)	Drainage area
1	<1	155,500.99
2	1-5	12,837.24
3	5-10	49,058.79
4	10-20	31,524.53
5	>20	2,614.14

Table 4. Area of different soil erosion intensities of pengyang county (hm²).

Erosion intensity	Minor	Slight	Moderate	Intense	Extremely intense-severe
Area	34,855.43	42,679.39	80,245.59	59,253.75	34,501.53

Table 5. Vegetation coverage (hm²).

Vegetation coverage	High coverage	Medium-high coverage	Medium coverage	Medium-low coverage	Low coverage	No coverage
Area	7,366.82	30,089.13	51,867.33	94,111.97	64,136.3	3,964.14

Table 6. Weight analysis.

Evaluation factor	Score of ecological sensitivity					Weight value
	Exposure	Gradient	Erosive channel	Soil erosion	egetation	
Exposure	1	1/2	1/3	1/3	1/4	0.077
Gradient	2	1	1	1	1/2	0.102
Erosive channel	3	1	1	1	1	0.187
Soil erosion	3	1	1	1	1	0.187
Vegetation	4	2	1	1	1	0.447

Table 7.   Areas & proportions of different ecological sensitive zones of pengyang county.

Sensitivity rating	Area/ hm^2	Proportion/%
Strong sensitive zone	82,956.47	32.98
Medium sensitive zone	104,613.69	41.59
Slight sensitive zone	63,965.53	25.43
Total	251,535 .69	100. 00

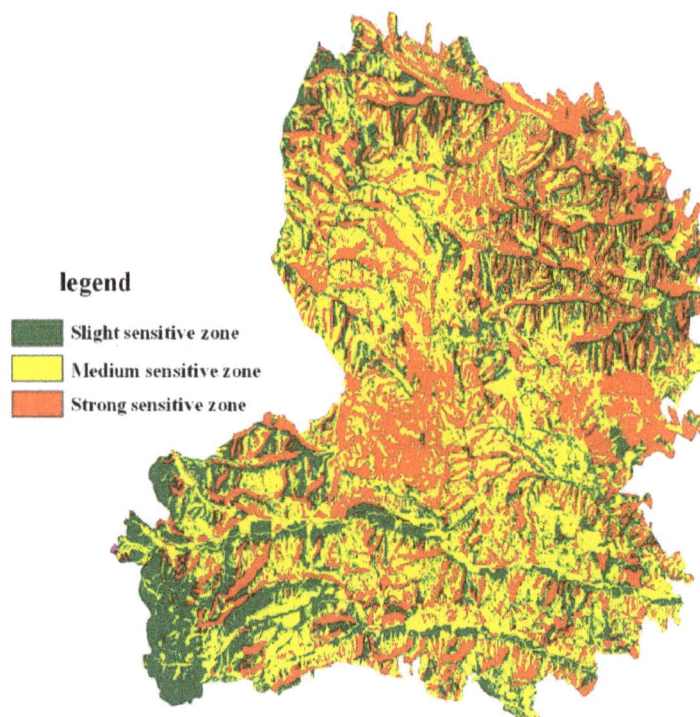

Fig. (1). The corresponding areas of different sensitive zones.

5.3. Multi-factor Weighted Stacking Ecological Sensitivity Analysis

Gradient, exposure, small watershed, soil erosion and vegetation are taken as evaluation factors in the ecological sensitivity analysis. After superposition of multiple factor analysis, final score according to the natural discontinuities classification method from 0 to 3.14, 3.14 and 4.87, 4.87 and 8.84 in Pengyang county was divided into slight sensitive zone, medium sensitive zone and strong sensitive zone, at the same time get the ecological sensitivity rating map of Pengyang County is then obtained through weight stacking, shown as Fig. (1). The corresponding areas of different sensitive zones are shown as Table 7.

As shown in the table7, the strong sensitive zone covers an area of 82,956.47 hm^2, accounting for 32.98% of the total, mainly distributed in the middle and north parts of of Pengyang County, involving loess ruin areas of Wangwa Town, Baiyang Town, Xiaocha Town, and Mengyuan Town; the medium sensitive zone covers an area of 104,613.69 hm^2, accounting for 41.59% of the total, mainly distributed at the south, east and southeast parts of the county; and the slight sensitive zone covers an area of 63,965.53 hm^2, accounting

for 25.43% of the total, mainly distributed in the southwest part of Pengyang County, and valleys of Ruhe River and Honghe River.

6. DISCUSSION

With factor weighted stacking method and fuzzy evaluation method as the theoretical supports, the paper establishes the ecological factor evaluation index system, determines the evaluation standard, and concludes that the multi-factor weighted stacking is the commonly used method for evaluating the ecological sensitivity. By selecting key factors, the research, different from previous ones, focuses on the gradient and soil erosion-especially channel factor-related to the evaluation of the ecological sensitivity of the study area. Through extracting data of erosive channel and small watersheds, and taking channel vertical gradient as referential index, the paper divides small watersheds into different ratings in terms of the possibility of water and soil loss, thus offering the scientific basis for evaluating the ecological sensitivity of the study area and obtaining the rating spatial distribution of the ecological sensitivity of

Pengyang County. After the analysis, the paper comes to the following conclusions:

1. The strong sensitive zone of the study area covers 82,956.47 hm^2, accounting for 32.98%. Landforms in the area usually include steep slopes with rich vegetation, and high ecological value, erosive channels, and zones affected by erosive channels. The zone is extremely sensitive to development and construction. Any disturbance or destruction to the zone would affect the complex ecological system of the whole region. Therefore, the strong sensitive zone is under the key protection.

2. The medium sensitive zone covers 104,613.69 hm^2, accounting for 41.59% of the total. Landforms in the area usually include forestlands of gentle relief and diversified plants. The lands are highly sensitive to human activities, and the ecology, which it is hard to restore, plays the important role in maintaining the function and environment of the strong sensitive zone. It should be considered whether it is feasible to develop and utilize the medium sensitive zone.

3. The slight sensitive zone covers 63,965.53 hm^2, accounting for 25.43%. Landforms in the area are usually low and flat, mainly rivers and forests with singular vegetation, areas with poor vegetation, and farmlands. The areas are resistant to human disturbance and able to go through development and construction of the specific intensity, and the lands can be developed for a variety of purposes. However, serious disturbance may lead to water and soil loss as well as related natural disasters, and the ecological restoration is slow.

ABOUT THE AUTHORS

First Author

Shi Yun, Associate professor, studying for PhDs in Ningxia University. The author's major is Grass science.

Second Author

Geng Sha, studying for master degree in Ningxia University. The author's major is GIS.

Third Author

Mi Wenbao, Professor of Ningxia University. The author's major is Human geography.

ACKNOWLEDGEMENTS

This work was financially supported by the Natural Science Foundation of China (41161081).

REFERENCES

[1] Z. Wu, M. Liu, Z. Wang, *GIS-based Sensitivity Evaluation of Water and Soil Loss in Anshan City,* Liaoning, 2009.

[2] Z. Ouyang, X. Wang, H. Miao, "China's eco-environmental sensitivity and its spatial heterogeneity", *Acta Ecologica Sinica*, vol. 20, pp. 9-12, 2000

[3] J. Pan, X. Dong, "GIS-based assessment and division on eco-environmental sensitivity in the heihe river basin", *Journal of Natural Resources.* vol. 21, pp. 267-272, 2006

[4] Compiling Committee of Annals of Pengyang County, *Annals of Pengyang County,* Yinchuan, 1996

[5] X. Tao, J. Zhang, Y. Wang, "Eco-sensitivity and its spatial distribution in hangzhou", *Journal of Hangzhou University*, vol. 32, pp. 27-30, 2006

[6] J. Mo, Y. Lu, L. Wei, "Evaluation & analysis on urban eco-sensitivity in nanning based on GIS", *Geomatics World*, vol. 1, pp. 33-38, 2007

[7] J. Hong, Y. Song, "application research of analytic hierarchy process (AHP) in water environmental planning", *Environmental Science & Technology*, vol. 1, pp. 35-35, 2000

[8] J. Liu, "Discussion on the method of making slope classification map with DEM in ArcGIS", *Geomatics & Spatial Information Technology*, vol. 34, pp. 140-141, 2011

[9] B. Ren, H. Tian, "Study on indicators and methods with regards to national eroded gully survey", *Yangtze River*, vol. 41, pp. 103-105, 2010

[10] J. Ren, A. Ding, W. Liu, "Study and realization of automatic extraction method for gully erosions in northwest loess plateau area", *Journal of Gansu Sciences*, vol. 24, pp. 16-19, 2012

[11] H. Feng, Y. Shang, X. Liu, J. Ma, "Conversion method of extracting ground gradient through different resolution dem in lanzhou", *Journal of Lanzhou University (Natural Sciences)*, vol. 1, pp. 80-84, 2010

[12] Y. Feng, L. Yuexiang, *Study on Returning Farmland to Forest Project in Ningxia,* Yinchuan, pp. 13-19, 2012.

[13] L. Jinzhao, G. Qingxi, G. Jianping, " DEM based automated extraction system of gully in the loess upland gully area", *Journal of Northwest Forestry University,* vol. 24, pp. 220-223, 2009.

9

Screening and Characterization of a Mutant Fungal Aspartic Proteinase from *Mucor pusillus*

Li Yuqiu, Tan Hua, Li Da, Li Zhoulin, Chi Yanping, Jiang Yuanyuan, Liu Xiangying, Wang Jinghui[*] and Li Qiyun[*]

Center of Agro-food Technology, Jilin Academy of Agricultural Sciences, Changchun, Jilin, 130033, P.R. China

Abstract: In this study, site-directed mutagenesis was carried out to alter properties of Mucor pusillus rennet (MPR) in order to find a potential substitution of commercial chymosin. Mutant G186D/E13D screened from thousands of mutants showed a significant milk-clotting activity (MCA). Mutant G186D/E13D rennet was purified and characterized. The molecular weight was estimated to be 44 kDa by SDS-PAGE. The maximum enzyme activity was at a wide range of pH (5.0-7.0) and 60°C. The enzyme was inhibited by metal ions (Fe^{2+}, Fe^{3+}, Cu^+ and Zn^{2+}), 1.10-Phenantrolin and pepstatin A. Further texture analysis of types of cheddar cheese made by non-mutant rennet, mutant (G186D/E13D) rennet and commercial rennet suggested that the soluble nitrogen content and hardness of cheddar cheese made by chimeric mutant rennet was decreased without any significant change in flavor between these cheeses. The result implicated that, to some extent, the mutant rennet could decrease hydrolysis of protein during ripening of cheese, probably as a candidate for a useful milk coagulant.

Keywords: Aspartic proteinase, *Mucor pusillus* rennin, Mutation, Thermostability, Proteolytic activity.

1. INTRODUCTION

Chymosin is an aspartic proteinase (EC 3.4.23.4) that is responsible for the coagulation of milk in the fourth stomach (abomasum) of unweaned calves in the form of an inactive precursor prochymosin [1], which is used extensively in cheese production because it cleaves κ-casein in a specific manner, at the Phe105-Met106 bond, with low proteolytic activity, and for the production of quality cheeses with good flavor and texture [2]. Unavailability of calf stomach and ethical problems associated with animal slaughtering has necessitated the finding of other alternatives to calf chymosin. In this regard, various plants and microbial proteases alternatives are used for chymosin production. Plant sources for milk-clotting enzymes have been identified from Cynara scolymus [3], Carica papay [4], Streblus asper [5], Centaurea calcitrapa [6] and Albizia [7]. Unfortunately, most of these sources are not suitable for production of quality cheese as they produce a bitter taste [2]. Proposed microbial substitutes for animal proteases include those from fungi and bacteria, such as Basidiomycete [8], Mucor pusillus [9], Bacillus sphaericus [10], Rhizomucor pusillus [11], Rhizopus oryzae [12] and Aspergillus [13]. At present, microbial rennet is used for one third of the entire cheese produced worldwide [14].

Mucor rennins are an aspartic proteinase produced by two closely related strains of Mucorales fungi, Mucor pusillus and Mucor miehei [15, 16]. These enzymes possess similarly characterized milk-clotting characteristics to those of calf chymosin, and they have been used as substituting enzymes for calf chymosin in the cheese industry. However, these enzymes are more proteolytic than bovine chymosin, thus leading to a lower yield in the production of cheese due to continued proteolysis following milk coagulation. Additionally, during cheese ripening, the curd-entrapped enzyme remains active and further degrades casein fractions by extensive non-selective peptide bond attack. This phenomenon may also lead to bitter flavor and structural deficiencies in ripe cheese, even after the heat-treatment step often present in its processing [17]. Moreover, the enzyme fraction in the whey could degrade proteins of economic value [18].

Site-directed mutagenesis is a good tool to research on the relationship of structure and function of proteins, especially for researching on chymosin, to vary systematically the sequence of peptide substrates and also to vary the specificity subsites using site-directed mutagenesis [19-23]. Meanwhile, site-directed mutagenesis is also a good strategy to alteration and modification of sequence. We have cloned the preproRMPP gene, and have developed efficient expression systems for the enzymes as zymogens in Pichia pastoris [24]. By using this system, site-directed mutagenesis of many milk-clotting enzymes was carried out to generate mutant enzymes with amino acid exchanges at position 13, 101 and 186. The mutant enzyme was purified and characterized biochemically, and cheeses made by the enzyme were analyzed for textural parameters and proteolysis after cheese ripening. In this case, replacement of Glu13/Gly186 was found to cause a marked decrease in the proteolytic activity.

*Address correspondence to these authors at the Center of Agro-food Technology, Jilin Academy of Agricultural Sciences, Changchun, Jilin, 130033, P.R. China;
E-mail: 846812862@qq.com

2. MATERIALS AND METHOD

2.1. Strains and Plasmids

The MPR gene encoding preproMPR, cloned from the genome of Mucor pusillus consists of a pre-sequence of 22 amino acids for secretion, a pro-sequence of 44 residues and a mature enzyme of 362 residues [24]. The plasmid pPICZα A, containing the MPR gene downstream of the AOX promoter, was introduced into Pichia pastoris GS115 (his4, lacZ) as a host to produce the wild-type and mutated MPR genes.

2.2. Media and Culture Conditions

YPD medium contained 2% Bacto-peptone (Difco, USA), 1% Bacto-yeast extract (Difco). The yeast transformants were pre-cultured in YPD medium at 30°C for 24 h. The cells were then harvested and re-suspended in 100 mL of BMMY medium containing 1.34% (w/v) bacto-yeast nitrogen base, 1% yeast extraction, 2% (w/v) peptone, 0.5% (w/v) biotin and 1% (v/v) methanol. Cultivation was continued at the same temperature for an additional 5 days [25].

2.3. Mutagenesis

Site-directed mutagenesis was carried out using a Polymerase Chain Reaction mediated method. Substitutions were made by overlap extension PCR mutagenesis [20]. Five mutants Ala101Thr (A101T), Glu186Asp (G186D), Glu13Asp (E13D), Glu13Gln (E13Q) and Glu13Ala (E13A) were obtained by reverse overlap primers as shown in Table 1. All the mutations were checked by nucleotide sequencing and introduced into the corresponding position of pPICZα A by forming single mutants (A101T, G186D, E13D, E13Q, E13A and E13P) and chimeric mutants (A101T/G186D, A101T/E13D, G186D/E13D and G186D/E13Q). Transformation of Pichia pastoris was carried out by the electroporation method of Becher et al. 1991 [26].

2.4. Screening

Clotting activity determination of all transformants by microplate assay according to [20] was carried out using 4% (w/v) skim milk tempered to 35°C. Clotting was measured as an increase in absorption at OD800, and the clotting activity calculated from a standard curve was obtained using serially diluted standardized unmutated MPR. Selected transformants were used for the shake flask experiments, which were transformant cultured in BMMY induced by methanol for 5 days at 30°C on orbital shaker at 220 rpm. The cells were harvested by centrifugation at 10,000g for 20 min at 4°C and the supernatant (crude enzyme) was used for purification experiments [10].

2.5. Purification of MPR

The crude enzyme solution was precipitated at 80% saturation of (NH4)2SO4. The active fraction with high milk clotting activity (MCA) was further purified by passing through Sephadex G-100 column (100cm ×1.2cm) pre-washed with 50 mM sodium phosphate buffer at pH 5.8. Fractions of 5 ml each were collected at room temperature at a flow rate of about 20 ml/h. The active fractions were dialyzed against distilled water and concentrated via lyophilization [9]. The concentrated enzyme was loaded on to a DEAE-52 (20cm×1.6cm) pre-equilibrated with 50 mM sodium phosphate buffer at pH 5.8. Elution of protein was then carried out by batch-wise addition of 50 ml portions of increasing molarities (0.0 -0.5 M) of NaCl in 50 mM sodium phosphate buffer at pH 5.8. Fractions of 5 ml each were collected at room temperature (25°C) at a flow rate of about 30 ml/h and analyzed for MCA and protein content [12]. The active enzyme fractions were pooled and stored at 4°C for further studies. The purified proteinase was examined for protein by electrophoresis under denaturing conditions in 12% polyacrylamide slab gels [17, 27].

2.6. Assay of Milk-Clotting Activity (MCA)

The MPR was assayed using the method described by Arima et al. [11]. A 10% solution of skim milk (Snow Brand

Table 1.   Nucleotide sequences used for site-directed mutagenesis of Ala101, Gly186 and Glu13.

Mutant Site	Primers	Sequence of Mutant Primer
101	A101T-F	5'CGGCGGTACGACCGTGAAG3'
	A101T-R	5'CTTCACGGTCGTACCGCCG3'
186	G186D-F	5'GTCTTTGGTGACGTCAACAACACC3'
	G186D-R	5'GGTGTTGTTGACGTCACCAAAGAC3'
13	E13Q-F	5' GACTTGGAGCAGTACGCCATTC3'
	E13Q-R	5' GAATGGCGTACTGCTCCAAGTC3'
	E13D-F	5' GACTTGGAGGACTACGCCATTC3'
	E13D-R	5' GAATGGCGTAGTCCTCCAAGTC3'
	E13P-F	5'GACTTGGAGCCGTACGCCATTC3'
	E13P-R	5'GAATGGCGTACGGCTCCAAGTC3'
	E13A-F	5' GACTTGGAGGCATACGCCATTC 3'
	E13A-R	5'GAATGGCGTATGCCTCCAAGTC3'

milk products Co.) containing 10mM CaCl2 was used as the substrate. Substrate solution (5ml) was added to the enzyme solution (0.5ml) at 35°C. The time required for curd particles to form was measured with a stop watch. Under the above assay condition, 1 unit of activity (Soxhlet Unit) was defined as the amount of enzyme that clotted the milk solution in 40 min [28].

2.7. Assay of Proteolytic Activity (PA)

The proteolytic activity was measured using a 1.0% solution of casein (M/10 50 mM sodium phosphate buffer, pH 5.8) as the substrate. Five milliliters of the substrate solution was incubated with 1ml of enzyme solution at 45oC for 30 min and the enzyme reaction was stopped with 5ml of trichloroacetic aci mixture solution. After 30 min of incubation, the reaction mixture was filtered using filter paper and 2 ml of the filtrate was added to 5 ml of 0.55M Na_2CO_3 and 1 ml of Folin's reagent. This mixture was measured. One unit of the activity was defined as the amount of enzyme, which released 1ug of amino acid expressed as the tyrosine concentration per min under the above condition [28].

2.8. Effect of Temperature on Enzyme

The effect of temperature on milk-clotting activity of MPR was studied by measuring the activity of the purified enzyme at different temperatures (30-75°C).

The thermal stability of the purified enzyme was studied by measuring the milk-clotting activity of the residual enzyme after incubation for 0, 10, 20, 30, 40, 50, 60, 90 and 120 min at 55°C [3].

2.9. Effect of pH on Enzyme

The effect of pH on milk-clotting activity of the purified enzyme was studied at pH range of 5.0-8.0. The buffers used were: 0.1M citrate-phosphate (pH5.0-6.0) and 0.1M sodium phosphate (pH 6.0-8.0) [11, 14].

The pH Stability of the enzyme was studied at pH values from 3.0 to 10.0. The buffers employed were 50mM sodium citrate for pH 3.0, 4.0 and 5.0; 50mM sodium phosphate for pH 6.0 and 7.0; 50mM Tris/HCl for pH 8.0, 9.0 and 10.0. After incubation for 16 h at 25°C, the residual enzyme activity was measured [29, 30].

2.10. Effect of Metal Ions and Inhibitors

The effect of some metal ions (Ni^{2+}, K^+, Zn^{2+}, Mg^{2+}, Mn^{2+}, Cu^+, Fe^{2+} and Fe^{3+}) at 5mM concentration and some inhibitors (o-Phenantrolin, Aprotinin, Leupeptin, phenyl-methylsulphonyl fluoride (PMSF), EDTA and pepstatin A on purified enzyme activity was tested. The concentrations of the inhibitors are listed in Table 3. The purified enzyme was incubated at room temperature for 30 minutes with metal ions or inhibitors and the residual milk-clotting activity was measured [31].

2.11. Cheese Manufacture and Analysis

Textural and proteolytic properties of cheese made with MPR were analyzed using a standard procedure [32]. Cheddar cheese was made with the mutant rennin (G186D/E13D)

The composition of the cheese was determined in triplicate. Fat content [33] and moisture content were determined [34]. The pH of the cheese was estimated [35].

The proteolysis of cheeses was assayed by determination of total nitrogen (TN), pH 4.6 phosphotungstic acid-soluble nitrogen (PTASN) according to methods described by Christensen et al. [36]. Total protein content was then obtained by multiplying the TN value by 6.38 [37].

Texture profile analysis (TPA) was carried out of the cheeses according to the methods described by Bhaskaracharya RK [38]. All analyses were carried out thrice.

2.12. Statistical Analysis

The SPSS package (SPSS 12.0 for Windows, SPSS Inc. Chicago, ILand USA) was used for statistical analysis of the results. Analysis of variance (ANOVA) was undertaken and the mean was established for $P < 0.05$. Mean comparisons were performed according to the Tukey's honest significant differences (HDS) test. Thus, a, b, c superscripts were employed to state significant differences between the lots for the exact same ripening time [37].

3. RESULT AND DISCUSSION

3.1. Screening of Mutant MPR

To test the effect of mutations at positions A101, G186 and E13 on the thermostability and proteolytic activity of MPR, one or two of the residues were substituted in MPR and the chimeric mutants in Pichia pastoris were expressed [11, 24]. The results of measurements for milk-clotting activity, proteolytic activity and thermostability of these mutants are presented in Table 2. The mutants' enzymes (A101T/G186D, G186D/E13D, G186D/E13Q, G186d/E13Q and A101T /G186) were shown to be more sensitive to thermos than that of the non-mutated enzyme, especially the double mutant A101T/G186D that lost milk-clotting by 50% when incubated for 40min at 55oC, but without any change in the proteolytic activity. When residues were exchanged in position E13, the proteolytic activity of all the mutants (E13P, E13Q, E13D and E13A) decreased remarkably to almost a half of that of non-mutant enzyme, but there was no change in thermostability. Chimeric mutants (G186D/E13D and G186D/E13Q) had a reduction in thermostability along with a sharp decrease in proteolytic activity. The mutant G186D/E13D was selected for further characterization as described below.

Generally, protein cores are typically hydrophobic. Hydrophobic interaction is considered as a dominant force in structural stability and increased packing efficiency is often correlated with increased hydrophobicity [39-41], therefore hydrophobic amino acids always exist in protein cores to keep the proteins stable. Hydrophobic amino acids (Ala and

Table 2.   Milk clotting and proteolytic activities of the non-mutant and mutant MPRs.

Enzyme	Clotting Activity (U/μg)	Proteolytic Activity (U/μg)	C/P ratio	Thermostability Relative Activity (%) at 55°C 40 min
Non-mutation	10.19	2.83	3.6	88.6
A101T	13.6	2.72	4.8	75.8
G186D	16.63	2.64	6.3	58.6
E13P	8.63	1.87	4.6	86.2
E13Q	7.97	1.19	6.7	88.2
E13D	8.38	1.18	7.1	89.1
E13A	8.26.	1.23	6.7	85.5
A101T/ G186D	16.24	2.62	6.2	53.5
G186D/E13D	10.47	1.36	7.7	70.2
G186D/E13Q	9.53	1.27	7.5	73.6

Gly) at positions 101 and 186 were substituted by charged amino acids (Thr and Asp) respectively, which made it difficult to maintain hydrogen cores of rennet [42], finally leading to unstable or flexible conformation of MPR.

Site E13 in RMPP plays a critical role in forming the correct hydrogen bond network around the active center and influence catalytic rate of RMPP as reported by Aikawa [12]. Substitutions at position 13 exchanged residue from Glu to Asp to alter space conformation of MPR resulting in decreased affinity of the MPR to substrate. Simulation analysis and molecular docking analyzed by software found possibility that position 13 may be related to increased substrate specificity (Analysis by our colleague, data is not shown), which was later proved in this experiment and also in RMPP Aikawa in 2001.

3.2. Purification of Mutated MPR

The results pertaining to purification of mutated MPRs using a combination of different purification techniques are summarized in Table **3**. After fermentation, the mutated MPR produced by recombinant Pichia pastoris GS115 was fractioned at 70% ammonium-sulphate with 1.3-fold purification and 91.6% recovery. Passage through Sephadex G-75 column resulted in about 5.7-fold purification of the enzyme with specific activity of 17203 U/mg. Finally, the concentrated active fractions passed through DEAE -52 column and the enzyme was purified about 7.1-fold with 5.2% recovery. Figs. (**1, 2**) show the elution diagrams of the mutant G186D/E13D using Sephadex G-75 and DEAE-52 columns, respectively. Purified MPRs were separated into two peaks of proteins with one activity peak on Sephadex G-75 and one eluted activity peak (elution peak I) with 0.2 mol/L NaCl sodium phosphate buffer, pH 6.0 on DEAE-52. These peaks with MCA of proteins appeared as one band with molecular mass of 44 kDa on SDS-PAGE (shown in Fig. **3**).

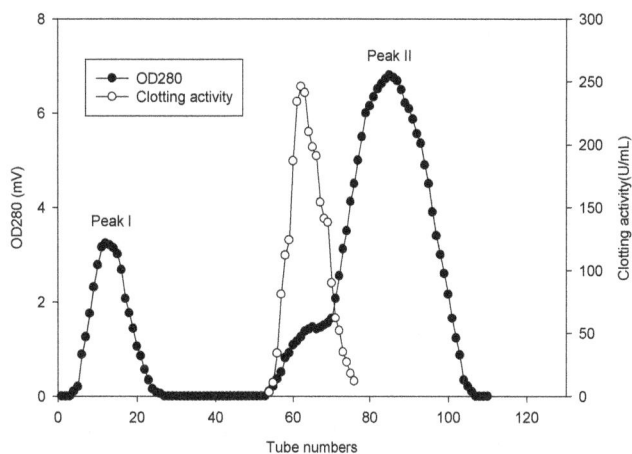

Fig. (1). Elution profile of mutant MPR on sephadex G-75 chromatography.

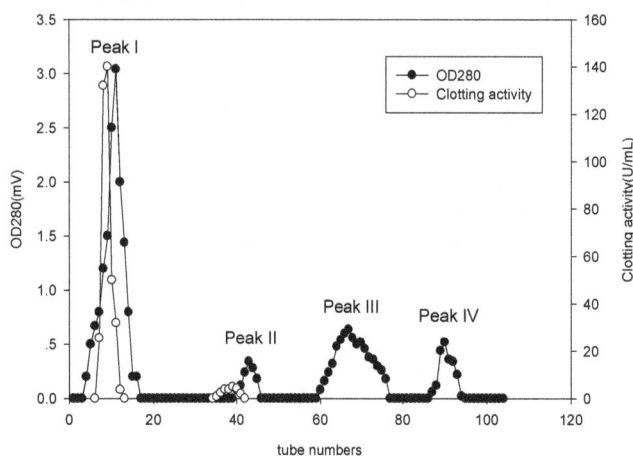

Fig. (2). Elution profile of mutant MPR on DEAE-52 chromatography.

Table 3.   Purification scheme of Mutant MPRs produced by *Pichia pastoris*.

	Volume (mL)	Clotting Activity (U/mL)	Total Clotting Activity (U)	Protein Content (ug/mL)	Total Protein Content (mg)	Specific Activity (U/mg)	Yield (%)	Purification Fold
Crude extract	100	83	8300	27.5	2.75	3018	100	1
(NH4)₂SO4	10.0	760	7603	193.7	1.94	3922	91.6	1.3
G-75	3.0	462	1386	26.9	0.081	17203	16.7	5.7
DE-52	3.0	144	432	6.7	0.022	21428	5.2	7.1

3.3. Effect of Temperature

Purified mutant G186D/E13D and non-mutant enzymes acted optimally at 60°C and later, they started losing their activity rapidly leading to complete inactivation at 75°C (Fig. **5**). The thermal inactivation experiments indicated that the mutant enzymes were slightly sensitive to heat than the non-mutant enzymes (Fig. **4**), despite that MPR is a relatively thermostable protein. The non-mutated enzyme at 55°C remained almost fully active even after 60min of incubation, but G186D/E13D lost 30% of its activity after 50 min at the same temperature.

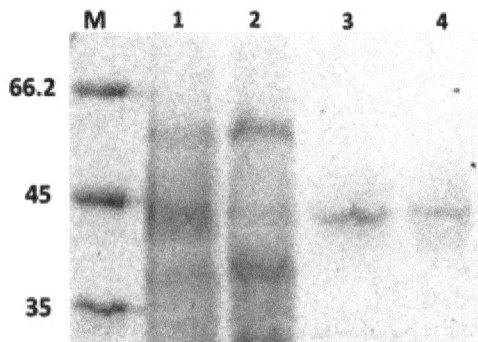

Fig. (3). Electrophoretogram of mutant MPR (G186D/E13D) after various steps of purification on SDS-PAGE of purification steps. Lane M: standard molecular weight markers; Lane1: Supernatant extract; Lane 2: (NH4)2SO4 fractionation; Lane 3: Sephadex G-75, Lane 4: DEAE-52.

Previously, the optimal temperature of MCA produced by Enterococcus faecalis TUA2495L [28] was 70°C. The maximum MCA of purified enzyme produced by Rhizopus oryza [12] and Bacillus sphaericus [10] was at 60°C and 55°C respectively. The crude enzyme from Yeast Extracellular [29] showed maximum activity at 65°C with 10% of clotting activity lost at 60 min incubation at 45°C.

3.4. Effect of pH

Fig. (**6**) shows that the MCA of all of the mutant enzymes decreased as the pH increased from 5.0 to 7.5, and no activity was observed at pH 8.0. The optimum pH for both the non-mutant and mutant enzymes was 5.5 and the same result has been reported by Ashwani for milk clotting protease from Capra hircus. The pH optimum of purified APs from C. calcitrapa cell suspensions was detected at pH 5.1 [43]. The milk-clotting enzyme from glutinous rice wine mash liquor exhibited maximal MCA in milk at a pH of 5.5 [14]. Initial enzyme activity remained stable after treatment at pH 6.0, but there was an almost 50% loss in the activity at

pH 5.0 or 7.0 (Fig. **7**). Sushil found that extracellular acid protease from Rhizopus oryzae [12] retained 96% of its activity at pH 5.5-7.5.

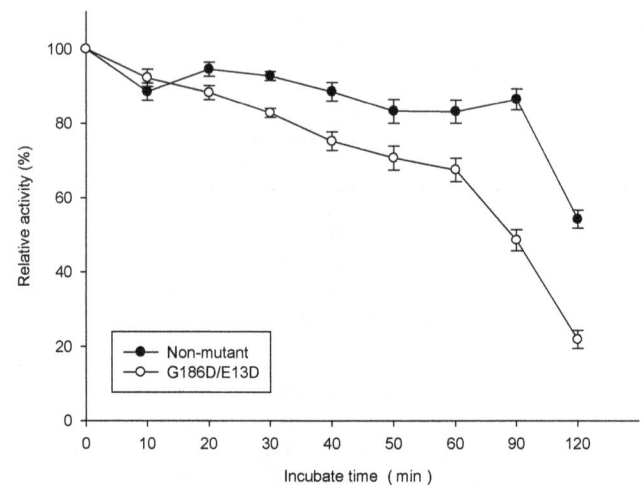

Fig. (4). Thermostability of purified non-mutant and mutant MPRs. Relative milk-clotting activities (%) were determined between 10-120 min at 55°C for non-mutant MPR and mutant MPR G186D/E13D produced by recombinant yeasts.

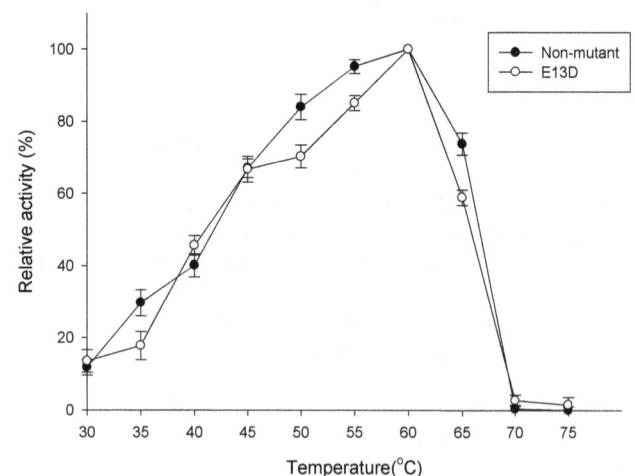

Fig. (5). Effect of temperature on purified proteinase activity. Relative milk-clotting activities (%) were measured using skim milk as substrate from 30°C to 75°C for non-mutant MPR and mutant MPR G186D/E13D produced by recombinant yeasts.

3.5. Effect of Metal Ions and Inhibitors

The effect of various metal ions on the percent residual activity is shown in Table **4**. As shown, Ca^{2+}, Mn^{2+}, and Mg^{2+} were activators, whereas Ni^{2+}, Fe^{3+}, Fe^{2+}, and Zn^{2+} were inhibitors of the MCA for these enzymes. However, Cu^+, K^+ and Na^+ had no effect on the enzyme activity. Contrary to our results, Wang and others [14] reported that K+ was an inhibitor of the MCA.

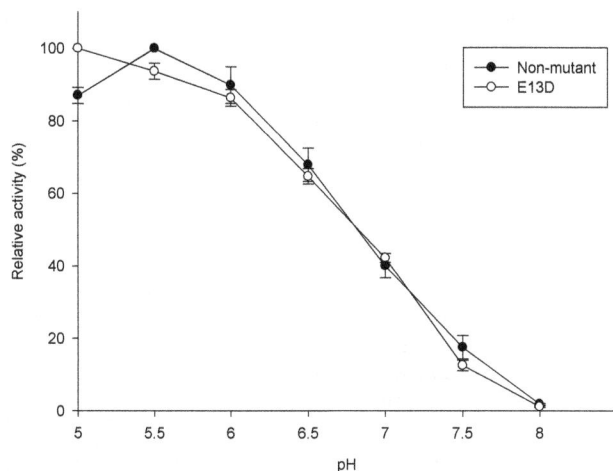

Fig. (6). Effect of pH on purified proteinase activity. Relative milk-clotting activities (%) were measured using skim milk as substrate from 5.0 to 7.0 for non-mutant MPR and mutant MPRsG186D/E13D produced by recombinant yeasts.

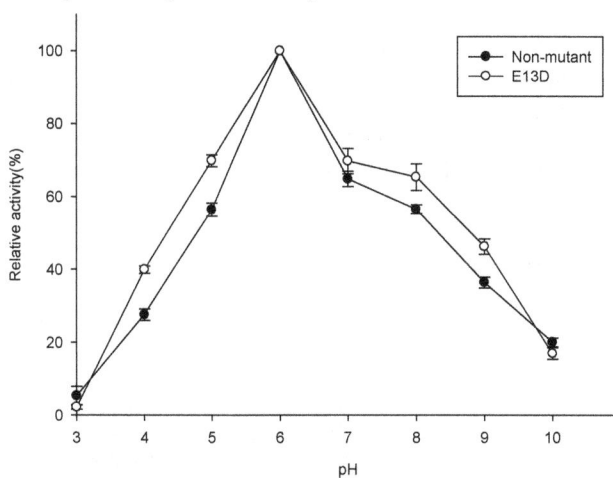

Fig. (7). Stability of proteinase to pH. The non-mutant and mutant MPR were incubated at various pH values for 4h at 25°C, the residual activities were measured.

Mutant G186D/E13D retained above 90% of relative activity in the presence of PMSF or aprotin (Table **5**), hence serine participation at the active site of the enzyme has also been ruled out. EDTA and aprotinin had little effect on the activity of these enzymes. The protease activity was found to be sensitive to pepstatin, thereby confirming the aspartic protease nature of the activity. But the mutant G186D/E13D was found to be sensitive to 1.10-Phenantrolin, which suggested that these enzymes may be metalloenzyme. However, this result was totally different from other MCPs as reported by Sara [43], Magda [10], Arun [44] and Sushil [12].

3.6. Textural and Proteolytic Properties of Cheese Made with MPR the Mutants

The composition and pH of cheddar cheese made with commercial rennet, non-mutant and mutant rennet (G186D/E13D) with ripening 180d is shown in Table **6**. The values for pH, moisture, protein and fat content were in the range 5.16 to 5.21, 36.68 to 38.37%, 23.25 to 26.72% and 29.25 to 31.53%, respectively. The protein content of non-mutant cheese was marked (P<0.05) lower than that of mutant rennet cheese whose protein contenthad no significant (P<0.05) difference with commercial rennet cheese. Similar levels of moisture, fat and pH were shown between the three cheeses.

Table 4. Effect of metal ions on MCA of purified non-mutant and mutant enzymes.

Metal Ion (5mM)	Relative Activity (%)	
	G186D/E13D	Non-Mutant
Control	100	100
Ca^{2+}	227.6	218.8
Mn^{2+}	148.3	120.7
Ni^{2+}	28.7	47.8
Mg^{2+}	119.5	106.8
Cu^+	96.4	100
K^+	99.8	117.3
Na^+	97.4	98.9
Fe^{3+}	79.8	85.8
Fe^{2+}	75.4	87.7
Zn^{2+}	85.9	85.5

Control: Enzymes without adding metal ions.

Table 5. Effect of inhibitors on MCA of purified non-mutant and mutant enzyme.

Inhibitor	Concentration (mmol/L)	Relative Activity (%)	
		G186D/E13D	Non-Mutant
PMSF	1	93.4	93.1
1.10-Phenantrolin	1	0	0
Aprotinin	1	73.6	72.6
Leupeptin	0.2	95.2	96.5
Pepstatin A	0.05	15.2	24.5
EDTA	50	77.2	71.1
control		100	100

The level of pH4.6-SN increased with the increase in ripening time in each of the three cheeses. The extent of the increase in pH4.6-SN level was greatest (P≤0.05) in the non-mutant cheese at 180 d of ripening (Table **6**), and there were no obvious differences (P< 0.05) between mutant cheese and

Table 6. Composition, pH and pH 4.6 SN (soluble nitrogen) of Cheddar cheeses made by non-mutant, mutant rennet (G186D/E13D) and commercial rennet at 180 d of ripening[1].

Enzymes	pH	Moisture%	Protein%	Fat%	pH4.6 SN%
Commercial	5.16(0.03)[a]	38.37(0.82)[a]	26.72(1.42)[a]	31.53(1.17)[a]	5.32(0.09)[a]
Non-mutant	5.16(0.11)[a]	37.54(1.12)[a]	23.25 (0.59)[b]	30.92(0.84)[a]	6.81(0.24)[b]
G186D/E13D	5.21 (0.02)[a]	36.68(0.93)[a]	25.29(1.14)[ab]	29.25(0.63)[a]	5.19(0.18)[a]

[a,b]Means within a row with different superscripts are significantly different (Tukey's HSD; $P \leq 0.05$).
[1]Values represent means (SD; n = 3).

Table 7. Texture profile analysis parameters hardness, cohesiveness, springiness, and chewiness for Cheddar cheeses made with commercial rennet (control), non-mutant rennet, and mutants rennet (G186D/E13D) at 180 d of ripening[1].

Enzymes	Hardness (N)	Springiness	Cohesiveness	Chewiness (N)
Commercial	1134.44(12.34)[a]	0.39(0.01)[a]	0.25(0.01)[a]	93.44(5.11)[a]
Non-mutant	982.11(14.59)[b]	0.37(0.01)[a]	0.25(0.01)[a]	72.13(5.38)[a]
G186D/E13D	1160.71(16.24)[a]	0.36(0.07)[a]	0.22(0.01)[b]	88.90(6.01)[a]

a, b Means within a row with different superscripts are significantly different (Tukey's HSD; P <0.05).
1 Values represent means (SD; n = 3).

commercial cheese. The level of the increased pH4.6-SN was greatest in non-mutant cheese, which supports the earlier results showing that the protein content was lowest in non-mutant cheese of 180d ripening. Table 7 summarizes the mean values for the parameters obtained in the instrumental texture evaluation. Hardness (p<0.05) of non-mutant cheese was marked lower than that of the mutant cheese. The softening of cheeses was related to the hydration of the protein matrix and the effects of proteolysis [38]. This is in agreement with previous observations in which the level of the increased pH4.6-SN resulted in cheese softening. With respect to springiness, cohesiveness and chewiness, there was no significant difference (p<0.05) between the three cheeses, except for a slight lower inclination of non-mutant rennet cheese with respect to springiness and chewiness.

CONCLUSION

Site 186 and 13 of MPR are key positions, which relate to temperature sensitivity and substrate specificity. Themostability of MPR decreases remarkably when G186 is substituted by Asp and light reduction of proteolytic activity in MPR happens to be E13, substituted by Glu. Site 186 and 13 belonged to different core regions of MPR. They did not have any influence on each other, which is also proved by mutant G186D/E13D due to the reduction in the thermostability and proteolytic activity simultaneously.

Except for reduction in thermostability and proteolytic activity, the other properties of enzyme of mutant rennet were similar to that of the non-mutated rennet. Most of the textures of cheese made by non-mutated MPR were no different from the other two cheeses, except in protein content, pH4.6-SN, hardness and chewiness. Above all, these properties may assure the practical use of this double mutant as an improved milk coagulant. The effective expression system of Pichia pastoris allowing extracellular secretion used in this study will provide a possibility to produce the improved enzyme on large scale.

ACKNOWLEDGEMENTS

The financial support for this work from the Science & Technology Innovation Project of Jilin Academy of Agricultural Science (ZYCX201315), Science & Technology Development Project Agreement in Jilin Province (20150307019NY), Jilin Postdoctoral Science Foundation Funded Project (RB201305) is gratefully acknowledged.

REFERENCES

[1]   Maria CVH, Alicia GC, Jesús V, Félix CM. Molecular cloning and expression in yeast of caprine prochymosin. J Biotech 2004; 114: 69-79.

[2]   Kumar A, Sharma J, Mohanty AK, Grover S, Batish VK. Purification and characterization of milk clotting enzyme from goat (Capra hircus). Comp Biochem Physiol 2006; B 145: 108–13.

[3]   Llorente BE, Brutti CB, Caffini NO. Purification and characterization of a milk-clotting aspartic proteinase from globe artichoke (Cynara scolymus L.). J Agri Food Chem 2004; 52(26): 8182-9.

[4]   O'Hara BP, Hemmings AM, Buttle DJ, Pearl LH. Crystal structure of glycyl endopeptidase from Carica papaya: a cysteine endopeptidase of unusual substrate specificity. Biochemistry 1995; 34(40): 13190-5.

[5]   Senthilkumar S, Ramasamy D, Subramanian S. Isolation and Partial Characterisation of Milk-clotting Aspartic Protease from Streblus asper. Food Sci Technol Int 2006; 12(2): 103-9.

[6]   Domingos A, Cardoso PC, Xue ZT, Clemente A, Brodelius PE, Pais MS. Purification, cloning and autoproteolytic processing of an aspartic proteinase from Centaurea calcitrapa. J Fed Europ Biochem Soci 2000; 267(23): 6824-31.

[7]   Egito A, Girardet J, Laguna L, et al. Milk-clotting activity of enzyme extracts from sunflower and albizia seeds and specific hydrolysis of bovine κ-casein. Int J Technol 2007; 17(7): 816-85.

[8]   Kobayashi H, Kim H. Characterization of Aapartic Proteinase from basidiomycete, laetiporus sulphureus. Food Sci Technol Res 2003; 9(1): 30-4

[9]   Khan MR, Blain JA, Patterson JDE. Extracellular Proteases of Mucor pusillus. Appl Environ Microbiol 1979; 37(4): 719-24.

[10]   Magda AE, Maysa EM, Thanaa HA. Purification and Characterization of Milk Clotting Enzyme Produced by Bacillus sphaericus. J Appl Sci Res 2007; 3(8): 695-9.

[11] Aikawa J, Park YN, Sugiyama M, Nishiyama M, Horinouchi S, Beppu T. Replacements of Amino Acid Residues at Subsites and Their Effects on the Catalytic Properties of Rhizomucor pusillus Pepsin, an Aspartic Proteinase from Rhizomucor pusillus. J Biochem 2001; 129: 791-4.

[12] Sushil K, Neeru SS, Mukh RS, Randhir S. Process Biochem 2005; 40: 1701-1705.

[13] Dunn-Coleman NS, Bloebaum P, Berka RM, et al. Commercial levels of chymosin production by Aspergillus. Biotechnology (NY) 1991; 9: 976-81.

[14] Wang YP, Cheng QL, Ahmed Z, Jiang XX, Bai XJ. Purification and partial characterization of milk-clotting enzyme extracted from glutinous rice wine mash liquor. Korean J Chem English 2009; 26(5): 1313-1318.

[15] Hiramatsu R, Aikawa J, Horinouchi S, Beppu T. Secretion by Yeast of the Zymogen Form of Mucor Rennin, an Aspartic Proteinase of Mucor pusillus, and Its Conversion to the Mature Form. J Biolog Chem 1989; 264(28): 16862-6.

[16] Beldarraín A, Acosta N, Montesinos Rl, Mata M, Cremata J. Characterization of Mucor pusillus rennin expressed in Pichia pastoris: enzymic, spectroscopic and calorimetric studies. Biotechnolog Appl Biochem 2000; 31: 77-84.

[17] Yamashita T, Higashi S, Higashi T, et al. Mutation of a fungal aspartic proteinase, Mucor pusillus rennin, to decrease thermostability for use as a milk coagulant. J Biotechnol 1994; 32: 17-28.

[18] Fernandez-Lahore HM, Auday RM, Fraile ER, de Jimenez Bonino MB, Pirpignani L, Machalinski C, Cascone O. Purification and characterization of an acid proteinase from mesophilic Mucor sp. solid-state cultures. J Peptide Res 1999; 53(6): 599-605.

[19] Mark GW, Julie W, Phil N, et al. Mutagenesis, biochemical characterization and X-ray structural analysis of point mutants of bovine chymosin. Protein Eng 1997; 10(9): 991-7.

[20] Young NP, Aikawa J, Nishiyama M, Horinouchi S, Beppu T. Involvement of a residue at position 75 in the catalytic mechanism of a fungal aspartic proteinase, Rhizomucor pusulus pepsin. Replacement of tyrosine 75 on the flap by asparagine enhances catalytic efficiency. Protein Eng 1996; 9(10): 869-75.

[21] Chitpinityol S, Goode D, Crabbet MJC. Site-specific mutations of calf chymosin B which influence milk-clotting activity. Food Chem 1998; 62(2): 133-9.

[22] Elena G, Lev R, Lev G, Pavel M. A Natalia. Post X-ray crystallographic studies of chymosin: The existence of two structural forms and the regulation of activity by the interaction with the histidine-proline cluster of κ-casein. FEBS lett 1996; 379: 60-2.

[23] Lowther WT, Majer P, Dunn BM. Engineering the substrate specificity of rhizopuspepsin: The role of Asp 77 of fungal aspartic proteinases min facilitating the cleavage of oligopeptide substrates with lysine in PI. Protein Sci 1995; 4: 689-702.

[24] Jiang YY, Wang JH, Li YQ, Li D, Yang ZN. Cloning and expression of a rennet gene from mucor pusillus. Dairy Indust China 2010; 38(2): 7-9.

[25] Young NP, Aikawa J, Nishiyama M, Horinouchi S, Beppu T. Site-Directed Mutagenesis of Conserved Trp39 in Rhizomucor pusillus Pepsin: Possible Role of Trp39 in Maintaining Tyr75 in the Correct Orientation for Maximizing Catalytic Activity. J Biochem 1997; 121: 118-21.

[26] Becher DM, Guarente L. High-efficiency transformation of yeast by electroporation. Meth Enzymol 1991; 194: 182-7.

[27] Laemmli UK. Cleavage of structural proteins during the assembly of the head of bacteriophage T. Nature 1970; 227: 680-7.

[28] Sato S, Tokuda H, Koizumi T, Nakanishi K. Purification and characterization of an extracellular proteinase having milk-clotting activity from Enterococcus faecalis TUA2495L. Food Sci Technol Res 2004; 10(1): 44-50.

[29] Alessandro M, Federico F. Partial Purification and Characterization of a Yeast Extracellular Acid Protease. J Dairy Sci 1980; 63: 1397-1400.

[30] Chazarra S, Sidrach L, Lopez-Molina D, Rodrıguez-Lopez JN. Characterization of the milk-clotting properties of extracts from artichoke (Cynara scolymus, L.) flowers. J Int Dairy 2007; 17: 1393-1400.

[31] Nuala MR, Thomas PB, Alan LK, Timothy PG. Effect of milk pasteurization temperature and in situ whey protein denaturation on the composition, texture and heat-induced functionality of half-fat Cheddar cheese. J Int Dairy 2004; 14: 989-1001.

[32] O'Mahony JA, Lucey JA, McSweeney PLH. Chymosin-Mediated Proteolysis, Calcium Solubilization, and Texture Development During the Ripening of Cheddar Cheese. J Dairy Sci 2005; 88: 3101-14.

[33] AOAC, Association of Official Analytical Chemists International, Official methods of analyses (16th ed), Gaithersburg, Maryland (1997).

[34] AOAC, Association of Official Analytical Chemists International, Official Methods of Analysis (15th ed), Arlington, VA (1990).

[35] Muir DD, Banks JM, Hunter EA. A comparison of the Flavour and Texture of Cheddar Cheese of factory of farmhouse origin. J Int Dairy 1997; 7: 485- 97.

[36] St-Gelais D, Lessard J, Champagne CP, Vuillemard JC. Production of fresh Cheddar cheese curds with controlled postacidification and enhanced flavor. J Dairy Sci 2008; 92: 1856-63.

[37] Calvo MV, Castillo I, Diaz-Barcos V, Requena T, Fontech J. Effect of a hygienized rennet paste and a defined strain starter on proteolysis, texture and sensory properties of semi-hard goat cheese. Food Chem 2007; 102: 917-24.

[38] Bhaskaracharya RK, Shah NP. Texture evaluation of commercial mozzarella cheese. Dairy Indust Associ Aus 1999; 54(1): 36-40.

[39] Chirakkal H, Ford GC, Moir A. Analysis of a conserved hydrophobic pocket important for the thermostability of Bacillus pumilus chloramphenicol acetyltransferase (CAT-86). Protein Eng 2001; 14(3): 161-6.

[40] Alsop E, Silver M, Livesay DR. Optimized electrostatic surfaces parallel increased thermostability: a structural bioinformatic analysis. Protein Eng 2003; 16(12): 871-4.

[41] Xiao ZZ, Bergeron H, Grosse S, et al. Improvement of the Thermostability and Activity of a Pectate Lyase by Single Amino Acid Substitutions, Using a Strategy Based on Melting-Temperature-Guided Sequence Alignment. Appl Environ Microbiol 2008; 74(4): 1183-9.

[42] Querol E, Perez-Pons JA, Mozo-Villarias A. Analysis of protein conformational characteristics related to thermostability", Protein Eng 1996; 9(3): 265-71.

[43] Raposo S, Domingos A. Purification and characterization milk-clotting aspartic proteinases from Centaurea calcitrapa cell suspension cultures. Process Biochem 2008; 43: 139-44.

[44] A Sharma, A Eapen, SK Subbarao. Purification and characterization of a hemoglobin degrading aspartic protease from the malarial parasite plasmodium vivax. J Biochem 2005; 138: 71-78.

10

Consumer's Risk Perception of Genetically Modified Food and its Influencing Factors: Based on the Survey in Jiangsu Province, China

Ruixin Liu[1,3], Linhai Wu[1,2,*], Lijie Shan[1] and and Hua Li[2]

[1]*Food Safety Research Base of Jiangsu Province, School of Business, Jiangnan University, Wuxi 214122, China*

[2]*Synergetic Innovation Center of Food Safety and Nutrition, Wuxi 214122, China*

[3]*School of Tourism, Yangzhou University, Yangzhou 225127, China*

Abstract: Safety has always been the focus of debate on genetically modified food (GMF). To understand consumers' risk perception of GMF and its influencing factors, this study investigated 300 consumers from 6 cities in Jiangsu province, China by questionnare. The data were analyzed by using independent sample t-test, one-way ANOVA and principal component analysis. The results showed that most consumers worried about the safety of GMF and hoped GMF to be labeled with identity; meanwhile, their purchase intention of GMF was not high. Some of consumer characteristics including gender, education background, personal annual income, and with at least one child under 18, significantly influenced their risk perception of GMF. Furthermore, the main factors significantly influencing the consumers' risk perception of GMF were as follows in order of influence degree: health risks (0.386), ecological risk (0.187), and social risk (0.163). Consequently, the government should strengthen the popularization of scientific knowledge on genetically modified technology as well as GMF and simultaneously reinforce standardized management of GMF label.

Keywords: Consumers, genetically smodified food, risk perception, factor analysis.

1. INTRODUCTION

In recent years, with the rapid increase of urban population, continuous decrease of arable land and ever-growing material demand of people, genetically modified organism (GMO), especially genetically modified crop, has been developing fast due to its incomparable advantages such as disease and insect resistance, high yield, endurable storage, freedom from seasonal and climatic restriction, and quality improvement when compared with traditional one. Genetically modified food (GMF) is food made from GMO [1]. Because there's still no clear conclusion on GMF safety, many negative reports on GMF safety emerge constantly and GMF are even demonized on the internet. Thus, the public may resist and dread it, which has great negative influence on the promotion of genetically modified crop and the development of its relevant food processing industry.

Consumers are the direct recipients of GMF. Their consuming intention may guide market development, and factors affecting their choices may promote market transformation. According to the risk perception theory of food safety, it is not the actual risk of food safety itself but consumers' subjective risk perception of food safety that governs their consumption behavior [2]. Therefore, by making an investigation on consumers' risk perception of food safety in Jiangsu Province, China, this study aimed to have an understanding of Jiangsu consumers' risk perception and attitude towards GMF, and the main factors influencing their risk perception. The results of this study may enrich the achievement of consumers' risk perception of GMF and provide theoretical support for reducing the risk. Moreover, it may provide foundation for enterprises concerning GMF to ascertain target market and make corresponding marketing strategies, thereby promoting GMF market to develop in a faster and healthier way.

2. QUESTIONNAIRE DESIGN AND DATA ACQUISITION

2.1. Questionnaire Design

By combining Dong's [3] measurement scale about risk perception of GMF with consumers' actual situation in Jiangsu province, we developed a questionnaire about Jiangsu consumers' risk perception of GMF and its influencing factors. The questionnaire includes three parts. The first part examines consumers' cognition and attitude to GMF; the second part investigates the main factors affecting consumers' risk perception of GMF; the third part shows consumers' individual characteristics.

2.2. Data Acquisition

This study employed questionnaire to collect data for its operability. Stratified and random sampling were used to ensure that the samples are universal and representative.

*Address correspondence to this author at 2Synergetic Innovation Center of Food Safety and Nutrition, 214122, China;
E-mail: wlh6799@126.com

First, 13 cities in Jiangsu province were divided into three regions according to their geographic location, which are southern, northern, and central Jiangsu. Then two cities in each region were selected as representatives, which are Suzhou and Wuxi, Yangzhou and Nanjing, and Huai'an and Suqian, respectively. A total of 300 questionnaires, 50 per city, were distributed. The investigation was conducted in the supermarkets, farmers' markets, and squares from April 30 to May 5, 2014. The questionnaires were completed on the spot by communicating face to face between the trained investigators and respondents. Finally, 300 questionnaires were collected. The data were analyzed by SPSS19.0.

2.3. Characteristics of Samples

The total of 152 males and 148 females participated in this investigation, accounting for 50.7% and 49.3% of all participants, respectively. In terms of age, 107 participants were less than 30, 132 participants between 31 and 55, and 61 participants over 55, which accounted for 35.7%, 44.0%, and 20.3%, respectively. As to education background, participants with junior high school, senior high school, and college degree accounted for 30.7%, 40.3%, and 29.0%, respectively. In regard to personal annual income in 2014, the income of 67 participants were less than 30 000 yuan, that of 135 participants between 30 000 and 60 000 yuan, and that of 98 participants more than 60 000 yuan, accounting for 22.3%, 45.0%, and 32.7%, respectively. In addition, 65.3% of the participants owned at least one child under 18 in their families.

3. CONSUMERS' RISK PERCEPTION OF GMF

3.1. Cognition of GMF

As a newborn high-tech food, GMF has become an integral part in our life in only more than 10 years. The investigation showed that 5.4% of the respondents were very familiar with GMF, 13.6% knew it, 29.0% had a little knowledge of it, 34.3% have heard of it but did not know it, and only 17.7% have never heard of it. Compared with the result of Zhong [4] who investigated Nanjing consumers' attitudes towards GMF in 2004 and found only 43.3% of the responders had heard of GMF, the percentage of the respondents having heard of GMF increased to 82.3%. This increase indicates that GMF has developed rapidly in Jiangsu province, one of the most developed provinces in China, during the past 10 years.

Although 48.0% of the respondents said that they knew GMF more or less, their cognition was relatively vague in distinguishing specific GMF. They were more familiar with soybean, corn, and tomato which are common in daily life, but had a little knowledge of papaya, sweet pepper, and oilseed rape. Moreover, even 19.5% of the respondents considered non-genetically modified wheat and peanut as GMF.

3.2. Channels to Obtain Information on GMF

The public has many channels to obtain information about GMF in modern society. The results showed that 51.5%, 38.2%, and 31.7% of the respondents got to know GMF through TV, internet, and newspapers and magazines, respectively; in addition, some respondents obtained the information through salespersons' introduction in supermarket (24.1%), books (23.5%), and their relatives and friends (17.6%). Therefore, TV, internet, and newspapers and magazines are the main approaches for consumers to obtain information.

As to the most trusted channel, 33.6% of the respondents chose relatives and friends, 31.3% TV, and 22.7% newspapers and magazines, while only 7.4% internet. It suggests that TV, newspapers and magazines play a significant role in guiding public opinion, while information obtained by internet is low credible.

3.3. Risk Perception of GMF

The safety of GMF is the bone of contention, and to date, no final decision has been reached on this matter. Consequently, consumers' opinions upon the matter may depend on the available information. Among the 238 respondents who have heard of GMF, 23.2% considered GMF to be safe, 39.3% unsafe, and 37.5% were uncertain.

Further analysis showed that consumers who had only a little knowledge or have never heard of GMF were inclined to be uncertain about GMF safety. In contrast, consumers who were very familiar with or knew GMF tended to regard it as safe. Moreover, the number of males who deemed GMF safe was much more than that of females; the proportion of the former was 53.4%, while the latter 32.3%.

3.4. Attention to GMF Label

Due to the particularity of GMF, GMF for sale is required to put on special label in many countries. As regards the controversial matter whether GMF should put on special label, consumers' attitudes are highly consistent, because most of them hope to distinguish GMF from traditional foods by the label. In this study, up to 91.6% of the respondents wished for label. This result is similar to Ruan's [5] investigation which examined consumers' cognition of GMF label in Shenzhen.

The investigation also showed that label played an important role for consumers in choosing goods. When they were purchasing foods, 44.3% of the respondents read component description every time and 36.9% read often. Therefore, to safeguard consumers' rights and benefits, 86.1% of the respondents held the opinion that the government should adopt mandatory labeling system of GMF.

3.5. Purchasing Intention of GMF

To make it easier for consumers to understand, this investigation took genetically modified soybean oil, which is common in daily life, as an example to inspect consumers' purchasing intention of GMF. The result revealed that if the price of GMF was same as that of the traditional one, only 9.8% of the respondents would choose the former and 63.7% the latter; if the two prices were different, 55.0% would still buy non-GMF even though it was more expensive, while 45.0% would choose the cheaper one even though it was GMF. Hence, we can see that consumers hold prudent attitu-

Table 1. Influence of gender and a child under 18 on consumers' risk perception level of GMF.

Test variable	Levene's test		Descriptive statistical analysis				Independent Sample t-test	
	F-value	P-value	Grouping Variable	Number	Mean	SD	T-value	P-value
Gender	0.003	0.354	Male	152	3.52	0.33	5.604	0.047
			Female	148	3.94	0.55		
Child under 18	0.160	0.289	with	196	4.12	0.32	-4.822	0.039
			without	104	3.62	0.66		

Table 2. Influence of age, income, and education background on consumers' risk perception level of GMF.

Test Variable	Levene's Test		Descriptive Statistical Analysis				One-way AVOVA	
	F-value	P-value	Grouping Variable	Number	Mean	SD	F-value	P-value
Age	7.354	0.058	<30	107	2.47	0.51	3.341	0.274
			30-55	132	3.95	0.70		
			>55	61	3.52	0.39		
Personal annual income	0.152	0.061	<30 000	67	3.46	0.46	3.054	0.018
			30 000-60 000	135	3.98	0.74		
			>60 000	98	4.21	0.78		
Education back-ground	5.973	0.547	Low degree	92	2.81	0.32	5.255	0.002
			Medium degree	121	3.63	0.53		
			High degree	87	4.05	0.80		

de towards purchasing GMF which safety is still controversil, and they tend to buy GMF with a lower price than traditional food.

4. INFLUENCE OF CONSUMER CHARACTERISTICS ON GMF RISK PERCEPTION

Taking respondents' risk perception level of GMF as dependent variable, independent sample t-test and one-way ANOVA were conducted to examine whether there were significant effects of respondent characteristics on risk perception level of GMF. The Levene's test showed that all variables conform to homoscedasticity (Tables 1 and Table 2).

Independent sample t-test was employed to determine whether consumers' gender and possession of at least one child under 18 had significant effects on risk perception of GMF. As shown in Table 1, both gender and possession of at least one child under 18 had remarkable influence on consumers' risk perception of GMF (p<0.05). The risk perception level of GMF was higher by females than by males. One reason for this could be that females take more responsibilities than males for food purchasing. Meanwhile, consumers with a child under 18 have remarkably higher risk perception

level of GMF than those without, and it may be due to stronger requirements for food safety and nutrition out of the consideration of their children's health.

One-way ANOVA was employed to determine whether consumers' age, personal annual income and education background had significant effects on risk perception of GMF. As shown in Table 2, only education background (p<0.01) and personal annual income (p<0.05) influenced risk perception of GMF significantly.

The mean of risk perception level for consumers with senior high school and college degree were 3.63% and 4.05%, respectively, clearly higher than those with lower degree. The possible reasons may be that the information which highly educated consumers receive is much more comprehensive, so they may obtain more negative reports about GMF to cause their higher level of risk perception.

Higher-income consumers had significantly higher risk perception level than lower-income ones. A possible reason for it may be that the former pursue higher-quality life and have higher requirement for food quality and safety, so they hold cautious attitude to GMF whose safety is still pending.5. Main factors influencing consumers' risk perception of GMF

The main factors influencing consumers' risk perceptive level of GMF were examined by 14 questions, and principal

Table 3. Main factors influencing on consumers' risk perception of GMF.

Index	Health Risk	Ecological Risk	Social Risk	Function Risk
Worry about harm to family's health	0.846	0.038	-0.401	0.061
Worry about influence on children's growth	0.820	0.165	0.306	0.036
Worry about influence on human reproduction	0.701	0.046	0.128	0.095
Worry about emergence of super-weed or super-pest	-0.142	0.769	0.095	0.292
Worry about destruction of biological diversity	0.099	0.675	-0.392	0.393
Worry about destruction of ecological balance	0.367	0.613	0.221	0.035
Worry about environmental pollution	0.330	0.592	0.114	0.114
Worry about influence on food security in china	0.400	0.096	0.879	0.110
Worry about control of breeding area by foreign countries	-0.343	0.055	0.726	0.255
Worry about excessive intervention in biological evolution process	0.152	-0.110	0.693	0.009
Worry about poor nutrition	0.059	0.134	0.120	0.731
Worry about inferior taste	0.102	0.020	0.248	0.754
Eigenvalue	2.351	1.871	1.401	1.027
Total variance explained (%)	35.778	18.392	12.775	9.315
Cumulative proportion of variance (%)	35.778	54.170	66.945	76.260

components were extracted. First, KMO and Bartlett's test of spherical were conducted. The KMO value was 0.815, which indicates that the 14 questions are closely related. In addition, chi-square value of Bartlett's test of spherical was 832.619 (p=0.000), which indicates that correlation matrix is significant different from identity matrix. Therefore, the data were appropriate to be analyzed by factor analysis.

Then, factor loading matrix was rotated by varimax rotation. Index 2, which indicates worry about food allergy or intoxication, was double loaded; index 12, which indicates worry about destructing natural selection, had a lower load. Therefore, these two indexes were excluded and other twelve indexes left.

On the basis of eigenvalue greater than 1, four common factors were extracted. Factor loading after rotation and four common factors were shown in Table **3**. Cumulative proportion of variance of the common factors was 76.260%, which indicates that these factors can adequately express the information of these indexes.

Eigenvalue of the first common factor was 2.351 and total variance explained was 35.778%. It included three indexes: worry about harm to family health, children's growth and human reproduction, which can be summarized as harm of GMF to consumers' and their families' health, so it may be named as health risk.

Eigenvalue of the second common factor was 1.873 and total variance explained was 14.392%. It mainly included four indexes: worry about emergence of super-weed or super-pest, destruction of biological diversity, destruction of ecological balance and environmental pollution. Because

these four indexes are mainly about destruction of ecological environment, it may be named as ecological risk.

Eigenvalue of the third common factor was 1.401 and total variance explained was 10.775%. It mainly included three indexes: worry about the influence on Chinese food security, control of breeding area by foreign countries and excessive intervention in biological evolution process. Thus, it may be named as social risk.

Eigenvalue of the fourth common factor was 1.211 and total variance explained was 9.315%. It mainly included two indexes: worry about poor nutrition and inferior taste. Nutrition and taste are two major elements taken into consideration in purchasing food, so it may be named as function risk.

On the basis of extraction of four common factors, multivariate linear regression analysis was adopted to further inspect whether these common factors had influence on consumers' risk perception level of GMF and their influencing degree. The model is expressed as follows:

$$Y = \beta_0 + \beta_1 X_1 + \beta_2 X_2 + \beta_3 X_3 + \beta_4 X_4 + \varepsilon \qquad (1)$$

Where Y refers to consumers' risk perception level of GMF; $\beta 0$ means constant term; four common factors extracted, which are health risk, ecological risk, social risk, and function risk, are independent variables X1, X2, X3, and X4; $\beta 1$, $\beta 2$, $\beta 3$, and $\beta 4$ represent unstandardized regression coefficients; ε is a random disturbance term.

The results of regressive analysis showed that the adjusted R2 was 0.413, p=0.000, which indicates that the model passes test. According to Table **4**, the multiple linear

Table 4. Regression analysis of factors influencing consumers' risk perception level of GMF.

Variable	Beta	S.E.	t	Sig.
(Constant)	2.475	0.277	55.015	0.000
Health risk (X1)	0.386	0.041	-5.865	0.001
Ecological risk (X2)	0.187	0.020	3.216	0.007
Social risk (X3)	0.163	0.041	2.844	0.049
Function risk (X4)	0.145	0.036	-0.589	0.281

regression equation that influences consumers' risk perception level of GMF can be expressed as follows:

$$Y = 2.475 + 0.386X_1 + 0.187X_2 + 0.163X_3 \qquad (2)$$

The regression equation indicates that different factors have different influencing degrees on consumers' risk perception level of GMF. Among them, health risk that consumers can perceive has the deepest influencing degree ($p<0.01$), which conforms to reality. What consumers concern most about food is its influence on health. With the increased awareness of safety and health, consumers become prudent about GMF and tend to avoid purchasing them when faced with much negative information about GMF.

The common factor, ecological risk, also had a significant influence on consumers' risk perception of GMF ($p<0.01$). That is, the greater destruction of transgenosis to ecological environment consumers considered, the higher risk perception level of GMF they had. This result may be related to the environmental protection idea advocated in current society. There are some opinions of harm to environment due to genetic modification, such as producing super-weed or super-pest, polluting environment and destructing ecology.

The common factor, social risk, also had a significant influence on consumers' risk perception of GMF ($p<0.05$). On the one hand, consumers worry that spreading GMF will cause Chinese breeding area under the control of transnational corporations with dominant position in genetically modified technology development and commercial operation, which will pose threat to native food security. On the other hand, to change organism's biological features, transfer some gene from one species to another may affect normal evolution process of plants. This also contributes to consumers' worry about GMF to some extent, and then intensifying their risk perception of GMF.

By contrast, the common factor, function risk, had no significant influence on consumers' risk perception of GMF. Compared to food nutrition and taste, food safety is much more important for the public. For newborn high-tech GMF, consumers pay more attention to its safety. For example, whether it will give rise to food allergy or intoxication. Meanwhile, for vegetarians or religious persons, the most important isn't its nutrition or taste, but its accordance with doctrines and ethics. Consequently, compared with the safety and ethics of GMF, its food functions such as nutrition and taste have less influence on consumers' risk perception of GMF. This result is similar to that of Dong [3].

6. RESEARCH CONCLUSIONS AND POLICY SUGGESTIONS

Focusing on consumers' risk perception of GMF and its influencing factors in Jiangsu province, this paper comes to the following conclusions.

First, although up to 82.3% of the respondents have heard of GMF, their congnition of GMF is vague. Consumers gain information about GMF mainly by TV, internet, and newspapers and magazines, but they most distrust the information from internet. Moreover, they most trust the information from friends and relatives.

Second, most consumers worry about GMF safety and they hope the government set up mandatory labeling system of GMF to help them to differentiate GMF from traditional food.

Third, consumers are more likely to buy GMF with a lower price than traditional food. If there is no difference in price, most consumers are reluctant to buy GMF.

Fourth, some characteristics of consumer, including age, education background, personal annual income, and whether has a child under 18, have significant influence on their risk perception of GMF. Females and those with children under 18 have obviously higher level of risk perception of GMF than males and those without children under 18. Furthermore, compared with consumers with lower education degree or lower income, higher-educated or higher-income consumers have much higher level of risk perception of GMF.

Fifth, different factors impose different influencing degrees on consumers' risk perception of GMF. According to influencing degree, they can be listed from the highest to the lowest as health risk (0.368), ecological risk (0.187) and social risk (0.163). However, function risk has no significant influence on consumers' risk perception of GMF.

According to the conclusions, policy suggestions can be summarized as follows.

First, the government should strengthen the propaganda of genetically modified technology as well as GMF through

TV, newspapers and magazines, and ensure that the information access to consumers is scientific and understandable. Popularization of the knowledge of GMF may help the consumers get rid of confusion about GMF, so may effectively reduce consumers' risk perception of GMF.

Second, the government should perfect supervision policy on GMF and enhance consumer confidence in GMF. Because GMF label plays an important part in consumers' purchasing decision, the government need to reinforce standardized management of GMF label and supervision of enterprises that manufacture GMF. Meanwhile, the government should guide consumers to have a correct understanding of GMF and its labels, thus guarantee consumer right to the truth and the options sufficiently, which is conducive to improving consumer welfare.

ACKNOWLEDGEMENTS

The authors are grateful for the financial support of the social science project of Jiangsu Education Department of China No. 2013SJD630063, Yangzhou university humanity and social science project No. xjj2014-67, Key Projects of National Social Science Foundation of China No. 14ZDA069, the National Natural Science Foundation of China No. 71273117, Central University Basic Research Funds No. JUSRP51325A & JUSRP51416B, the Project of the Six Top Talents in Jiangsu Province No. 2012-JY-002 and the project of college Innovation Team of Jiangsu Province social science No. 2013-011, Soft Science Research Project of Yangzhou No. YZ2014250.

REFERENCES

[1] S. Liu, J. Huang, and J. Bai, "Consumer's willingness to pay genetically modified foods in China", *Journal of International Food and Agribusiness Marketing*, vol. 43, pp. 571-584, 2006.

[2] J. Scully, "Genetic engineering and perceived levels of risk", *British Food Journal*, vol. 105, no. 1-2, pp. 59-77, 2003.

[3] Y. Y. Dong, Z. H. Qi, and D. M. Zhang, "Effect of perceived risks of genetically modified food on consumers' purchase intention: Based on a survey conducted in Wuhan", *Journal of China Agricultural University*, vol. 19, no. 3, pp. 27-33, 2014.

[4] F. N. Zhong and Y. L. Ding, "Consumer awareness and response to genetically modified food in Nanjing", *China Rural Survey*, no. 1, pp. 22-27, 2004.

[5] J. L. Ruan, C. Chen, and L. H. Chen, "Investigation and analysis of consumer recognition of genetically modified foods and transgenic labelling: a case study of Shenzhen city", *Modern Food Science and Technology*, vol. 29, no. 4, pp. 848-852, 2013.

11

Influence of Ionic Liquid 1- butyl-3-methylimidazolium Chloride on the Soil Micro-Ecological System

Yanjie Tong[a], Qijun Wang[a,b], Yafan Bi[a], Mingke Lei[a], Yezi Lv[a], Yangyang Liu[a], Jiali Liu[a], Lili Lu[a], Yali Ma[a], Yuanxin Wu[a] and Shengdong Zhu*,[a]

[a]*Key Laboratory for Green Chemical Process of Ministry of Education, Hubei Key Laboratory of Novel Chemical Reactor and Green Chemical Technology, School of Chemical Engineering and Pharmacy, Wuhan Institute of Technology, Wuhan 430073, P.R. China*

[b]*College of Horticulture and Landscape Architecture, Key Laboratory of Horticulture Science for Southern Mountainous Regions, Ministry of Education, Southwest University, Chongqing 400715, P.R. China*

Abstract: In order to evaluate the influence of ionic liquid 1- butyl-3-methylimidazolium chloride ([Bmim]Cl) on the soil micro-ecological system, the toxicity of [Bmim]Cl to soil microorganisms and its impact on soil physico-chemical properties were investigated. Three soil samples, which were taken from the rape land, nursery land and the broad bean land respectively, were used for this study. The toxicity test results show that the [Bmim]Cl inhibited the growth of soil microorganisms including bacteria and actinomycetes. This inhibition became stronger with the [Bmim]Cl concentration increasing. The EC50 of soil bacteria was close to that of the *Vibrio fischeri,* and the EC50 of soil actinomycetes was near to that of the *Pseudokirchneriella subcapitata.* The soil physico-chemical properties test results indicate that the organic mass and the soluble salts in soil increased with the increase of the [Bmim]Cl concentration. The [Bmim]Cl also caused the pH change in the soil micro-ecological system. It suggests that the ionic liquid [Bmim]Cl would influence the soil micro-ecological system by inhibiting the growth of soil microorganisms and altering the soil physico-chemical properties when it contaminated the soil system.

Keywords: Ionic liquid, [Bmim]Cl, soil micro-ecological system, microorganism, physico-chemical property.

1. INTRODUCTION

Ionic liquids (ILs) are a group of new organic salts that exist as liquids at a relatively low temperature (<100 ℃). They are composed entirely of ions, typically large organic cations and small inorganic anions [1, 2]. Interests in ILs have steadily grown in recent years because they have many attractive properties, such as chemical and thermal stability, non-flammability and immeasurably low vapor pressure, which provide the possibility for clean manufacturing in chemical-related industry [3]. Researches on the applications of ILs have become one of the most active areas in green chemistry, the ILs have been widely used in organic synthesis, separation, biotransformation, new material preparation and renewable resource utilization [4-8]. Although most of these applications are still in a laboratory scale, some of them have been coming into pilot or commercial stage since the BASF first successfully used the IL in an industrial scale in 2003 [9, 10]. Compared using the traditional volatile organic solvents, the industrial application of ILs can help to reduce air pollution because of their immeas-

urably low vapor pressure, but it is still possible to release them into environment by accidental spills or effluents, which might cause water or soil pollution [11-13]. Therefore, much attention should be paid on their influence on the environment and ecology when they are used in an industrial scale. The impact of ILs on the environment and ecology is closely related to their bioaccumulation, toxicity and degradability. Based on the reported data on their bioaccumulation, toxicity and degradability in recent years, the ILs might cause water or soil pollution as the commonly used chemicals [11-13]. However, as we know, there are no reports on the influence of ILs on their contaminated soil microecological system. This work is to deal with how the ILs will affect their contaminated soil micro-ecological system. In this work, the most commonly used IL, 1- butyl-3-methylimidazolium chloride ([Bmim]Cl) was selected as the model IL. Three soil samples, which were taken from the rape land, nursery land and the broad bean land respectively, were used as the model soil micro-ecological system. The toxicity of [Bmim]Cl to soil microorganisms and its impact on soil physico-chemical properties were investigated to evaluate its influence on the soil micro-ecological system.

2. MATERIALS AND METHODS

All experiments were carried out three times. The data reported are expressed as the mean values ± standard deviation.

---

*Address correspondence to this author at the Key Laboratory for Green Chemical Process of Ministry of Education, Hubei Key Laboratory of Novel Chemical Reactor and Green Chemical Technology, School of Chemical Engineering and Pharmacy, Wuhan Institute of Technology, Wuhan 430073, P.R. China; E-mail: zhusd2003@21cn.con

2.1. Materials and Chemicals

Soil samples were taken from different sites (rape land, nursery land and the broad bean land) in Wuhan, Hubei province, China. All soil samples were collected in the beginning of April in 2010. The soil samples from the rape land, nursery land and the broad bean land were chosen to evaluate the influence of [Bmim]Cl on the soil micro-ecological system with different biological and physico-chemical characteristics.

Sampling was carried out at 5 randomly chosen points from each site. Samples were collected at a depth of 1~15 cm, after removing the top layer. All samples from one site were mixed, then were air-dried, homogenized by sieving to less than 2 mm to separate roots and large objects, and stored in a polyethylene bag at room temperature for later use.

Each soil sample from one site was divided eight portions, one portion was used as the control, and the remaining seven portions were added the suitable amount of [Bmim]Cl to their required [Bmim]Cl concentration respectively. In this study, their [Bmim]Cl concentration were set at 10000 ppm, 1000 ppm, 100 ppm, 10 ppm, 1 ppm, 0.1 ppm and 0.01 ppm respectively. These soil samples with different [Bmim]Cl concentrations were stored in polyethylene bags for 15 days at room temperature, and then they were used to determine their biological and physico-chemical properties.

The [Bmim]Cl used in this study was obtained from Henan Lihua Pharmaceutical Co. Ltd., China. All other chemicals employed in this study were of reagent grade and purchased from Wuhan Chemicals & Reageng Corp., China.

2.2. Analytical Methods

2.2.1. Determination of Biological Properties

2.2.1.1. Determination of the Number of Living Microorganisms

The number of living microorganisms (bacteria or actinomycetes) in soil samples was estimated by viable count on serial spread plates [14]. A series of 10–fold dilution for the sample was prepared starting with 9 ml of sterilized phosphate buffered saline added to 1.0 g soil sample. The flask was then closed and the contents were stirred for 30 min, and 1.0 ml suspension of sample was added to 9.0 ml of sterilized phosphate buffered saline. The dilutions were repeated to 7 continuous dilutions. Finally, 200 µl from each serial dilutions of the sample suspension was spread over an agar plate with beef extract peptone medium for bacteria and an agar plate with Gause's No.1 synthetic medium for actinomycetes. All plates were in triplicates and each soil sample was diluted for three independent measurements. Finally, the plates were incubated two or three days at 28℃ until colonies appeared (2 days for bacteria and 3 days for actinomycetes), and colony forming units (CFU) were count which varied from 30 to 300.

2.2.1.2. Calculation of the EC50 of Microorganisms

The concentration of added [Bmim]Cl which inhibits 50% of the growth of microorganisms (bacteria and actinomycetes) comparing with the control culture was de-fined as EC50, and it was calculated by a regression method [15].

2.2.2. Determination of Physico-Chemical Properties

The organic matter in each soil sample was determined using a $K_2Cr_2O_7$-H_2SO_4 method as described in Chinese National Standard GB9834--88. The soluble salts in each soil sample was measured following the procedures described in Chinese Agricultural Standard NY/T1121.16-2006. The pH value in each soil sample was determined according to the method described in ISO10390-2005.

3. RESULTS AND DISCUSSION

3.1. Toxicity of [Bmim]Cl to the Soil Microorganisms

The soil microorganisms are an important element in soil micro-ecological system and play a vital role in soil micro-ecological system. The toxicity of [Bmim]Cl to the soil microorganisms is an important part of its influence on the soil micro-ecological system. Table 1 listed the number of microorganisms in different soil samples at different [Bmim]Cl concentrations. As indicated in Table 1, the number of microorganisms including bacteria and actinomycetes decreased with the increase of [Bmim]Cl concentration for all soil samples. It demonstrated that the [Bmim]Cl could inhibit the growth of such soil microorganisms as bacteria and actinomycetes. Table 2 listed the EC50 of microorganisms including bacteria and actinomycetes of different soil samples. As shown in Table 2, the EC50 values were at relatively lower concentrations for both microorganisms. It indicated that the inhibition of the [Bmim]Cl to the growth of both microorganisms was not just simply due to the pH shift because the [Bmim]Cl concentration changed, but it came from its toxicity to these microorganisms. For all soil samples, the EC50 of bacteria was greater than that of actinomycetes. It indicated that actinomycetes in soil was more sensitive to [Bmim]Cl than bacteria, that is, the [Bmim]Cl had a stronger inhibition on the growth of actinomycetes. For different soil samples, the EC50 of microorganisms including bacteria and actinomycetes existed difference and it showed that the [Bmim]Cl had a different extent influence on the growth of the soil microorganisms which come from the soil micro-ecological system with different biological and physico-chemical characteristics. Compared with the reported EC50 of pure cultures [16-19], the EC50 of soil bacteria was close to that of the *Vibrio fischeri*, and the EC50 of soil actinomycetes was near to that of the *Pseudokirchneriella subcapitata*. It seems that the toxicity of ILs to soil microorganisms is close to that of some well-investigated organisms, such as *Vibrio fischeri* and *Pseudokirchneriella subcapitata*. Whether the ILs toxicity data to these well-investigated organisms can be used to predict the toxicity of ILs to soil microorganisms, further research work are needed because the soil microbial community are far more complicated than the pure cultures. Anyway, the [Bmim]Cl would influence the soil micro-ecological system by inhibiting the growth of soil microorganisms including bacteria and actinomycetes. The well-established IL toxicity data of some pure cultures could provide useful information in the prediction of its toxicity to soil microorganisms.

Table 1.   The Number of Microorganisms in Different Soil Samples at Different [Bmim]Cl Concentrations

	Soil type	[Bmim]Cl concentration (ppm)							
		10000	1000	100	10	1	0.1	0.01	control
Bacteria number (10^7CFU.g^{-1})	Rape land	1.30±0.08	3.20±0.17	7.65±0.54	11.06±1.03	11.51±0.87	11.22±0.79	11.75±0.86	11.80±0.92
	Broad bean land	1.20±0.23	1.25±0.35	3.85±0.045	4.35±0.46	4.85±0.63	5.05±0.73	5.29±0.39	5.45±0.47
	Nursery land	0.13±0.01	0.95±0.02	2.45±0.39	4.50±0.81	5.45±0.43	5.37±0.54	5.53±0.61	5.50±0.79
Actinomycetes number (10^6CFU.g^{-1})	Rape land	0.30±0.02	0.65±0.06	1.80±0.02	3.00±0.31	3.35±0.54	3.69±0.43	3.87±0.49	4.05±0.75
	Broad bean land	1.00±0.13	2.5±0.83	3.00±0.38	5.00±0.58	9.00±0.87	9.45±0.48	10.80±0.70	11.51±0.82
	Nursery land	0.10±0.02	0.15±0.02	0.50±0.07	0.85±0.13	1.33±0.19	1.42±0.21	1.73±0.19	1.95±0.16

Table 2.   The EC50 of Microorganisms of Different Soil Samples

Microorganism type	Land type	EC50 (ppm)
Bacteria	Rape land	453.93±20.70
	Broad bean land	489.42±14.47
	Nursery land	868.29±34.31
Actinomycetes	Rape land	81.23±4.27
	Broad bean land	83.13±2.96
	Nursery land	7.50±1.34

3.2. Influence of [Bmim]Cl on the Soil Physico-Chemical Properties

The impact of [Bmim]Cl on the soil physico-chemical properties is another important part of its on the soil micro-ecological system besides its toxicity to soil microorganisms. Table 3 listed some important physico-chemical properties of different soil samples at different [Bmim]Cl concentra-tions. As indicated in Table 3, the organic mass and the soluble salts in soil samples increased with the increase of the [Bmim]Cl concentration. This is because the [Bmim]Cl itself was an organic compound and it had strong solubility to salts in the soil samples. The [Bmim]Cl also caused the pH change in the soil micro-ecological system. When the [Bmim]Cl concentration was greater than 100 ppm, the soil

Table 3.   The Physico-Chemical Properties of Different Soil Samples at Different [Bmim]Cl Concentrations

	Soil type	[Bmim]Cl concentration (ppm)							
		10000	1000	100	10	1	0.1	0.01	control
Organic matter (mg.g^{-1})	Rape land	31.78±1.59	25.30±1.42	22.11±1.15	20.69±1.43	19.29±0.97	17.85±1.02	15.20±1.07	15.06±1.35
	Broad bean land	22.11±2.13	18.01±1.59	15.47±1.67	13.96±2.49	13.05±1.71	12.69±1.63	11.96±1.56	10.97±1.55
	Nursery land	75.40±1.77	69.14±2.37	67.45±1.97	65.70±1.29	62.24±2.11	60.59±2.23	58.70±1.43	58.23±2.12
Soluble salts (mg.g^{-1})	Rape land	1.57±0.08	1.23±0.06	1.19±0.06	0.97±0.05	0.85±0.05	0.76±0.04	0.73±0.04	0.74±0.04
	Broad bean land	1.8±0.07	1.44±0.09	1.32±0.07	1.17±0.05	1.05±0.04	0.92±0.05	0.79±0.03	0.74±0.03
	Nursery land	2.84±0.09	1.65±0.08	1.46±0.07	1.13±0.05	0.91±0.05	0.78±0.03	0.67±0.04	0.62±0.03
pH	Rape land	5.82±0.02	6.51±0.02	6.93±0.03	6.88±0.05	6.85±0.03	6.89±0.06	6.91±0.02	6.96±0.03
	Broad bean land	6.29±0.04	6.62±0.03	6.81±0.03	6.90±0.04	6.88±0.04	6.96±0.05	6.93±0.05	7.04±0.05
	Nursery land	6.08±0.04	6.44±0.05	7.05±0.05	7.03±0.03	7.07±0.05	7.03±0.04	7.07±0.05	7.02±0.05

pH decreased with the [Bmim]Cl concentration increasing. When the [Bmim]Cl concentration was less than 100 ppm, the soil pH was almost the same as the control. This is the result of interaction between the [Bmim]Cl and the soil microorganisms. At lower [Bmim]Cl concentration, although the [Bmim]Cl itself cause the pH to decrease because of its acidity, the soil microorganisms had the ability to adjust the soil pH and the soil pH could remain almost unchanged. However, at higher [Bmim]Cl concentration, the [Bmim]Cl cause the pH to decrease and, at the same time, the soil microorganisms lost the ability to adjust the soil pH because of its toxicity, the final result was that the soil pH decreased.

CONCLUSIONS

This work investigated the toxicity of [Bmim]Cl to soil microorganisms and its impact on soil physico-chemical properties to evaluate its influence on the soil micro-ecological system. The main conclusions are as follows:

1) The [Bmim]Cl inhibited the growth of soil microorganisms including bacteria and actinomycetes. This inhibition became stronger with the [Bmim]Cl concentration increasing. Based on the EC50 of the soil samples, the actinomycetes were more sensitive to the [Bmim]Cl than the bacteria.

2) The organic mass and the soluble salts in soil increased with the increase of the [Bmim]Cl concentration. The [Bmim]Cl also caused the pH change in the soil micro-ecological system.

3) The [Bmim]Cl would influence the soil micro-ecological system by inhibiting the growth of soil microorganisms and altering the soil physico-chemical properties when it contaminated the soil system.

ACKNOWLEDGEMENTS

This work was support by Hubei Key Laboratory of Novel Chemical Reactor and Green Chemical Technology (RGCT201005) and the National Natural Science Foundation of China (No. 21176196). We thank Dr Shixue Zheng (Huazhong Agricultural University, Wuhan, China) for his assistance in experiment work.

REFERENCES

[1]   Wasserscheid P, Stark. Handbook of Green Chemistry: Green Solvents, Vol.6-Ionic Liquid. Wiley-VCH, 2010.

[2]   Zhang S, Xu C, Lv X, Zhou Q. Ionic liquid & green chemistry. China Science Press, Beijing, 2009.

[3]   Rogers RD, Seddon KR. Ionic liquids: Industrial applications for green chemistry. Oxford University Press, England, UK, 2002.

[4]   Wasserscheid P, Welton T. *Ionic Liquid in Synthesis* (2nd edn). Wiley-VCH, 2007.

[5]   Zhao H, Xia SQ, Ma PS. Use of ionic liquids as green solvents for extractions. J Chem Technol Biotechnol 2005; 80: 1089-96.

[6]   Rantwijk FV, Sheldon RA. Biocatalysis in ionic liquids. Chem Rev2007; 107: 2757-85.

[7]   Zhu S, Wu Y, Chen Q, *et al.* Dissolution of cellulose with ionic liquids and its applications: a mini-review. Green Chem 2006; 8: 325-7.

[8]   Zhu S. Use of ionic liquids for the efficient utilization of lignocellulosic materials. J Chem Technol Biotechnol 2008; 83: 777-9.

[9]   Freemantle M. BASF's smart ionic liquid. Chem Eng News 2003; 81: 9.

[10]   Plechkova NV, Seddon KR. Applications of ionic liquids in chemical industry. Chem Soc Rev 2008; 37: 123-50.

[11]   Zhu S, Wu Y, Chen Q, Chi R, Shen X, Yu Z. A mini-review on greenness of ionic liquids. Chem Biochem Eng Q 2009; 23: 207-11.

[12]   Pham TPT, Cho CW, Yun YS. Environmental fate and toxicity of ionic liquids: a review. Water Res 2010; 44: 352-72.

[13]   Zhao D, Liao Y, Zhang Z. Toxicity of ionic liquids. Clean-Soil, Air, Water, 2007; 35: 42-8.

[14]   Zheng S, Yao J, Zhao B, Yu Z. Influence of agriculture practice on soil microbial activity measured by microcalorimetry. Eur J Soil Biol 2007; 43: 151-7

[15]   Lee SM, Chang WJ, Choi AR, Koo YM. Influence of ionic liquids on the growth of *Escherichia coli*. Korean J Chem Eng 2005; 22: 687-90.

[16]   Couling DJ, Bernot RJ, Docherty KM, Dixon JK, Maginn EJ. Assessing the factors responsible for ionic liquid toxicity to aquatic organisms *via* quantitative structure-property relationship modeling. Green Chem 2006; 8: 82-90.

[17]   Garcia MT, Gathergood N, Scammells PJ. Biodegradable ionic liquids. Part II: effect of the anion and toxicology. Green Chem 2005; 7: 9-14.

[18]   Docherty KM, Kulpa CF. Toxicity and antimicrobial activity of imidazolium and pyridinium ionic liquids. Green Chem 2005; 7: 185-9.

[19]   Wells AS, Coombe VT. On the freshwater ecotoxicity and biodegradation properties of some common ionic liquids. Org Pro Res Dev 2006; 10: 794-8.

12

Responses of *Sargassum thunbergii* Germlings to Acute Environmental Stress

Tang Yongzheng[*], Chu Shaohua, Lu Zhicheng, Yu Yongqiang and Li Xuemeng

Ocean School, Yantai University, Yantai 264005, PR China

Abstract: The responses of *Sargassum thunbergii* germlings to high temperature, low salinity, desiccation, combined thermal and osmotic stress (35 °C combined with 12 psu), anthracene, and eutrophication were examined. Probit regression analysis results showed that the median lethal time (LT50) values of high temperature decreased with the increase in temperature. The 24 h median lethal temperature was 36.9 °C. For salinity treatment, the LT50 value of fresh water was 47.6 h. Survival rates of germlings were over 60% when germlings were exposed to salinities ranging from 27 psu to 7 psu at a time interval of 108 h post-treatment. The LT50 values of desiccation and combined thermal and osmotic stress (35 °C combined with 12 psu) were 7.0 h and 9.8 h, respectively. Anovas showed that germlings were inhibited by high concentrations of anthracene (5 mg L^{-1} and 10 mg L^{-1}) with low survival rates of below 50% and low relative growth rates of below 1% after 25 days of culture; however, low concentrations (0.01-1 mg L^{-1}) had no significant effects. In addition, neither severe eutrophication nor disproportionality of N/P showed any significant effect on the survival and growth of germlings. Of the environmental stresses tested, possible occurrence of high temperature of 40 °C and combined thermal and osmotic stress directly impacted the survival of germlings, suggesting that the deterioration of S. thunbergii bed may be related more to increasing extreme climatic events.

Keywords: Environmental stress, germling, median lethal value, *Sargassum thunbergii*.

1. INTRODUCTION

As a member of *Sargassum* beds, *Sargassum thunbergii* is ecologically important in the maintenance of a healthy coastal ecosystem [1], such as serving as a primary producer; in spawning, nursery and feeding ground for marine organisms; and as nutrient cycling controller [2]. However, the natural populations of *S. thunbergii* along the coast of China have evidently deteriorated in recent years [3]. It is necessary to explore the reason for this phenomenon. Mostly, deterioration of seaweed beds is caused by various anthropogenic interferences and climate changes which cause increasing extreme events and physical stresses [4]. For example, increased temperature is generally thought to have negative effects on spore production, germination, sporophyte growth and recruitment of seaweeds [5]. Particularly, when they experience periods of temperature change, which are sufficiently high to result in disruptive stress, such damage and any reallocation of resources for protection and repair can cause slow growth, delay development and also lead to mortality [6]. In addition, seaweed growths have declined due to water pollution such as polycyclic aromatic hydrocarbons (PAHs) and eutrophication. Anthracene is a PAH with higher solubility than most other PAHs due to its low molecular capacity and may prove a threat to the environment if widely distributed [7]. It acts as a photosensitizer causing an oxidative damage of algal cells [8]. The eutrophication of coastal zones resulted from land run-off, river inflow and sewage discharges with an imbalanced N/P ratio can cause harmful algal blooms [9] and deterioration of macroalgal communities [10]. Deeper in water, total algal abundance and abundance of perennial algae decrease along a eutrophication gradient [11]. In brief, both physical and chemical stresses may have adverse effects on the survival of seaweed.

The role of recruitment from germlings in the *S. thunbergii* populations has not been investigated. In other *Sargassum* species, the role of recruitment from propagules in local persistence and stabilising densities of populations remains controversial [12]. In any case, early stage is a bottleneck for algae [9]. Therefore, it is necessary to test the tolerance of *S. thunbergii* germlings to environmental stresses to search for the reasons of deterioration of *S. thunbergii* bed.

Our previous studies investigated the responses of *S. thunbergii* germlings to combined physical stresses [13]. However, the effects of individual physical and chemical stress are still unknown. Therefore, the present study focuses on the chemical stress and the median lethal values of germlings responding to the individual physical stress. As a common experimental measure of stress tolerance, the acute LD50 (median lethal dose) is used for evaluating tolerance to physical stress due to its high accuracy and flexibility [14]. The lower the value of LD50, the higher is the damage of the stress. The median lethal time was also estimated for the combined thermal and osmotic stress (35 °C combined with 12 psu) which previously resulted as extreme for germlings [13]. The objective of this study was to investigate if the

*Address correspondence to this author at the Ocean School, Yantai University, yantai, shandong, 264005, P.R. China;
E-mail: yongzht_cn@sina.com

physical and chemical parameters emerge as environmental stresses for *S. thunbergii* germlings; and if the environmental stresses caused by local (pollution) and overall (climate change) anthropogenic disturbances can explain the deterioration of *S. thunbergii* beds.

2. MATERIAL AND METHODS

2.1. Collection of Germlings

Fertile female and male specimens of *S. thunbergii* were collected on June 25, 2011, in the intertidal zone of Zhanqiao (37°31'N, 121°26'E), Yantai. Selected thalli (about 25 cm) were healthy and yellowish-brown in appearance with intact and inflated receptacles which had no obvious shedding. Once transported to the laboratory, they were placed in two 15L plastic tanks (30cm × 20cm × 25cm) filled with filtered seawater which was continuously aerated. Tanks were kept at 25 °C, 60 µmol photons m^{-2}s^{-1} and with a 10L:14D (light: dark cycle) photoperiod. The irradiance was provided with fluorescent illumination and measured on the water surface.

Released germlings sank to the bottom of tanks within 24 h after fertilization. Germlings were then collected by filtering the remaining seawater with an 80-mesh nylon sieve, followed by filtration with a 200-mesh nylon sieve. A total of about 3.0 × 10^5 germlings were obtained. They were subsequently transferred to a 3 L glass tank and even stirred to produce a homogeneous suspension.They were then immediately poured into each Petri dish (60mm × 15mm). A total of 109 Petri dishes with about 700 germlings in each were used for the experiment.

2.2. Stress Treatments

After 24 h, germlings attached to the Petri dishes were cultured in seven light and temperature-controlled incubators on the basis of the following experimental design. Three independent replicates were used for each treatment. For desiccation treatment, samples were cultured in a total of five durations of desiccation: 3 h, 6 h, 9 h, 10 h and 12 h. The desiccation treatment was conducted at relative humidity of about 80%. The chosen level of relative humidity was in accordance with that in the sample plot. The relative humidity was measured with a hygrometer and adjusted by placing soaked filter paper or calcium sulfate in three incubators. After desiccation, germlings were fully immersed for 12 h prior to the record of survival rate.

For hyposalinity treatment, samples were cultured at five levels of salinity: 27, 20, 13, 7 and 0. Solutions of different salinities were prepared from seawater diluted with distilled water, measured with a salinity hydrometer (GM Manufacturing Co.) and changed on a daily basis.

For anthracene treatment, samples were cultured at seven concentrations of anthracene: 0.01 mg L^{-1}, 0.05 mg L^{-1}, 0.1 mg L^{-1}, 0.5 mg L^{-1}, 1 mg L^{-1}, 5 mg L^{-1} and 10 mg L^{-1}, respectively. The tests of anthracene were performed on the basis of two controls (germlings in culture with and without 0.5% DMSO).

A two-way factorial experimental design was used to test the effects of severe eutrophication and disproportionality of N/P on germlings, along-with the concentrations of N and P

as fixed factors [15]. The concentrations of N and P sources were designed as 0.072 mg L^{-1}, 0.36 mg L^{-1} and 2 mg L^{-1} for N; and 0.011 mg L^{-1}, 0.03 mg L^{-1} and 0.3 mg L^{-1} for P. NaNO$_3$ and KH$_2$PO$_4$ were added to the seawater as respective N and P sources for a final concentration. Therefore, high N/P ratios (0.36/0.011, 0.36/0.03, 2/0.011 and 2/0.03 mg L^{-1}) and low N/P ratios (0.072/0.03, 0.072/0.3 and 0.36/0.3 mg L^{-1}) were obtained. In addition, severe eutrophication with normal N/P ratio (2/0.3 mg L^{-1}) was also included.

The treatments of desiccation, reduced salinity, anthracene and eutrophication listed above were conducted at 25 °C. For high temperature treatment, samples were cultured at six temperature conditions: 35 °C, 36 °C, 37 °C, 38 °C, 39 °C and 40 °C. The extreme combined condition (35 °C combined with salinity of 12) was also conducted to estimate the median lethal time. All treatments were cultured at irradiance of 60 µmol photons m^{-2} s^{-1} with a 10 L: 14 D light: dark cycle. The irradiance within the wavelength range 400-700 nm was measured using a Li-Cor LI-250 light meter equipped with a LI-190SA quantum sensor.

2.3. Measurements of Survival and Growth

Prior to the stress treatment, the mean survival rate of germlings in four dishes was measured as the initial survival rate. The survival rate used in the present study was calculated as a ratio of the actual survival rate in the treatment to the initial survival rate. In each dish, germlings were counted in five areas arranged in a cross pattern (the upper, lower, left and right peripheries and the center) under stereoscopy microscope. A total of about 200 germlings were used to calculate the survival rate of each dish. Germlings were classified as dead if they collapsed or were structurally fragmented.

Considering that anthracene and eutrophication may have no evident effects on the survival of germlings, for the treatments of anthracene and eutrophication, growth of germlings was measured in terms of changes in lengths excluding rhizoids at the end of the experiment. Lengths were measured using a microscope with an ocular micrometer in five areas in each Petri dish as mentioned above. The initial mean length of germlings in four dishes was also measured before the stress treatment. A total of 20 germlings in each dish were used to estimate the relative growth rate (RGR, % d^{-1}). RGR was calculated as 100 (ln (L_1) − ln (L_0))/t, from initial versus final values, where L_0 and L_1 are germlings lengths at the start and at the end of treatment, respectively, and t is the length of treatment period calculation in days [16].

2.4. Statistical Analysis

All the data were analyzed using SPSS 13.0 for Windows. Median lethal values of temperature, salinity, desiccation and extreme combined condition with 95% confidence limits (95% CL) were determined using the probit regression analysis [17]. Probit regression equations were estimated as Y = a + b X; where Y is the percent of mortality in probit units, a and b are the intercept and slope constants, and X is the log time or log dose of stress. The effects of DMSO and anthracene on the survival of germlings were tested by repeated measures ANOVA. One-way ANOVAs were used to test the effect of DMSO and anthracene on the growth of germlings. Main effects and interactions of N and P on the survival of germlings were analyzed by two-way repeated measures ANOVA. A two-way univariate analysis of variance was performed to test the significance of the main effects and interactions on the growth of germlings, with N and P as fixed factors. For repeated measured ANOVA, the Huynh-Feldt correction was used to adjust the degrees of freedom when the sphericity assumption was violated (*i.e.* if Mauchly's test of sphericity was statistically significant at $p < 0.05$). Tukey's tests were used for post-hoc comparisons. The differences were considered to be statistically significant if the probability value was less than 5% ($p < 0.05$).

3. RESULTS

3.1. Acute Physical Stress for Germlings

For high temperature treatment, the survival rates of germlings decreased with time (Fig. 1). In comparison to the gradual decline of survival rates at 35 °C and 36 °C, the survival rates decreased sharply to 0% within 40 h when exposed to 37 °C to 40 °C. Especially at 40 °C, the survival rate of 0% was even recorded at about 10 h after treatment (Fig. 1). The results of acute thermal stress tests are listed in Table 1. Variation in LT_{50} values at different temperatures was evident. 50% mortalities occurred after exposure to 35 °C and 36 °C at over 50 h and 70 h, respectively (Table 1). However, LT_{50} values decreased sharply from 37 °C to 40 °C (Table 1). When exposed to 40 °C, the LT_{50} value was even lower than 5 h (Table 1). By the end of 24 h thermal tolerance experiment, the median lethal temperature was estimated as 36.9 °C ($\chi^2 = 73.8$, df = 13, $p < 0.001$, 95% CL, 36.6 – 37.1 °C).

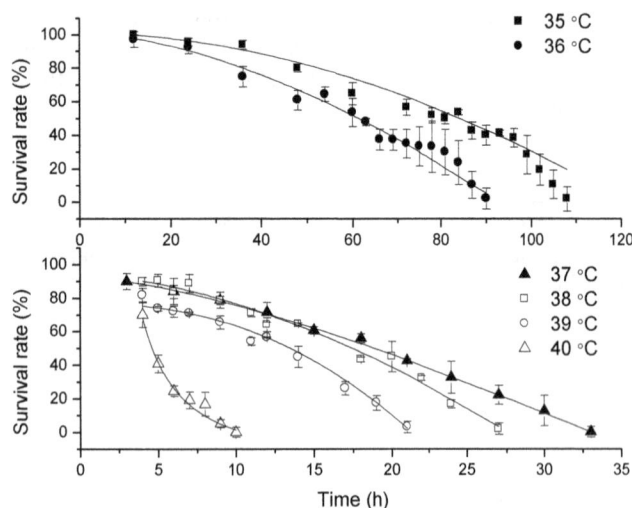

Fig. (1). Survival rates of germlings exposed to various temperatures (35 °C, 36 °C, 37 °C, 38 °C, 39 °C and 40 °C). Solid lines are logistic fits (all *p* values < 0.01 and all R^2 values > 0.97). Values are means ± SE (n = 3).

For hyposalinity treatment, germlings maintained high survival rates at over 90% when exposed to 27 psu, 20 psu and 13 psu for 108 h (Fig. 2). Over 60% survival rate was obtained after 108 h exposure to 7 psu (Fig. 2). However, in fresh water, survival rate decreased sharply to about 0% at a time period of 60 h (Fig. 2). Probit analysis estimated an LT_{50} value of 47.6 h in fresh water (Table 1).

For desiccation treatment, the survival rates decreased quickly in a short time span and all germlings died at 12 h (Fig. 3). The LT_{50} value of desiccation was estimated as 7.0 h (Table 1).

For combined thermal and osmotic stress treatment, the survival rates decreased to about 0% at 21 h (Fig. 3). The LT_{50} value of combined thermal and osmotic stress was estimated as 9.8 h (Table 1).

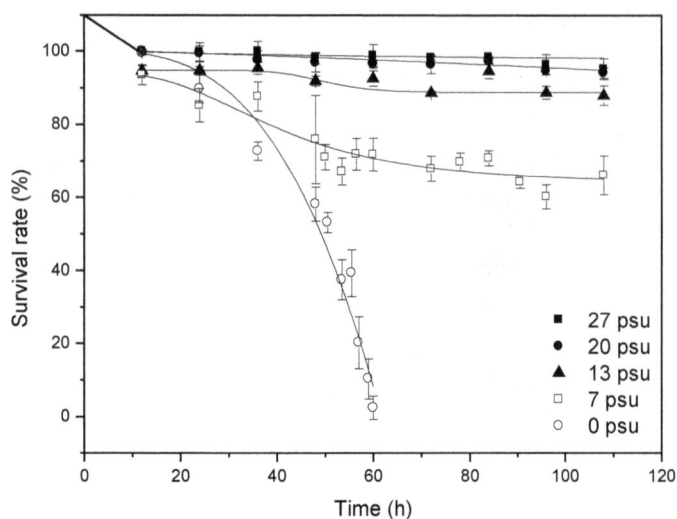

Fig. (2). Survival rates of germlings exposed to various salinities (27 psu, 20 psu, 13 psu, 7 psu and 0 psu). Solid lines are logistic fits (all p values < 0.01 and all R^2 values > 0.70). Values are means ± SE (n = 3).

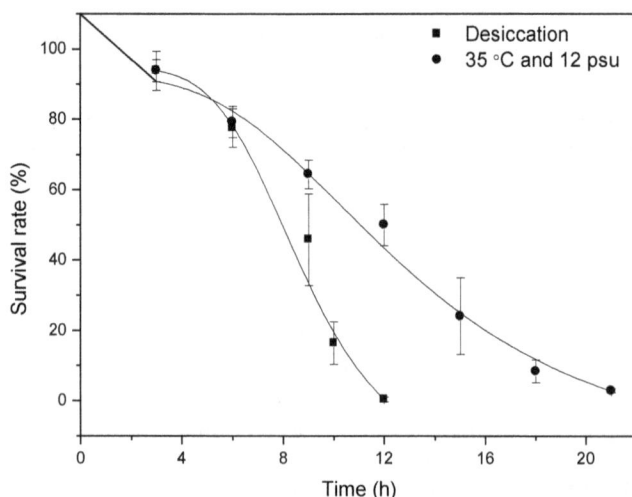

Fig. (3). Survival rates of germlings exposed to desiccation and 35 °C combined with 12 psu. Solid lines are logistic fits (all p values < 0.01 and all R^2 values > 0.99). Values are means ± SE (n = 3).

3.2. Effect of Anthracene on Germlings

Results of one-way ANOVA indicated that the difference between seawater and DMSO (0.5% v/v) for the growth of germlings was not significant ($F = 0.053$, $p = 0.830$). DMSO also had no significant effect on the survival of germlings

Table 1. Median lethal time of germlings exposed to various stresses (n = 3).

Variable	df	χ^2	Regression Equation	LT_{50} (h)	95% Confidence Limit (h)	
					Lower	Upper
35 °C	49	123.7*	Y = - 7.9 + 4.2 X	77.2	74.6	12.0
36 °C	40	189.0*	Y = - 6.1 + 3.5 X	57.9	80.0	3.5
37 °C	28	155.1*	Y = - 3.0 + 2.5 X	16.4	54.3	4.8
38 °C	34	98.5*	Y = - 3.3 + 2.8 X	15.8	61.5	45.0
39 °C	28	70.5*	Y = - 2.5 + 2.4 X	11.0	14.6	50.4
40 °C	13	63.0*	Y = - 2.5 + 4.0 X	4.3	18.4	6.2
Fresh water	22	76.6*	Y = - 7.9 + 4.7 X	47.6	14.7	8.3
Desiccation	13	169.3*	Y = - 4.9 + 5.6 X	7.0	17.1	8.7
35 °C + 12 psu	19	137.6*	Y = - 4.0 + 4.0 X	9.8	10.2	10.9

* Since Goodness-of-Fit test is significant ($p < 0.05$), a heterogeneity factor is used in the calculation of confidence limits.

(Table 2). Therefore, the interference of 0.5% DMSO on the growth and survival of *S. thunbergii* germlings can be excluded. Although anthracene significantly affected the survival (Table 2), low concentrations (0.01 ~ 1 mg L^{-1}) had no significant effects (Tukey's tests: $p= 0.433$, $p = 0.996$, $p = 0.830$ and $p = 0.340$, respectively). Survival rates of over 70% were obtained after 25 days of exposure to concentrations rangeing from 0.01 mg L^{-1} to 1 mg L^{-1} (Fig. 4). Similar to the survival effect, although a significant effect of anthracene on the growth of germlings was found ($F = 49.360$, $p < 0.001$), low concentrations of anthracene (0.01 mg L^{-1} – 1 mg L^{-1}) had no significant effects on growth (Tukey's tests, $p = 0.510$, $p = 0.982$, $p = 0.163$, $p = 0.676$ and $p = 0.234$, respectively). Germlings were severely inhibited by high concentrations of anthracene (5 mg L^{-1} and 10 mg L^{-1}) with low survival rates and RGRs of below 50% and 1%, respectively (Tukey's tests, $p < 0.001$) (Fig. 4). The initial mean length of germlings was measured as 127.6 μm. At the end of anthracene treatment, final lengths ranged from 129.3 to 461 μm.

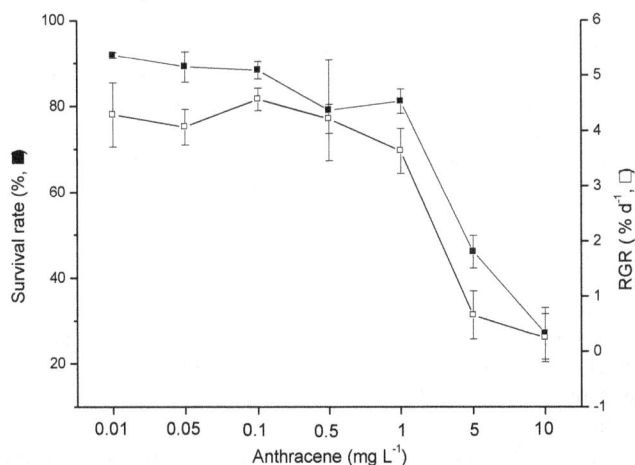

Fig. (4). Relative growth rates (RGR; % day^{-1}) and survival rates of germlings exposed to various concentrations of anthracene (0.01mg/L, 0.05 mg/L, 0.1 mg/L, 0.5 mg/L, 1 mg/L, 5mg/L and 10 mg/L) after 25 days of culture. Values are means ± SE (n = 3).

3.3. Effect of Eutrophication on Germlings

The results of repeated measures ANOVA showed that neither N nor P source had significant effect on the survival of germlings (Table 3). No significant effects of N or P on the growth were found (Table 4). Therefore, severe eutrophication had little effect on germlings. Tables 3 and 4 also showed that the interaction between N and P source had no significant effects on the survival and growth, indicating that various imbalanced N/P ratios also had no effect on germlings. By the end of eutrophication treatment, final lengths of germlings ranged from 227.115 to 344.93 μm.

4. DISCUSSION

Of the environmental stresses tested in the present study, high temperature of 40 °C and combined thermal and osmotic stress (35 °C combined with 12 psu) may directly impact the survival of germlings.

Although the LT_{50} value of 39 °C was evaluated as 11h, its duration in the mid-tidal zone occupied by *S. thunbergii* cannot exceed the LT_{50} value even with this temperature , because the temperature decreases due to immersion in seawater during the high tide. This suggests that the high temperature not over 39 °C causes no mass mortality of *S. thunbergii* germlings in the field. However, the LT_{50} value of 40 °C was only about 4 h. Although *S. thunbergii* germlings are rarely exposed to 40 °C for 4 h during low tide, this extreme thermal condition possibly occurs under extreme hot weather.

The LT_{50} value of combined thermal and osmotic stress (35 °C combined with 12 psu) was much lower than the separate LT_{50} values of 35 °C or 13 psu indicating that there is a synergistic effect between thermal and osmotic stress, that is, germlings living near the limit of one tolerance were more sensitive to additional stress, which is in accordance with the results reported for *Fucus vesiculosus* [18]. Low salinity usually results from strong rainfall which also causes a decrease in temperature. Hence, high temperature and low

Table 2.   Repeated measures ANOVA for effects of DMSO and anthracene on survival of germlings.

	Variable	df	Mean Square	F	p
DMSO	Within-subjects				
	Time	24	25.785	5.555	< 0.001
	Time × DMSO	24	7.264	1.565	0.066
	Error (Time)	96	4.642		
	Between-subjects				
	DMSO	1	29.695	0.343	0.59
	Error	4	86.532		
Anthracene	Within-subjects				
	Time	19.174	302.948	23.310	< 0.001
	Time × Anthracene	115.047	292.546	22.509	< 0.001

Table 3.   Repeated measure ANOVA for effects of $NaNO_3$ and KH_2PO_4 on survival of germlings. N: $NaNO_3$, P: KH_2PO_4.

Variable	df	Mean Square	F	p
Within-Subjects				
Time	4	91.923	14.428	< 0.001
Time × N	8	5.178	0.813	0.594
Time × P	8	15.936	2.501	0.019
Time × N × P	16	3.847	0.604	0.871
Error (Time)	72	6.371		
Between-Subjects				
N	2	12.68	0.394	0.680
P	2	27.278	0.849	0.444
N × P	4	10.557	0.328	0.855
Error	18	32.145		

Dependent variable (survival) was untransformed and the assumption of homogeneity met Levene's test ($F = 4.367$, $p = 0.212$).

salinity can generally not be concurrent with each other in the field. However, extreme climatic events, such as El Niño which can cause high temperatures and storms, may severely destroy *S. thunbergii* germlings. For example, on the Pacific Coast of California and Baja California, a reduction of the brown algae *Macrocystis pyrifera* bed size and biomass with up to 100% in some areas resulted from the influence of El Niño [4].

Table 4.   Univariate analysis of variance for effects of $NaNO_3$ and KH_2PO_4 on relative growth rate (RGR, % d^{-1}) of germlings. N: $NaNO_3$, P: KH_2PO_4.

Variable	df	Mean Square	F	p
N	2	< 0.001	1.779	0.197
P	2	< 0.001	2.824	0.086
N × P	4	< 0.001	1.887	0.156
Error	18	< 0.001		

Dependent variable (RGR) was untransformed and the assumption of homogeneity met Levene's test ($F = 0.843$, $p = 0.547$).

The capacity to tolerate desiccation is thought to be a major factor in determining the upper limits of distribution for intertidal seaweed [19]. In comparison to other stresses, germlings were more vulnerable to desiccation at a relative humidity of 80%, which is consistent with our previous study [13]. In contrast, small thallus pieces of *Codium fragile* (green alga) can survive long periods of emersion (90 days) when kept under high relative air humidity of 90% [20]. We suggest that desiccation tolerance is related to relative humidity. Therefore, lower relative humidity may have greater adverse effects on *S. thunbergii* germlings in a short span of time. During the low tide, adult canopy can act as a nurse thalli by buffering germlings beneath them from desiccation stress. It is inferred that in the upper tidal zone which is characterized by its long periods of emergence, germlings beneath the adult canopies are the main sources of *S. thunbergii* recruitments.

Intertidal seaweeds are subjected to low salinity stress when exposed to low tide or trapped in tide pools, where fresh water from rain may cause a decrease in salinity [21]. In seaweeds, hypo-osmotic stress causes increases in cell volume and turgor, resulting in the loss of ions and organic solutes as well as in damage to membranes and organelles, culminating in cell rupture [22]. However, in this study, *S. thunbergii* germlings exhibited high tolerance to reduced salinity, even in fresh water. Similar phenomena were reported in *Fucus*. Simulated rainfalls during low tides caused photosynthetic activity of *Fucus spiralis* to drop to 50% of initial Fv/Fm, independent of the length of the rain period. Treated thalli also fully recovered after 6 min re-submersion in seawater [23]. It suggested that even heavy rain during the low tide cannot result in mass deaths of *S. thunbergii* germlings. Furthermore, extremely low salinity cannot be maintained for a long time due to tidal motions. Therefore, the deterioration of *S. thunbergii* beds may not be related to hyposalinity resulting from heavy rain.

It has been reported that algae are more tolerant to PAHs than other aquatic organisms [24]. In the present study, *S. thunbergii* germlings also showed high tolerance to anthracene. Although the survival and growth of germlings were significantly affected by the concentration of over 5 mg L^{-1}, the solubility of anthracene in water (0.073 mg L^{-1}) wasmuch lower than its concentration [25]. Although anthracene at 0.25 mg L^{-1} reduced the growth of three *Scenedesmus* species [26], growth of *S. thungergii* germlings was not significantly affected at this concentration. It appears that direct anthracene toxicity is not important in assessing the environmental hazard posed to algae by anthracene and PAH contamination [27]. However, there was a significant interaction between anthracene and UV-A radiation, which, in combination, caused significant toxic effects on *Selenastrum capricornutum* [27]. Anthracene at nominal concentrations exceeding 0.05 mg L^{-1} inhibited the growth of the algae in a concentration- and irradiance-dependent manner [8]. Therefore, response of *S. thunbergii* germlings to the interaction between anthracene and irradiance needs further investigation.

Eutrophication can result in accelerated development of the early stages of some algal species [9]. However, in the present study, neither positive nor negative effects of eutrophication on *S. thunbergii* germlings were found. Effect of N/P ratio on marine environment has received great deal of attention. The appearance of different red tides was related to the N/P ratio [28]. Our results showed that N/P ratio had no significant effect on *S. thunbergii* germlings. However, indirect effects including increased sediment cover of substrata, scouring caused by wind-induced resuspension of sediments, and grazing, were also expected to be negative [9]. A deterioration of the light climate due to increased phytoplankton biomass, suspended matter and overgrowing (shading) by epiphytes are likely causes for the decline of *Fucus* spp. in Kiel Bay [29], and for a decrease in macrophyte numbers in general [30]. Therefore, indirect effects of eutrophication may be partially responsible for the deterioration of *S. thunbergii* beds.

Since the young stages of seaweeds are a sensitive link in species life cycle [18], recovery of populations from anthropogenic stress is likely to depend upon recruitment of these early stages [9]. Pollution of PAHs and eutrophication seems to have little effect on *S. thunbergii* germlings. Although high temperature of 40 °C and combined thermal and osmotic stress (35 °C combined with 12 psu) directly impacted the survival of germlings, such extreme conditions rarely exist in the middle of the latitude region unless extreme events occur. It is suggested that increasing extreme climatic events caused by various anthropogenic interferences are more responsible for the deterioration of *S. thunbergii* beds and other seaweed beds. These results will be useful in the search for evidence regarding the reasons of deterioration of *S. thunbergii* beds.

CONCLUSION

As an ecologically important member of *Sargassum* beds, the natural populations of *S. thunbergii* along the coast of China have evidently deteriorated in recent years. This paper represents the responses of *S. thunbergii* germlings to acute environmental stress. The median lethal time (LT50) values of high temperature, low salinity and combined thermal and osmotic stress (35 °C combined with 12 psu) by Probit regression analysis, and the effects on the survival and growth of germlings of high concentrations of anthracene (5 mg L^{-1} and 10 mg L^{-1}), low concentrations (0.01-1 mg L^{-1}), severe eutrophication and disproportionality of N/P by anovas, were examined. Results indicated that the deterioration of *S. thunbergii* beds may be related more to the increasing extreme climatic events.

ACKNOWLEDGEMENTS

This work was financially supported by the National Natural Science Foundation of China (NO.41376154) and a Project of Shandong Province Higher Educational Science and Technology Program (J10LC22).

REFERENCES

[1] Zhang QS, Li W, Liu S, *et al.* Size-dependence of reproductive allocation of *Sargassum thunbergii* (Sargassaceae, Phaeophyta) in Bohai Bay, China. Aquat Botany 2009; 91: 194-8.

[2] Zhang SY, Sun HC. Research progress on seaweed bed ecosystem and its engineering. Chinese. J Appl Ecol 2007; 18: 1647-53 (in Chinese with English abstract).

[3] Zhang QS, Tang YZ, Liu SK, *et al.* Zygote-derived seedling production of Sargassum thunbergii: focus on two frequently experienced constraints in tank culture of seaweed. J Appl Phycol 2012; 24: 707-14.

[4] Terawaki T, Yoshikawa K, Yoshida G, *et al.* Ecology and restoration techniques for *Sargassum* beds in the Seto Inland Sea, Japan. Marine Pollut Bull 2003; 47: 198-201.

[5] Buschmann AH, VaAquez J, Osorio P, *et al.* The effect of water movement, temperature and salinity on abundance and reproductive patterns of *Macrocystis* spp. (Phaeophyta) at different latitudes in Chile. Marine Biol 2004; 145: 849-62.

[6] Davison I R, Pearson G A. Stress tolerance in intertidal seaweeds. J Phycol 1996; 32: 197-211.

[7] Bonnet J L, Guiraud P, Dusser M, *et al.* Assessment of anthracene toxicity toward environmental eukaryotic microoganisms: *Tetrahymena Pyriformis* and selected micromycetes. Ecotoxicol Environ Safety 2005; 60: 87-100.

[8] Aksmann A, Tukaj Z. The effect of anthracene and phenanthrene on the growth, photosynthesis, and SOD activity of the green alga *Scenedesmus armatus* depends on the PAR Irradiance and CO sub (2) level. Arch Environ Contaminat Toxicol 2004; 47: 177-84.

[9] Coelho SM, Rijstenbil JW, Brown RM. Impacts of anthropogenic

stresses on the early development stages of seaweeds. J Aquat Ecosyst Stress Recovery 2000; 7: 317-33.

[10] Whitaker SG, Smith JR, Murray SN. Reestablishment of the Southern California rocky intertidal brown alga, *Silvetia compressa*: an experimental investigation of techniques and abiotic and biotic factors that affect restoration success. Restorat Ecol 2010; 18: 18-26.

[11] Krause-Jensen D, Sagert S, Schubert H, *et al.* Empirical relationships linking distribution and abundance of marine vegetation to eutrophication. Ecolog Indicat 2008; 8: 515-29.

[12] Ang P, De Wreede R. Matrix models for algal life history stages. Marine Ecol- Progress Series 1990; 59: 171-81.

[13] Chu SH, Zhang QS, Liu SK, *et al.* Tolerance of *Sargassum thunbergii* germlings to thermal, osmotic and desiccation stress. Aquatic Botany 2012; 96: 1-6.

[14] Fowler DB, Gusta LV, Tayler NJ. Selection for winter-hardiness in wheat: Screening methods. Crop Science 1981; 21: 896-901.

[15] Wang HS, Lei K, Li ZC, *et al.* Fuzzy comprehensive evaluation of w ater eutrophlcation in Liaodong Bay. Res Environ Sci 2010; 23: 413-9 (in Chinese with English abstract).

[16] Hunt R. Plant Growth Analysis. London: Edward Arnold 1978.

[17] Finney DJ. Probit analysis. Cambridge University Press, Cambridge 1973.

[18] Andersson S, Kautsky L, Kaystdy N. Effects of salinity and bromine on zygotes and embryos of *Fucus vesiculosus* from the Baltic Sea. Marine Biol 1992; 114: 661-5.

[19] Harker M, Berkaloff C, Lemoine Y, *et al.* Effects of high light and desiccation on the operation of the xanthophylls cycle in two marine brown algae. Europ J Phycol 1999; 34: 35-42.

[20] Schaffelke B, Deane D. Desiccation tolerance of the introduced marine green alga *Codium fragile* ssp. *tomentosoides* – clues for

[21] Macler BA. Salinity effects on photosynthesis, carbon allocation, and nitrogen assimilation in the red alga, *Gelidium coulteri*. Plant Physiol 1988; 88: 690-4.

[22] Lobban CS, Harrison PJ. Seaweed Ecology and Physiology. Cambridge: Cambridge University Press 1994.

[23] Schagerl M, Möstl M. Drought stress, rain and recovery of the intertidal seaweed *Fucus spiralis*. Marine Biol 2011; 158: 2471-9.

[24] Cody TE, Radike MJ, Warshawsky D. The phototoxicity of benzo[a]pyrene in the green alga *Selenastrum capricornutum*. Environ Res 1984; 35: 122-32.

[25] Johannes C, Majcherczyk A, Hüttermann A. Degradation of anthracene by laccase of *Trametes versicolor* in the presence of different mediator compounds. Appl Microbiol Biotechnol 1996; 46: 313-7.

[26] Zbigniew T, Wojciech P. Individual and combined effect of anthracene, cadmium, and chloridazone on growth and activity of SOD izoformes in three *Scenedesmus* species. Ecotoxicol Environ Safety 2006; 65: 323-31.

[27] Gala WR, Giesy JP. Flow cytometric determination of the photoinduced toxicity of anthracene to the green alga *Selenastrum capricornutum*. Environ Toxicol Chem 1994; 13: 831-40.

[28] Tseng C K. The past, present and future of phycology in China. Hydrobiologia 2004; 512: 11-20.

[29] Vogt H, Schramm W. Conspicuous decline of *Fucus* in Kiel Bay (Western Baltic): What are the causes? Marine Ecol Prog Seri 1991; 69: 189-94.

[30] Phillips G, Eminson D, Moss B. A mechanism to account for macrophyte decline in progressively eutrophicated freshwaters. Aquat Botany 1978; 4: 103-26.

likely transport vectors? Biol Invasions 2005; 7: 557-65.

13

Quantification of Lovastatin Produced by *Monascus purpureus*

A. Seenivasan[a, #], Sathyanarayana N. Gummadi[b], Tapobrata Panda[a,]* and Thomas Théodore[c]

[a]*Biochemical Engineering Laboratory, Department of Chemical Engineering, Indian Institute of Technology, Madras, Chennai 600036, Tamil Nadu, India*

[b]*Department of Biotechnology, Indian Institute of Technology, Madras, Chennai – 600036, Tamil Nadu, India*

[c]*Department of Chemical Engineering, Siddaganga Institute of Technology, Tumkur - 572103, Karnataka, India*

[#]*Present address: Department of Chemical Engineering, SSN College of Engineering, Chennai-603110, Tamil Nadu, India*

Abstract: Development of a novel method for the quantification of lovastatin is an interesting problem in the analytical field. In the literature, many reports use spectrophotometric method for the quantification of lovastatin. However, the analysis of fermentation broth containing lovastatin appears to be inaccurate using spectrophotometric method. Hence, the estimation of lovastatin produced by *Monascus purpureus* and pure lovastatin was attempted by UV-visible spectrophotometer as well as HPLC. It was observed that the analogues and/or intermediates of lovastatin synthesized in the fermentation broth and the products of fermentation caused superimposition effect on the absorption spectrum. Phosphate is a medium constituent for the production of lovastatin by the organism which contributed significantly to the superimposition of absorption spectrum. On the other hand, HPLC analysis consistently gave reliable results for the estimation of lovastatin under all the experimental conditions studied.

Keywords: Lovastatin, *Monascus purpureus*, spectrophotometry, HPLC.

1. INTRODUCTION

The conversion of 3-hydroxy-3-methyl-glutaryl coenzyme A (HMG-CoA) into mevalonate is the rate- limiting step in the cholesterol cascade mediated by HMG-CoA reductase (E.C 1.1.1.88). Statins are structural analogues of HMG-CoA and bind with the enzyme at its active site. This alters the conformation of the enzyme. The reversible competitive binding between statin and the enzyme prevents the access of the substrate HMG-CoA to the HMG-CoA reductase [1-6]. Among all statins, lovastatin is the first approved by USFDA for the treatment of hypercholesterolemia in 1987. Lovastatin also serves as a precursor for other important statins like simvastatin and wuxistatin [7, 8].

Development of analytical methods for the estimation of lovastatin is an interesting topic due to the existing problems in the analytical techniques. Pure and fermentation-derived lovastatin were quantified mostly using UV-visible spectrophotometer [9-16], HPLC [17-20], and capillary electrophoresis [21, 22].

The spectrophotometric method of analysis of lovastatin is simple, faster, eco-friendly, and less laborious method of analysis than other analytical techniques. However, the spectrophotometric quantification of lovastatin is not accurate in fermentation broth as well as for the system having competitive molecules which also absorbs light at 238 nm [2, 3]. Lovastatin has been produced commercially through fermentation route only [1, 6]. Hence, it is important to measure the content of lovastatin accurately in the fermentation broth. Many researchers have used HPLC technique for the quantification of lovastatin in biological samples, but the reason for shifting/moving from the UV-visible spectrophotometer to HPLC has not been explained so far for the estimation of fermentation-derived lovastatin. In the present study, we have used UV-visible spectrophotometer and HPLC techniques for the quantification of lovastatin produced in the fermentation broth. Our objective was to highlight the various forms of interferences in the quantification of lovastatin with necessary experimental proof, which has not been reported so far.

2. EXPERIMENTAL

2.1. Chemicals

Lovastatin (lactone form) was procured from Sigma-Aldrich. All other chemicals were of analytical grade and procured from Sisco Research Laboratories Pvt. Ltd., Mumbai, India. For the assay and preparation of reagents, Milli-Q water was used throughout the study. Lovastatin was dissolved in acetonitrile to get a stock solution of 1 mg/mL. The hydroxy acid form of lovastatin was prepared by hydrolysis as described by Yang and Hwang [23].

*Address correspondence to this author at the Department of Chemical Engineering, Indian Institute of Technology, Madras, Chennai 600036, Tamil Nadu, India; E-mail: panda@iitm.ac.in

2.2. Organism and Culture Conditions

Monascus purpureus MTCC 369 was obtained from the Institute of Microbial Technology, Chandigarh, India. A suspension containing 10^5-10^6 spores/mL was used to inoculate a basal medium having the following composition (in g/L): dextrose, 100; peptone, 10; KNO_3, 2; $NH_4(H_2PO_4)$, 2; $MgSO_4 \cdot 7H_2O$, 0.5; $CaCl_2$, 0.1. The pH was adjusted to 6 using 0.1 N NaOH solutions [16]. The culture was incubated at 30 ^{0}C for 48 h in an orbital shaker maintained at 120 rpm.

The defined production medium had the following composition (in g/L): dextrose, 29.59; NH_4Cl, 3.86; KH_2PO_4, 1.73; $MgSO_4 \cdot 7H_2O$, 0.86; $MnSO_4 \cdot H_2O$, 0.19. The pH was adjusted to 6 [16]. The complex production medium contained (in g/L): rice powder, 34.4; peptone, 10.8; glycerol, 33.2; glucose, 129.2; $MgSO_4 \cdot 7H_2O$, 1; KNO_3, 2. The pH was adjusted to 6 [24]. A 500-mL Erlenmeyer flask containing 100 mL of sterile medium was inoculated with 10% (v/v) seed culture and incubated at 30 ^{0}C for 15 days on a rotary shaker maintained at 120 rpm [18]. Care was taken to restrict evaporation loss during fermentation. Suitable correction factor has been used in this regard.

2.3. Analysis of Products

2.3.1. Pure Lovastatin

Working standards of lactone- and hydroxy acid- forms of lovastatin were prepared and the absorbance was read at 238 nm using a UV-visible spectrophotometer (JASCO V-630, Essex, UK). The zero-order absorption spectra were obtained over the wavelength range of 200-600 nm using the appropriate dispersion medium in a quartz cuvette (1 cm path length) at 1.5 nm slit width ($\Delta\lambda$).

The limit of quantification (QL) and detection limit (DL) was found using the equation given below,

$$DL; QL = \frac{F \times (SD)}{a}$$

where F value for DL and QL are 3.3 µg and 10 µg, respectively. SD is the standard deviation of the blank and 'a' is the slope of the standard plot [25].

Quantitative HPLC was performed using a gradient HPLC (Shimadzu, Prominence HPLC, Kyoto, Japan) equipped with a photodiode array detector and Luna C18 column (250×4.6 mm, 5µm). The mobile phase was composed of acetonitrile and acidified water (0.1% H_3PO_4) (60:40, v/v) which was filtered through a 0.22-µm membrane filter. The flow rate of solvent and column temperature was maintained at 1 mL/min and 40 ^{0}C, respectively. A 20 µL sample was injected into the column for analysis. The working standards of lactone- and hydroxy acid- forms of lovastatin were analyzed by HPLC at 238 nm.

2.3.2. Lovastatin Produced by Monascus Purpureus

The sample (cell suspension) was homogenized by ultrasonication for 20 min followed by extraction with an equal volume of ethyl acetate at 60 ^{0}C for 30 min with intermittent shaking. The upper organic layer was isolated from the aqueous layer and dried under vacuum in a rotary evaporator. The dry residue was resuspended in pure acetonitrile. This suspension was filtered through a 0.22-µm nylon filter paper (Pall Trincor®, USA) before analysis by UV-visible spectrophotometer and HPLC [17].

Thin layer chromatography (TLC) (using dichloromethane/ethyl acetate (70:30, v/v)) was done to confirm the presence of lovastatin [26].

Inductively-coupled plasma optical emission spectrometer (ICP-OES, Perkin-Elmer Optima 5300 DV, Shelton, CT) was used to measure potassium (λ_{max} = 766.5 nm) and phosphorus (λ_{max} = 213.62 nm) present in the suspension and also released during ultrasonication. The flow rates of the auxiliary, plasma, and nebulizer were maintained at 0.2, 15, and 0.8 L/min and that of the sample at 1.5 mL/min, respectively. A radiation source of 40 MHz and RF power of 1300 were used in the system. Potassium and phosphorus were measured using the axially-viewed plasma set-up.

3. RESULT AND DISCUSSION

3.1. Characterization of Pure Lovastatin Using Separately UV-visible Spectrophotometer and HPLC

Pure lovastatin has three different absorption maxima at 232, 238, and 247 nm in UV-visible spectrophotometry (*cf.* Fig. 1A), which suggests better identification of lovastatin from other compounds. The characteristics of the peaks could be due to the presence of diene [27]. The absorption spectra of both lactone form and hydroxy acid form of lovastatin in a mixture appear similar (*cf.* Fig. 1). The regression analysis of the individual forms of lovastatin and molar extinction coefficient (ε) are given in Table 1. The molar extinction coefficients for the lactone and hydroxy acid forms of lovastatin are $4.89 \times 10^4 \pm 21.8$ and $5.99 \times 10^4 \pm 36$ $M^{-1}cm^{-1}$, respectively. This is valid for the estimation range of lovastatin between 2 µg and 20 µg. Values of DL and QL for lovastatin are 0.39 µg and 1.16 µg, respectively.

Results from the spectrophotometric analysis gave the total amount of lovastatin (hydroxy acid plus lactone forms) rather than the amount of the individual forms of lovastatin, while distinct peaks were obtained for the hydroxy acid and lactone forms of lovastatin by HPLC technique (*cf.* Fig. **1B**). Hence, the individual forms of lovastatin cannot be distinguished in both pure and fermentation derived samples by UV-visible spectrophotometer. Therefore, the spectrophotometric method of quantification of lovastatin could lead to a confusion in clinical studies, since the individual forms of lovastatin have specific properties, viz., pharmacological activity, solubility, lipophilic nature, and transportation characteristics.

The lactone form of lovastatin is more lipophilic than the hydroxy acid form. For this reason, the lactone form is accepted by the hepatic cells. This behavior determines the bioavailability of lovastatin. In liver, the diffused lactone form of lovastatin is converted into hydrolyzed forms including the hydroxy acid form by cytochrome P450 3A4 [28]. Hence, the quantification of the individual forms of lovastatin as well as hydrolyzed/metabolized products of

Table 1.   Assay parameters of UV-visible spectrophotometric measurements.

Nature of Compound Teated	Test Sample	Regression equation, $Y = aX + b$		Correlation Coefficient (r^2)	Molar absorption coefficient ($\varepsilon_{\lambda_{238}}$), $M^{-1}cm^{-1}$
		Slope, μg^{-1} (a)	Intercept (b)		
For a pure compound	Lactone form of lovastatin standard	0.0404	-0.0032	0.999	$4.89 \times 10^4 \pm 21.8$
	Hydroxy acid form of lovastatin	0.0426	0.0218	0.9985	$5.99 \times 10^4 \pm 36$
In the presence of competitive molecules	Lovastatin + KH$_2$PO$_4$	0.0182	-0.0576	0.849	$1.55 \times 10^4 \pm 61.5$
	Lovastatin + Unautoclaved chemically defined medium	0.0208	-0.0539	0.9343	$1.80 \times 10^4 \pm 53$
	Lovastatin + Autoclaved chemically defined medium	0.0146	-0.0372	0.8971	$1.34 \times 10^4 \pm 36$
	Lovastatin + Unautoclaved complex medium	0.0156	-0.0098	0.9215	$1.71 \times 10^4 \pm 43$
	Lovastatin + Autoclaved complex medium	0.0139	0.0275	0.8227	$2.1 \times 10^4 \pm 55$

where Y-is the absorbance value at 238 nm and X is the amount of lovastatin (µg). The molar absorption coefficients are expressed as mean ± standard deviation for the sample size of n=5. The detection limit (DL) and quantification (DQ) are found 0.39 µg and 1.16 µg, respectively from the equation, $DL; QL = \dfrac{F \times (SD)}{a}$, where F value for DL and QL are 3.3 µg and 10 µg and SD is the standard deviation of the blank. 'a' is the slope of the standard plot [25]. The regression model for HPLC data is $y = 2 \times 10^6 x$ for all the cases (remained same), 'y' is the detector response (AU) and 'x' is the amount of lovastatin (µg).

lovastatin is necessary to determine the metabolic rate in controlled delivery systems. Several reports use the spectrophotometric method of quantification of *in vitro* and *in vivo* drug release characteristics [29-35]. A few studies have reported the controlled release characteristics of lovastatin *in vitro* by measuring lovastatin using UV-visible spectrophotometer and *in vivo* release characteristics of lovastatin measured by HPLC technique [29, 36].

3.2. Effect of Phosphate and Potassium on Lovastatin Estimation

The phosphate and potassium contents were estimated in the culture sample of *M. purpureus* grown in defined as well as in complex production medium, using ICP-OES. The samples were subjected to ultrasonication before estimation. Samples were collected every 24 h for 15 days. Significant amounts of phosphorus were released compared to

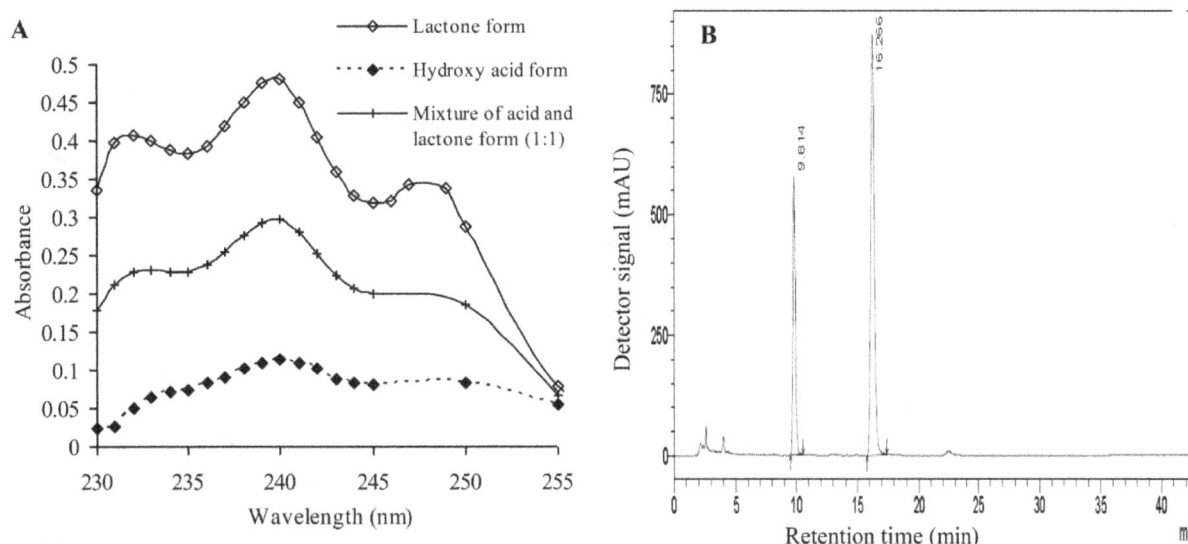

Fig. (1). UV-visible spectrophotometric and HPLC methods of estimation of pure lovastatin. (**A**) The lactone and hydroxy acid forms of lovastatin exhibited absorbance maxima (λ_{max}) at 232, 238, and 247 nm individually as well as in mixture (1:1) by UV-Visible spectrophotometer. (**B**) Both the forms of lovastatin quantified separately using HPLC. The retention times for hydroxy acid and lactone forms were 9.8 and 16.3 min, respectively.

potassium during the ultrasonication of the culture broth. The total amount of phosphorus in the defined medium was more than that in the complex medium. The maximum total amount of phosphorus (equivalent to KH_2PO_4 concentration) already present in the broth and released during ultrasonication was found to be 1.807 g/l and 0.0175 g/l for the defined and complex production medium, respectively. The effect of phosphate (equivalent to KH_2PO_4 concentration) in the estimation of a known quantity of pure lovastatin (lactone form) is shown in Figs. (2A) and (2C).

Jerotskaja *et al.* [37] measured some of the constituents in the spent human blood samples, after dialysis, using UV-visible spectrophotometer and quantified the levels of potassium and phosphate at 227 and 237 nm, respectively. It

suggests that the phosphate and potassium already present in the medium as well as that released from cells after sonication led to errors in lovastatin estimation by UV-visible spectrophotometer. The presence of diene groups in the phosphate and its related compounds might absorb at 238 nm. This could contribute to inaccurate measurement of this drug by the spectrophotometric method [27, 38].

It is pertinent to mention here that this study is relevant to a problem associated in the assay of drug delivery system, where the drug is dispersed in the phosphate buffer. The study is with respect to *in vitro* and *in vivo* release of lovastatin. The phosphate alone appears to be a potential interfering component as observed from the present study. Several reports suggested spectrophotometric measurement

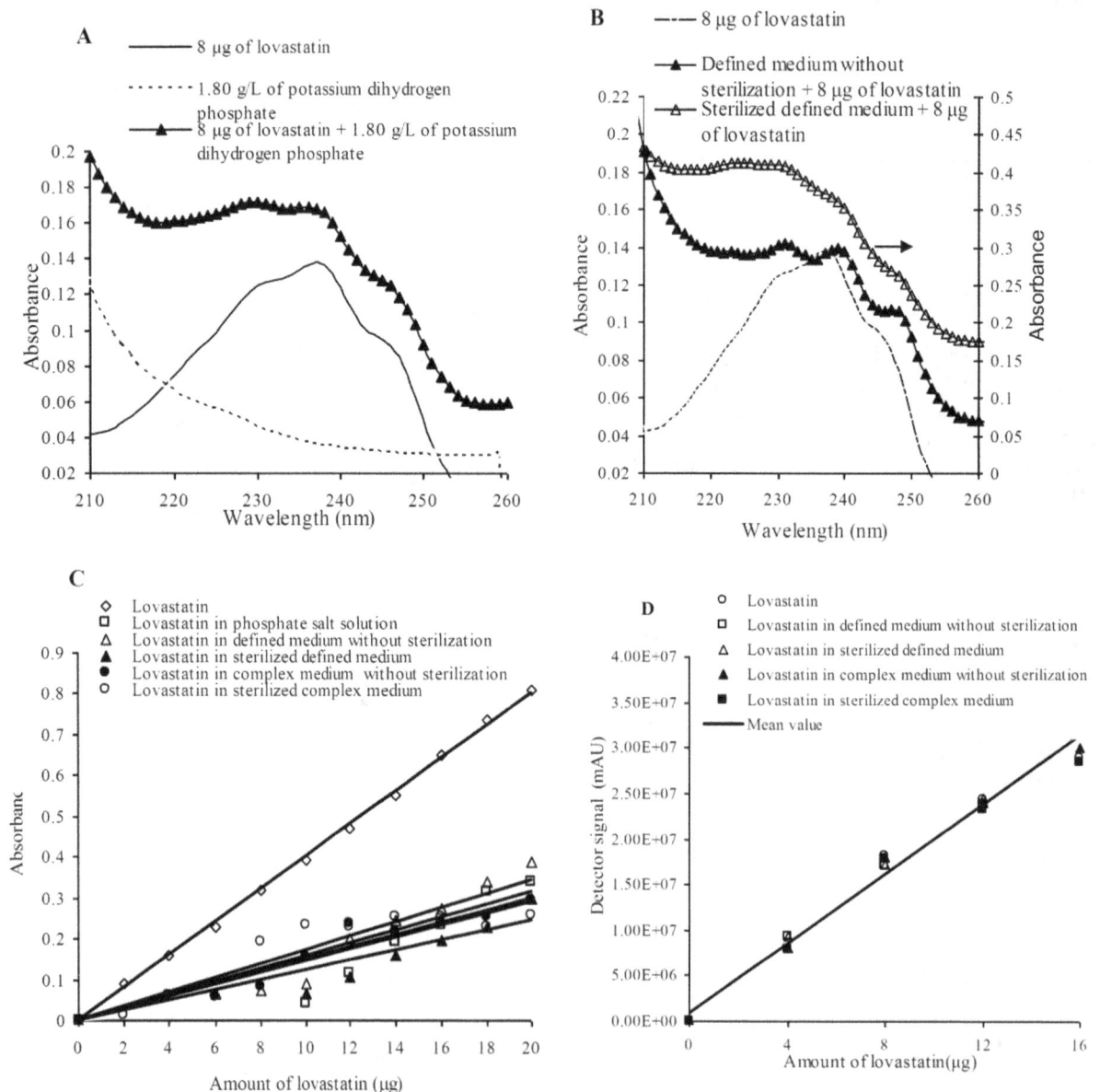

Fig. (2). Inaccuracy in lovastatin estimation by UV-visible spectrophotometry caused by phosphate and medium constituents. (A) Effect of maximum equivalent phosphate present in the medium as well as released from cells during sonication in defined medium (1.80 g/L) on the lovastatin absorption spectrum. (B) Effect of sterilization of defined medium constituents on lovastatin absorption spectrum. (C) Effect of medium constituents and phosphate on the absorbance versus amount of lovastatin plots measured by UV- visible spectrophotometry. (D) Effect of medium constituents on the standard plot of lovastatin obtained by HPLC technique.

for those *in vitro* and *in vivo* release characteristics of the drug [29-35]. Also, a few articles reported the release characteristics of the drug *in vitro* by UV-visible spectrophotometry and *in vivo* release characteristics using HPLC analysis, considering the competitive molecules in the serum [29, 36].

In the development of drug delivery systems, it is important to know the stability of lovastatin and its metabolism. The lactone form of lovastatin is metabolized or converted into hydrolyzed product including the acid form of lovastatin in the liver. Thus, the hydrolyzed product concentration is more important for the drug delivery and release characteristics of the drug. This change in concentration cannot be effectively determined by UV-visible spectrophotometer [39]. Jacobson *et al.* (1999) [40] also measured the hydrolyzed products of lovastatin in addition to lovastatin itself by HPLC at 239 nm. There are examples of lovastatin analogues and hydrolyzed products measured around 238 nm which cannot be distinguished in the spectrophotometric method.

3.3. Effect of Medium Constituents in Lovastatin Estimation

The constituents of the complex medium possibly contribute to error in lovastatin estimation and the corresponding absorbance values are higher than those obtained using the defined medium (*cf.* Fig. (**2B**)). The sterilized complex production medium showed a higher absorbance value than the same medium before sterilization. In the case of defined production medium, autoclaving of medium increased the absorbance which resulted in errors during lovastatin estimation (*cf.* Fig. (**2B**)). These results confirmed that the composition of the production medium (either defined or complex) and sterilization of the medium contribute to the error in the lovastatin assay (*cf.* Fig. (**2B**) and (**2C**)). The magnitude of error increased with the degree of complexity of the medium and the concentration of the

medium constituents. These results suggest that the spectrophotometric measurements are not suitable for quantification of lovastatin.

Pham *et al.* (2011) [41] reported a similar phenomenon for the estimation of xylose, where the availability of a rapid, inexpensive, and sensitive method like the spectrophotometric method is often limited by the complex nature of the medium. Some of the medium constituents, time of heating of medium, and trace elements in the medium were the reasons for inaccurate estimation of xylose.

3.4. Quantification of Lovastatin Produced by *Monascus sp.* Using UV-visible Spectrophotometric and HPLC Methods

The absorbances of aliquots of culture filtrate and of ethyl acetate-extract of culture filtrates obtained from the fermentation of *Monascus purpureus* and of *Monascus ruber* were separately measured against a suitable blank. Absorbance values were very high. There is no clear maximum peak value at 238 nm, which probably suggest that the presence of a number of compounds could possibly contribute to the absorbance value. Results described here exclusively for *Monascus purpureus.*

(a) Effect of Lovastatin Analogues and Its Intermediates

Batch fermentation was carried out for lovastatin production by *Monascus purpureus.* The presence of lovastatin was confirmed by TLC in the fermentation broth in the entire fermentation cycle. Lovastatin was measured in the ethyl acetate extract of culture filtrate separately by HPLC technique and on a spectrophotometer. Maximum production of lovastatin was between the 6th and 10th day of fermentation. The absorbance read on the spectrophotometer show equivalence to statin in the mg/l range, whereas the concentration of lovastatin measured by HPLC technique is

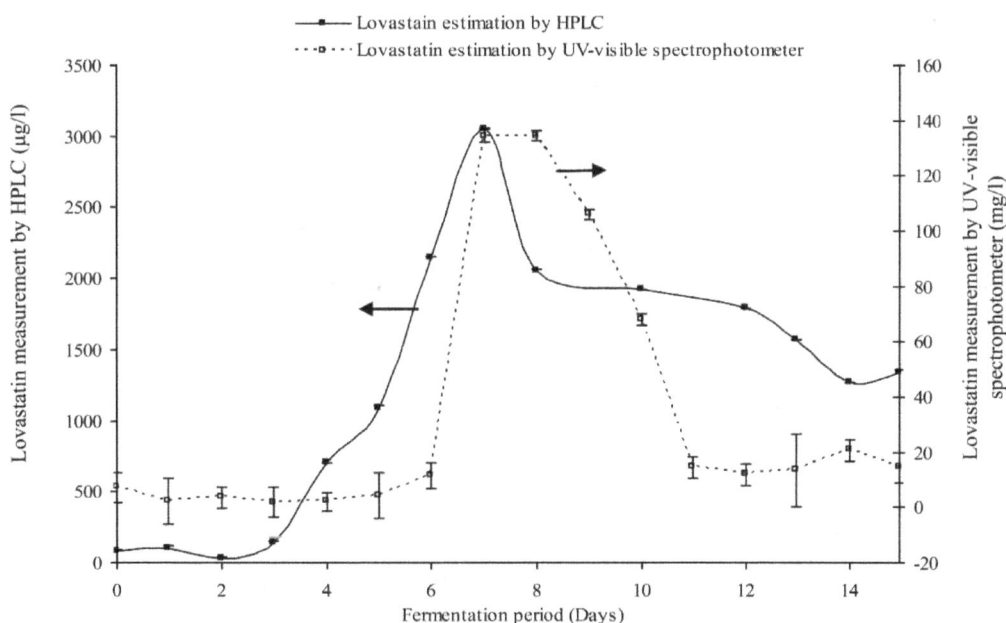

Fig. (3). Concentration profiles of lovastatin production analyzed by UV-visible spectrophotometry and HPLC.

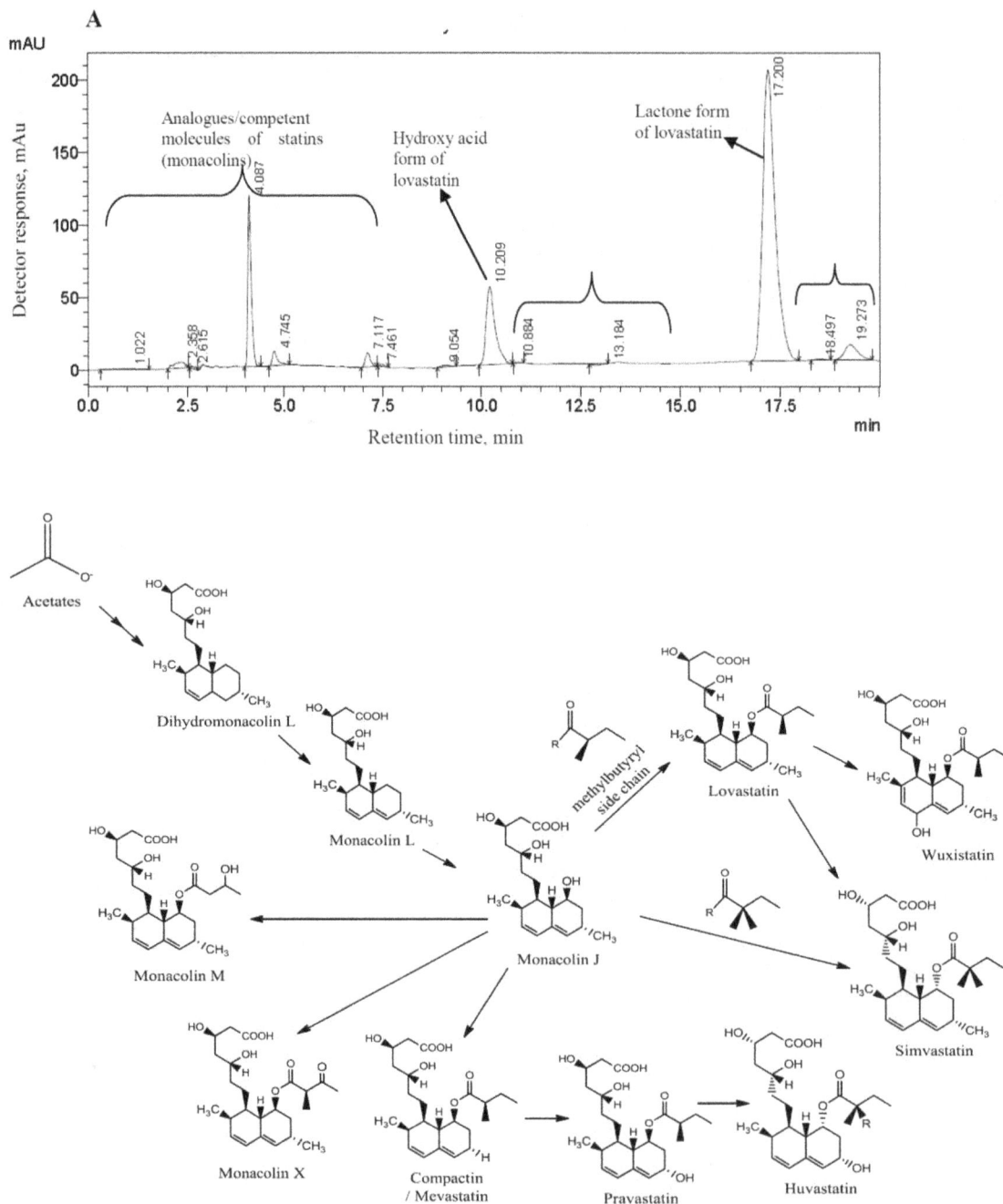

Fig. (4). Possible analogues/competent molecules present in the fermentation broth of *Monascus* sp. (**A**) Chromatogram (measured at 238 nm) of *Monascus purpureus* fermentation samples contains individual forms of lovastatin along with its competent/analogues (monacolins). (**B**) Biosynthesis pathway of lovastatin and its analogues from acetate [6, 7, 44-48]. The possible compounds might be accumulated in fermentation broth, which could showed HPLC chromatogram having different analogues eluted at different rentention time. These compounds has tri-substituted heteroannular diene chromophore which absorb at 238 nm [42, 44, 49].

actually in the range of µg/l (*cf.* Fig. (**3**)). These results also suggest that some intermediates produced by the organism during fermentation might absorb at 238 nm (*cf.* Fig. (**4**)).

It has been reported that the presence of diene groups in lovastatin and some of its intermediates were responsible for this typical interference. Some of the intermediates of lovastatin, *viz.*, monacolin J, X, L, and M were found to have a maximum absorbance at 238 nm [27, 42]. In particular, monacolin L, J, and X were found to exhibit three maximum

absorption peaks similar to the individual forms of pure lovastatin (*cf.* Fig. (**1A**)). However, some of the degraded compounds, *viz.*, methyl esters, anhydro, methoxy, and acetate ester forms of lovastatin and simvastatin could absorb at 238 nm due to the diene groups [43]. These compounds appear as probable interferences in the quantification of lovastatin using a spectrophotometer.

Also, the pleiotropic applications of statins vary among the statins as well as the intermediates [19]. On comparison between the fermentation-derived statins and the synthetic statins, monacolin J (hydrolyzed product of lovastatin) show better prevention of neurodegenerative condition due to its potential blood-barrier penetration and cholesterol lowering effects on neurons [19]. Thus, the determination of analogues of statins produced in the fermentation broth is necessary, especially, when the compound is used for the preparation of nutraceuticals, fermented food etc. By the spectrophotometric method, it could not be possible to estimate the intermediates and their analogues accurately. Using spectrophotometer one can measure monacolins with relative competitive molecules at 238 nm (*cf*. Fig. **4**). On the other hand, HPLC system can be used to measure the intermediates along with lovastatin with greater accuracy (*cf*. Fig. **4**). It is possible that more statins could be produced including lovastatin such as compactin, pravastatin, and monacolin J by *Monascus purpureus*. This speculation is almost equivalent to the observation by Manzoni *et al*. [44] for *Monascus sp*.

Jaivel and Marimuthu determined the concentration of lovastatin by spectrophotometer, showing higher estimate for lovastatin in the range of mg/l [9]. Studies on the applications of fermentation-derived lovastatin have used the spectrophotometric measurement for its quantification [15]. The quantification of lovastatin by spectrophotometer in the presence of intermediates also led to errors in the determination of the lethal dose concentration (LDC) in clinical studies. Osman *et al*. have reported the use of spectrophotometric assay method for lovastatin and found a maximum lovastatin production of 188.3 mg/l using *Aspergillus terreus* [10]. The high concentration of lovastatin produced could be the result of nutritional improvement of a complex medium. In this context, several reports are available for lovastatin biosynthesis using different organisms. Those reports have indicated the spectrophotometric method for the quantification of fermentation-derived lovastatin [9-13].

Xie *et al*. reported the accumulation of monacolins especially, monacolin J, inhibited lovastatin biosynthesis [45]. This suggests that there is a possibility of accumulation of monacolin J during the fermentation. Similarly, some of the intermediates of the lovastatin biosynthesis pathway and structural analogues of lovastatin could absorb light at 238 nm.

High absorbance values (> 0.8 after dilution) and the absence of a clear maximum at 238 nm could be attributed to the medium constituents, their complex nature, and the presence of secondary metabolites.

4. CONCLUSION

Lovastatin was estimated by UV-visible spectrophotometry and HPLC techniques. The interference of medium constituents, intermediates, and products of fermentation which led to inaccuracies during the estimation of lovastatin by spectro-photometry was discussed. The spectrophotometric method is suited for the quantification of pure lovastatin alone while HPLC analysis is the most suitable and reliable method for the estimation of lovastatin under all the experimental conditions heretofore studied.

ACKNOWLEDGEMENTS

The authors gratefully acknowledge the help rendered by Dr. V. Kesavan, members of Chemical Biology Laboratory, Mr. Z. Aslam Basha, and Mr. N. Arumugam of the Department of Biotechnology, IIT Madras and Mr. S. Venkatesan, Mr. S. Ravikumar, and members of Biochemical Engineering Laboratory of the Department of Chemical Engineering, IIT Madras during the different stages of this work.

REFERENCES

[1] Nigović B, Pavković I. Preconcentration of the lipid-lowering drug lovastatin at a hanging mercury drop electrode surface. J Anal Chem 2009; 64: 304-9.

[2] Abu-Nameh ESM, Shawabkeh RA, Ali A. High-performance liquid chromatographic determination of simvastatin in medical drugs. J Anal Chem 2006; 1: 63-6.

[3] Millership JS, Chin J. Determination of simvastatin in tablet formulations by derivative UV spectrophotometry. J Anal Chem 2010; 65(2): 164-8.

[4] Arayne MS, Sultana N, Hussain F, Ali SA. Validated spectrophotometric method for quantitative determination of simvastatin in pharmaceutical formulations and human serum. J Anal Chem 2007; 62: 536-41.

[5] Seenivasan A, Panda T, Théodore T. Lovastatin nanoparticle synthesis and characterization for better drug delivery. Open Biotechnol J 2011; 5: 28-32.

[6] Seenivasan A, Subhagar S, Aravindan R, Viruthagiri T. Microbial production and biomedical applications of lovastatin. Indian J Pharm Sci 2008; 70: 701-9.

[7] Zhuge B, Fang HY, Yu H, et al. Bioconversion of lovastatin to a novel statin by Amycolatopsis sp. Appl Microbiol Biotechnol 2008; 79: 209-16.

[8] Seenivasan A, Panda T, Théodore T. Characterization, modes of synthesis, and pleiotropic effects of hypocholesterolemic compounds–A review. Open Enzym Inhib J 2011; 4: 23-32.

[9] Jaivel N, Marimuthu P. Isolation and screening of lovastatin producing microorganisms. Int J Eng Sci Technol 2010; 2(7): 2607-11.

[10] Osman ME, Khattab OH, Zaghlol GM, Abd El-Hameed RM. Optimization of some physical and chemical factors for lovastatin productivity by local strain of Aspergillus terreus. Aust J Basic Appl Sci 2011; 5: 718-32.

[11] Sreedevi K, VenkateswaraRao J, Narasu L, Fareedullah Md. Strain improvement of Aspergillus terreus for the enhanced production of lovastatin, a HMG-COA reductase inhibitor. J Microbiol Biotech Res 2011; 1: 96-100.

[12] Latha MP, Chanakya P, Srikanth M. Lovastatin production by Aspergillus fischeri under solid state fermentation from coconut oil cake. Nepal J Biotechnol 2012; 2: 26-36.

[13] Reddy DSR, Latha PD, Latha HKPJ. Production of lovastatin by solid state fermentation by Penicillium funiculosum NCIM 1174. Drug Invention Today 2011; 3: 75-7.

[14] Prabhakar M, Lingappa K, Babu V, et al. Characterization of physical factors for optimum lovastatin production by Aspergillus terreus Klvb28mu21 under solid state fermentation. J Recent Adv Appl Sci 2012; 27: 1-5.

[15] Rajasekran A, Kalaivani M, Sabitha R. Anti-diabetic activity of aqueous extract of Monascus purpureus fermented rice in high cholesterol diet fed-streptozotocin-induced rats. Asian J Sci Inf 2009; 2: 180-9.

[16] Mielcarek J, Naskreni M, Grobelny P. Photochemical properties of simvastatin and lovastatin by radiation. J Therm Anal Calorim 2009; 96: 301-5.

[17] Lee C-L, Wang J-J, Pan T-M. Synchronous analysis method for detection of citrinin and the lactone and acid forms of monacolin K in red mold rice. J AOAC Int 2006; 89: 669-77.

[18] Sayyad SA, Panda BP, Javed S, Ali M. Optimization of nutrient parameters for lovastatin production by Monascus purpureus MTCC 369 under submerged fermentation using response surface

methodology. Appl Microbiol Biotechnol 2007; 73: 1054-8.

[19] Sierra S, Ramos MC, Molina P, *et al*. Statins as neuroprotectants: a comparative in vitro study of lipophilicity, blood-brain-barrier penetration, lowering of brain cholesterol, and decrease of neuron cell death. J Alzheimers Dis 2011; 23: 307-18.

[20] Yoshida MI, Oliveira MA, Gomes ECL *et al*. Thermal characterization of lovastatin in pharmaceutical formulations. J Therm Anal Calorim 2011; 106: 657-64.

[21] Li M, Fan L-Y, Zhang W, Sun J, Cao C-X. Quantitative analysis of lovastatin in capsule of Chinese medicine *Monascus* by capillary zone electrophoresis with UV-vis detector. J Pharm Biomed Anal 2007; 43: 387-92.

[22] Damić M, Nigović B. Analysis of statins in pharmaceuticals by MEKC. Chromatographia 2010; 71: 233-40.

[23] Yang D-J, Hwang LS. Study on the conversion of three natural statins from lactone forms to their corresponding hydroxy acid forms and their determination in Pu-Erh tea. J Chromatogr A 2006; 1119: 277-84.

[24] Chang Y-N, Huang J-C, Lee C-C, Shih I-L, Tzeng Y-M. Use of response surface methodology to optimize by *Monascus ruber*. Enzyme Microb Technol 2002; 30: 889-94.

[25] Sharaf El-Din MMK, Attia KAM, Nassar MWI, Kaddah MMY. Colorimetric determination of simvastatin and lovastatin in pure form and in pharmaceutical formulations. Spectrochim Acta A Mol Biomol Spectrosc 2010; 76: 423-8.

[26] Samiee SM, Moazami N, Haghighi S, *et al*. Screening of lovastatin production by filamentous fungi. Iran Biomed J 2003; 7: 29-33.

[27] Yang F, Weber TW, Gainer JL, Carta G. Synthesis of lovastatin with immobilized *Candida rugosa* lipase in organic solvents: effects of reaction conditions on initial rates. Biotechnol Bioeng 1997; 56: 671-80.

[28] Schachter M. Chemical, pharmacokinetic and pharmacodynamic properties of statins: an update. Fundam Clin Pharmacol 2004; 19: 117–25.

[29] Nanjwade BK, Derkar GK, Bechra HM, Nanjwade VK, Manvi FVJ. Design and characterization of nanocrystals of lovastatin for solubility and dissolution enhancement. Nanomed Nanotechnol 2011; 2: 107. doi:10.4172/2157- 7439.1000107.

[30] Shinde AJ, More HN. Design and evaluation of polylactic-co-glycolic acid nanoparticles containing simvastatin. Int J Drug Dev Res 2011; 3: 280-9.

[31] Patel MJ, Patel NM, Patel RB, Patel RP. Formulation and evaluation of self-microemulsifying drug delivery system of lovastatin. Asian J Pharm Sci 2010; 5: 266-75.

[32] Katare MK, Kohli S, Jain AP. Evaluation of dissolution enhancement of lovaststin by solid dispersion technique. Int J Pharm Life Sci 2011; 2: 894-8.

[33] Csempesz F, Söle A, Puskás I. Induced surface activity of supramolecular cyclodextrin–statin complexes: Relevance in drug delivery. Colloids and Surfaces A: Physicochem Eng Aspects 2010; 354: 308-13.

[34] Shinde AJ, Paithane MB, More HN. Development and in vitro evaluation of transdermal patches of lovastatin as a antilipidemic drug. Int Res J Pharm 2010; 1: 113-21.

[35] Ho M-H, Chiang C-P, Liu Y-F, *et al*. Highly efficient release of lovastatin from poly(lactic-co-glycolic acid) nanoparticles enhances bone repair in rats. J Orthop Res 2011; 29: 1504-10.

[36] Mandal S. Microemulsions drug delivery system: design and development for oral bioavailability enhancement of lovastatin. S Afr Pharm J 2011; 78: 44-50.

[37] Jerotskaja J, Lauri K, Tanner R, Luman M, Fridolin I. Optical dialysis adequacy sensor: wavelength dependence of the ultra-violet absorbance in the spent dialysate to the removed solutes. Conf Proc 29th IEEE Eng Med Biol Soc 2007: pp. 2960-3.

[38] Lin F, Ma QL. A novel photoproduct of thymine in phosphate-buffered saline under far UV irradiation. Chin Chem Lett 2003; 14: 1233-5.

[39] Wang RW, Kari PH, Lu AYH, *et al*. Biotransformation of lovastatin. IV. Identification of cytochrome P450 3A proteins as the major enzymes responsible for oxidative metabolism of lovastatin in rat and human liver microsomes. Arch Biochem Biophys 1991; 290: 355-61.

[40] Jacobsen W, Kirchner G, Hallensleben K, *et al*. Small intestinal metabolism of the 3-hydroxy-3-methylglutaryl-coenzyme A reductase inhibitor lovastatin and comparison with pravastatin. J Pharmacol Exp Ther 1999; 291: 131-9.

[41] Pham PJ, Hernandez R, French WT, Estill BG, Mondala AH. A spectrophotometric method for quantitative determination of xylose in fermentation medium. Biomass Bioenerg 2011; 35: 2814-21.

[42] Li Y-G, Zhang F, Wang Z-T, Hu Z-B. Identification and chemical profiling of monacolins in red yeast rice using high-performance liquid chromatography with photodiode array detector and mass spectrometry. J Pharm Biomed Anal 2004; 35: 1101-12.

[43] Krishna SR, Deshpande GR, Rao BM, Rao NS. A Stability-Indicating RP-LC method for the determination of related substances in simvastatin. J Chem Pharm Res 2010; 2: 91-9.

[44] Manzoni M, Bergomi S, Rollini M, Cavazzoni V. Production of statins by filamentous fungi. Biotechnol Lett 1999; 21: 253-7.

[45] Xie X, Watanabe K, Wojcicki WA, Wang CC, Tang Y. Biosynthesis of lovastatin analogs with a broadly specific acyltransferase. Chem Biol 2006; 13: 1161-9.

[46] Kimura K, Komagata D, Murakawa S, Endo A. Biosynthesis of monacolins: conversion of monacolin J to monacolin K (mevinolin). J Antibiot 1990; 43: 1621–2.

[47] Hutchinson CR, Kennedy J, Park C. Method of producing antihypercholesterolemic agents. US pat 6943017. 2005 Sep.

[48] Manzoni M, Rollini M. Biosynthesis and biotechnological production of statins by filamentous fungi and application of these cholesterol-lowering drugs. Appl Microbiol Biotechnol 2002; 58: 555-64.

[49] Alberts AW, Chen J, Kuron G, *et al*. Mevinolin: a highly potent competitive inhibitor of hydroxymethylglutaryl-coenzyme A reductase and a cholesterol-lowering agent. Proc Natl Acad Sci USA 1980; 77: 3957-61.

14

Modeling and Experimental Analysis of Cephalosporin C Acylase and Its Mutant

Ren Yu*

Key Laboratory of Ecology and Biological Resources in Yarkand Oasis at Universities under the Education Department of Xinjiang Uygur Autonomous Region, Department of Kashi Normal College, No.29. Xueyuan Road, Kashi Prefecture, Xinjiang Uygur Autonomous Region, China

Abstract: 7-amino cephalosporanic acid (7-ACA) is the crucial intermediate for the synthesis of semi-synthetic antibiotics, which is currently prepared by two-step biocatalysis using D-amino acid oxidase and glutaryl-7-amino cephalosporanic acid acylase (GL-7-ACA acylase) starting from cephalosporin C (CPC). Compared with the two-step enzymatic method, one-step method is more efficient and economical. But, the available Cephalosporin C acylase (CPC acylase) always take glutaryl-7-amino cephalosporanic acid (GL-7-ACA) as their primary substrate, and have low catalytic activities towards CPC to be used in industry. We investigated the catalytic mechanism of CPC acylase by the sequence alignment, homology modeling, and active site analysis to a series of CPC acylases from *Pseudomonas* where some effective mutations have been reported for activity enhancement. Two CPC acylases coded by the genes *acyII* and *S12* are studied intensively for the interaction between the amino acid residues in the activity region and the substrate CPC based upon the complex structure obtained from the homology modeling and molecular docking. Furthermore, the catalytic parameters of the two CPC acylases were measured experimentally in order to corroborate the modeling analysis and propose potential designing strategy for improvement of enzymic activity.

Keywords: 7-amino-cephalosporanic acid, Cephalosporin C, Cephalosporin C acylase, Homology modeling, Molecular docking, Protein designing.

1. INTRODUCTION

The semi-synthetic cephalosporins became the popular antibiotics due to their excellent characteristics such as broad spectrum, low toxicity, and resistance to the β-lactamase and made tremendous contribution to fight with bacterial infection [1]. The semi-synthetic cephalosporin is synthesized from the intermediate 7-amino cephalosporanic acid (7-ACA), which shares more than 40% of the global anti-infective market [2].

Currently, 7-ACA used for the semi-synthetic cephalosporin antibiotics is produced from Cephalosporin C (CPC) by either the chemical or the enzymatic methods. Among them, the two-step enzymatic method, is becoming dominant gradually [3, 4] because of its friendship to environment. However, this process is expensive and can't completely satisfy the industrial production. In comparison, the one-step enzymatic method is efficient and has been studied intensively [2]. Researchers separated CPC acylase from the micro-organisms which could convert CPC into 7-ACA directly [5-9]. But, the application of the wild strain was inconvenient [3] and the question was that CPC acylases

used glutaryl-7-amino cephalosporanic acid （GL-7-ACA） as their primary substrate normally and their specificity towards CPC was too low to be used in industry [10, 11]. Some mutations with improved activity towards CPC have been developed By protein engineering of the CPC acylases Oh *et al.* [12] found that the deacylation activity of the mutation Q50βM-Y149αK-F177βG toward CPC was improved by 790%. Pollegioni *et al.* [2] used the approach of the the homology modeling combined with the site-directed mutagenesis to produce the A215Y-H296S-H309S mutation which had slightly higher activity towards CPC (3.8U/mg protein) than to GL-7-ACA (2.7 U/mg protein). Ishii *et al.* [13] found that the mutation M269W caused the 1.6-fold increase of the specific activity against CPC and observed that the minor change of conformation induced by the mutation increased the stability of the enzyme-substrate complex. Saito *et al.* [14] suggested that Met164 was located in the binding region in the interior surface of the CPC acylase for recognition of the substrate and found that the mutantion M164L enhanced CPC acylase activity.

In this paper, the homology modeling and the structural analysis have been applied into a series of CPC acylases on which some mutations have been reported for improved activity (Table **1**). The two genes named *acy II* and *S12* coding the CPC acylase Acy II and its mutation named S12 respectively. *acy II* and *S12* were constructed into pET28a and expressed in the *E.coli* BL21(DE3) for experimental analysis.

*Address correspondence to this author at the Key Laboratory of Ecology and Biological Resources in Yarkand Oasis at Universities under the Education Department of Xinjiang Uygur Autonomous Region, Department of Biological Science, Kashi Normal College, No.29. Xueyuan Road, Kashi Prefecture, Xinjiang Uygur Autonomous Region, China;
E-mail: renyu1020@yahoo.com.cn

Table 1.   Reported CPC Acylases and Mutants from *Pseudomonas* sp.

Author, Year	Modified Amino Acids	Enzyme Activity
Ishii *et al.*, 1994	Y270F	Decreased activity
Nobbs *et al.*, 1994	Y270F	Decreased activity
Ishii *et al.*, 1995	M269Y or F	1.6-fold and 1.7-fold increase
Saito *et al*, 1996[b]	M164L	Enhance activity
Saito *et al.*, 1996[a]	A271Y	Increase 1.2-fold
Yamada *et al.*, 1996	Y270A or Y270F or Y270L or Y270S	decrease
	C199S,C277S,C305S,C391S,C496S	decrease
	C305S-M269S	1.6-fold increase

[a][14]; [b][26].

With the approach of the homology modeling and the experimental analysis, we established the preliminary knowledge about discovering the potentially efficient CPC acylase *in silico*, which played the important role for the one-step preparation of 7-ACA for CPC enzymatically.

2. MATERIALS AND METHODS

2.1. Homology Modeling

A series of CPC acylases originated from *Pseudomonas* sp. were chosen for modeling analysis in this work. Six protein templates from PDB [15] were selected to build the homology model of AcyII, i.e., Penicillin G acylase from *Escherichia coli* (PDB code 1e3a) sharing 36.3% sequence similarity with AcyII; Penicillin acylase complexed with 3, 4-dihydroxyphenylacetic acid (PDB code 1ai4) sharing 37.5% similarity; Cephalosporin acylase in complex with glutaryl-7-aminocephalosporanic acid (PDB code 1jvz) sharing 37.9% similarity; Penicillin amidohydrolase (PDB code

1pnm) sharing 36.5% similarity; Glutarylamidase (PDB code 1gk1) sharing 38.5% similarity, and Penicillin G acylase from *Alcaligence faecalis* (PDB code 3k3w) sharing 39.4% similarity with AcyII. The result of sequence alignment was obtained by using Discovery Studio 2.1 (Accelrys, v2.0, 2009).

The model of AcyII was constructed by using the Homology Modeling Module in Accelrys Discovery Studio 2.1. The quality of the predicted model was evaluated by the Discrete Optimized Protein Energy (DOPE) by running the Verify Protein module. By the CHARMm force field [16], the conformation of amino acid residues in AcyII structural model was further modified by a standard dynamics cascade created by joining a set of steps of minimization and equilibration including minimization with steepest descent, minimization with conjugate gradient, dynamics with heating, equilibration dynamics and production dynamics. Next, the potential binding region on which the CPC was docked were identified by the Dock Ligands Module in Discovery Studio.

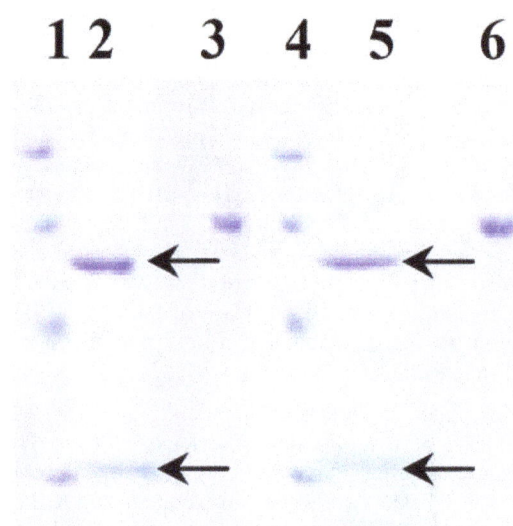

Fig. (1). SDS-PAGE of purified CPC acylase Acy II and S12
Lane 1, Molecular weight marker（from above to below: 97.2kDa, 66.4kDa, 44.3kDa, 29.0kDa）; lane2, Acy II; lane 3, BSA; lane 4, Marker; lane 5, S12; Lane 6, BSA.

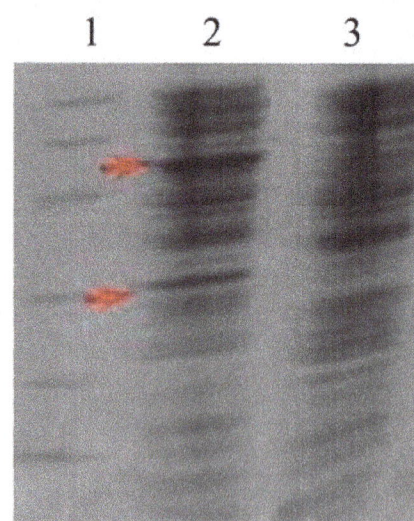

Fig. (2). Expression of CPC acylase Acy II and S12
Lane 1, Molecular weight marker（from above to below: 97.2kDa, 66.4kDa, 44.3kDa, 29.0kDa, 20.1kDa, 14,3kDa）; lane2, Acy II; lane 3, S12.

After obtaining the preliminary model of AcyII-CPC complex, the PRODA, a PROtein Design Algorithmic software [17, 18], was applied to place the CPC on the active region under the catalytic constraints between the CPC and the four catalytic residues, i.e., Ser β1, Hisβ23, Hisβ70 and Asnβ242.

2.2. Experimental Procedures

2.2.1. Mutagenesis

The gene *acy II* [19] was synthesized. Its mutant named *S12* was obtained by the overlapping primer PCR with the substituted amino acid residues V121αA-G139αS-F58βN-I75βT-I176βV-S471βC. The two genes were cloned into the pET28a(+) plasmids, sequenced and transformed into the *E.coli* BL21(DE3). For all primers, mutant positions were donoted in lowercase and the restriction sites are underlined. Backward primers designed with completely complementary role are marked with asterisk.

5'-GAGCTCATGACCATGGCGGCGAA-3'*CTCGAGTACTGGTA-CCGCCGCTT

5'-CAGGAACTGGTGCCGGCGCTCGAG-3'
*GTCCTTGACCACGGCCGCGAGCTC

5'-GCGTATgcgGCTGGAGTTAA-3'
*CGCATAcgcCGACCTCAATT

5'-CGAATATagcCTGCT-3'   *GCTTATAtcgGACGA

5'-GCTTTCCGCATaatGCGCA-3'
*CGAAAGGCGTAttaCGCGT

5'-CGTTTATGGATaccCAT-3'
*GCAAATACCTAtggGTA-3'

5'-GGCCTGgttGATCAT-3'   *CCGGACcaaCTAGTA

5'-CGCGCTGtgcCGTTAT-3'
*GCGCGACacgGCAATA

2.2.2. Expression and Purification of AcyII and S12

E. coli BL21(DE3) carrying the gene *acy II* was grown in LB medium containing 50 ug/mL kanamycin with shaking at 37°C overnight. A quantity of 100ml of fresh LB medium was inoculated with 1ml overnight culture and incubated with shaking to an O.D.$_{600nm}$ of 0.6. Then expression was induced by addition of 1mmol L^{-1}. The cell pellet of 100ml induced BL21(DE3) containing *acy II* was suspended in 100mM Tris-HCl buffer (pH8.0) and was sonicated for $20 \times 10s$ with 10s pause at 200-300w. The supernatant was loaded to 2ml Ni-NTA and eluted with 0mM inidazole, 50mM imidazole, 100mM imidazole, 200mM imidazole and 500mM imidazole in a succession. The target fractions were pooled and analyzed by SDS-PAGE.

2.2.3. Assay of AcyII and S12 Activity

Enzymic activity was determined for conversion of CPC to 7-ACA. 500 ul S12 (approximately 1 uM for CPC) was added to 500 ul CPC (20 mg/ml in 0.1 M Tris/HCl, pH 8.0), and the mixture was incubated at 37°C for 8 min. The reaction was stopped by addition of 5% acetic acid. After centrifugation (10,000 rpm, 5 min), the formed 7-ACA in the supernatant was determined by HPLC. One unit was defined as the amount of the enzyme liberating 1 umol 7-ACA/min.

3. RESULTS AND DISCUSSION

3.1. Analysis of AcyII Model

The Ramachandran diagram for the model of AcyII shows that there were 92.1% of residues falling in the allowed region, and 6.3% of residues in the marginal region. The remaining 1.6% of residues in the disallowed region was mostly far from the active region. In addition, the Verify Score of the model predicted by Accelrys Discovery Studio 2.1 was 178.74 while the Verify Expected High and Low Scores were 214.58 and 96.56, respectively. (The Verify Expected High Score is the score that would be expected for a correct structure having this sequence length, based on a statistical analysis of high-resolution structures in the Protein Data Bank. The Verify Expected Low Score is 45% of the first and is a score that is typical of grossly misfolded structures having this sequence length. The higher the Verify Score is and the more correct the structure is). These two types of data confirmed the reliability and correctness of the model of AcyII.

The homologous model and the active region of AcyII shown in Fig. (3a,b) and Fig. (4) were consistent with those of other acylases described by Fritz-Wolf *et al.* [20]. These acylases used conserved Ser1β as catalytic residue which was characteristic of the N-terminal hydrolase family. For Ser1β, its hydroxyl group was fixed by the conserved His23β and its NH group formed a hydrogen bond with His23β. The NH groups from the backbone of His70β and side chain of Asn242β formed the oxyanion hole for carboxyl group of CPC. The sequence alignment (Fig. **5a**) for different acylases, i.e., AcyII, CAD [21], and PGA [22] implied that those catalytic residues were conserved, just His70β was variable. The role of His70 β was to stabilize the hydrogen bond by using its backbone, which was consistent with the corresponding residues in PGA and CAD whose crystal structures were known. In the binding region, oxygen atoms from the carboxylate group of CPC interacted with Arg24β, Tyr32β and His57β. The amino adipyl moiety of CPC was stabilized

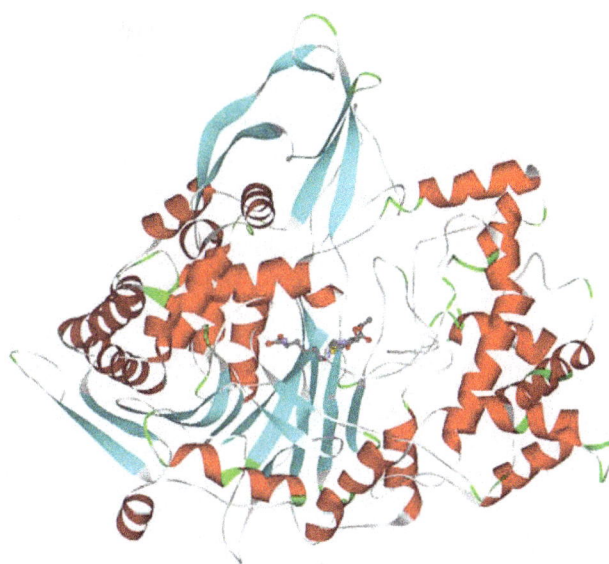

Fig. (3a). The model of AcyII, in which the substrate CPC is shown in ball and stick mode.

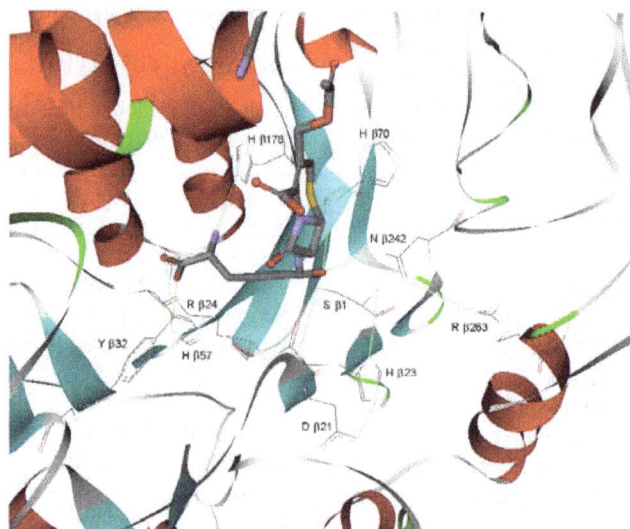

Fig. (3b). View of the active region with substrate for AcyII based on model.
The key catalytic and binding residues are shown in thin line model while the substrate CPC is shown in stick model. The hydrogen bonds are shown in green lines.

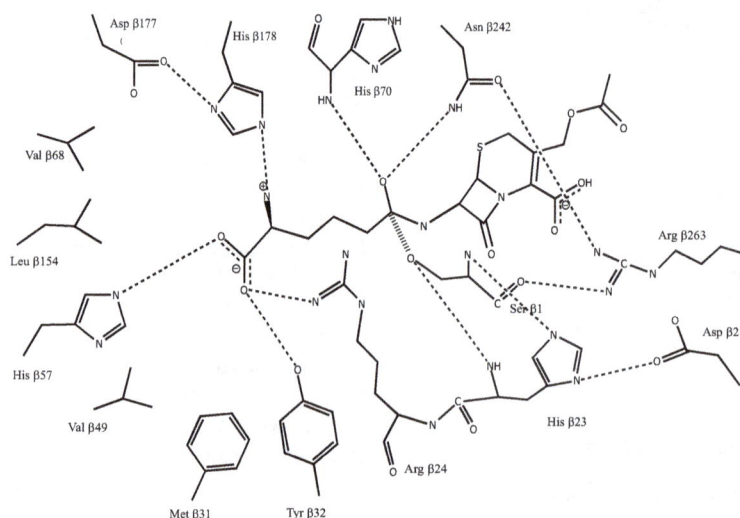

Fig. (4). Schematic drawing of the active amino acid residues for AcyII based on model.

by the formed hydrogen bond with His178β which simultaneously interacted with Asp177β stated by Fig. (4).

N176 was found active to CPC. Since AcyII and N176 shared a sequence similarity up to 93.5%, shown by Fig. (5b) for sequence alignment, it was reasonable to consider that N176 and S12 shared the high structural similarity. So the analysis of mutation of N176 could be based on the same model of AcyII.

3.2. Structural Analysis of Active Site Mutations of N176

Ishii *et al.* [23] reported that the mutation of W32βF led to the decreasing of k_{cat} to about 50%. Nobbs *et al.* [24] reported the mutation of W32βL whose k_{cat} was reduced by 32.2%. Another effort was made by Saito *et al.* [25] who changed Tyr32β to Ala/ Ser, which resulted in the decreasing of specific activity. These three studies confirmed the key binding effect of Tyr32β which interacted with the carboxyl group of CPC shown in Fig. (6a). Either of the mutations to

Phe32β or Leu32β demolished the original hydrogen bond with the carboxyl group on CPC, thus led to the decreasing of binding effect with CPC. The decreased activity from mutation W32βS was possibly that the distance between the polar groups on Ser32β and CPC was too short to form hydrogen bond.

The mutations from Met31β to Phe31β or Tyr31β reported by Ishii *et al.* [13] enhanced catalytic activity by 1.6-fold and 2.5-fold. The effect of these mutations is rational as the phenyl group from either phenylalanine or tyrosine could form π-π conjugation with that from the neighboring Tyr32β and this would help the neighboring residue Tyr32β to maintain the suitable conformation in order to form the hydrogen bond with CPC, as that shown by Fig. (6a).

3.3. Structural Analysis About the Mutations for AcyII

In this paper, the mutation V121αA-G139αS-F58βN-I75βT-I176βV-S471βC named S12 was investigated. F58βN,

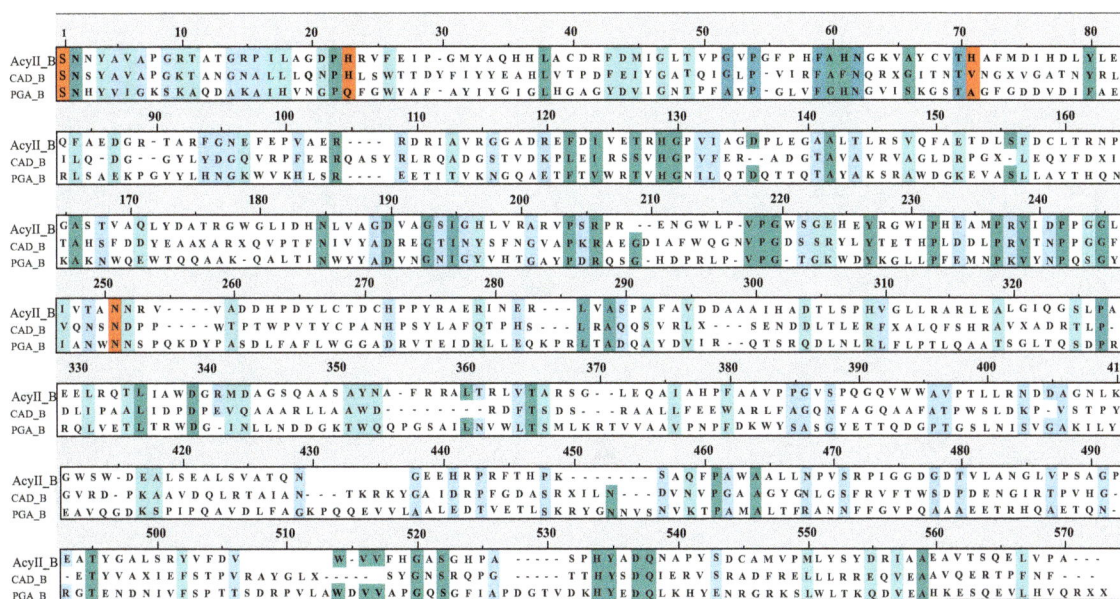

Fig. (5a). Sequences alignment of β-chains from AcyII, CAD and PGA

The four catalytic sites, i.e., Ser1β, His23β, His70β and Asn242β in AcyII, and their corresponding sites in CAD and PGA are shown in red.

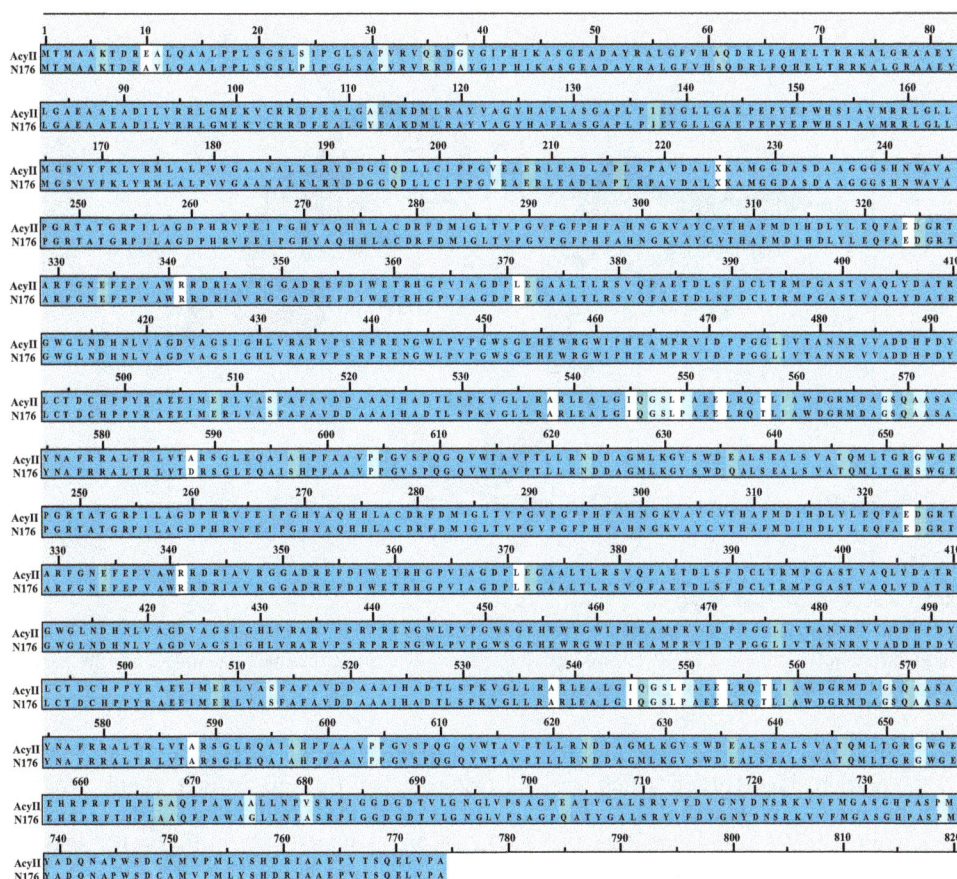

Fig. (5b). Sequence alignment of N176 and Acy II , where different residues are indicated.

I75βT and I176βV were close to the active region according to the model. For the mutation F58βN firstly, Asn58β had the polar carboxyl group which was different from the original hydrophobic Phe58β, thus could form hydrogen bond with the glyoxaline group on His 57β and stabilize the car- boxyl group of CPC by the N-O hydrogen bond, as that shown in Fig. (6a). Secondly, the phenyl group from Phe58β would clash with CPC because of its larger side chain and lead to increasing energy and decreasing stability compared with the mutation Asn58β.

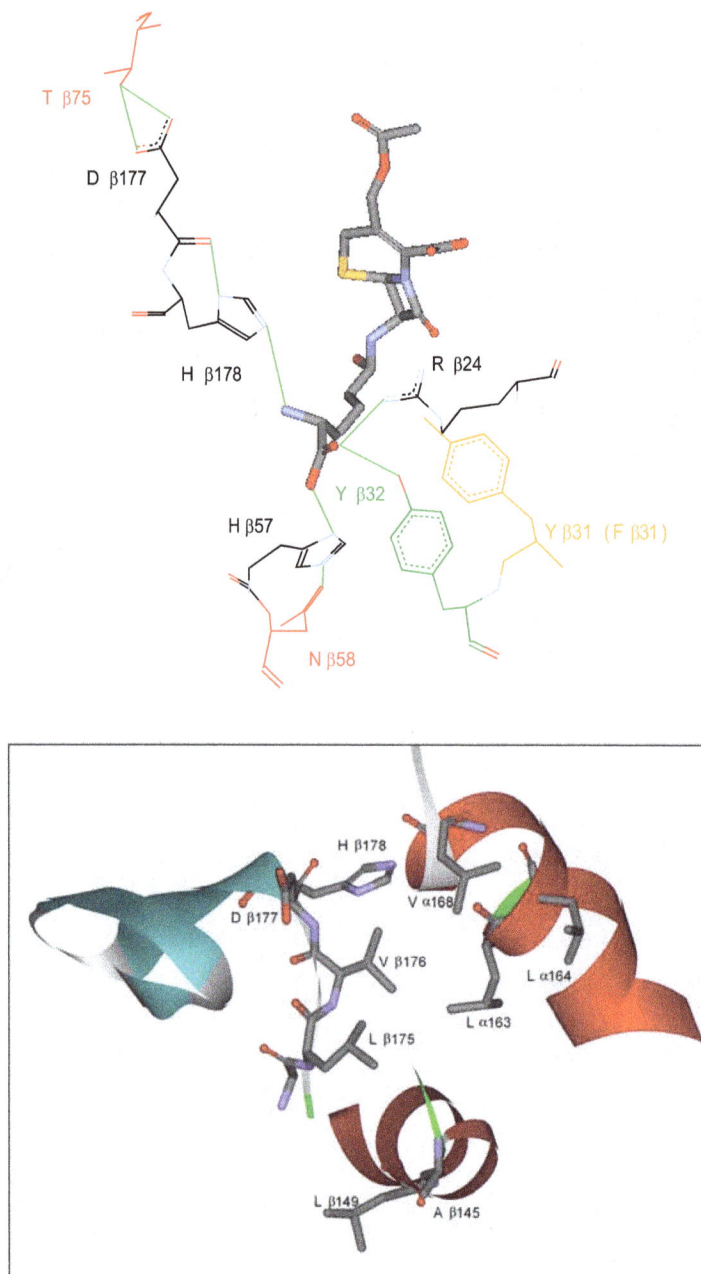

Fig. (6). View of the mutations at active sites of AcyII

(**a**). The two residues, i.e., Asn58β and Thr75β, are colored by red. The residues Tyr31β or Phe 31β is colored by orange. The important binding residue Tyr32β which interacts with the CPC is colored by dark green. The other residues are shown in line mode while CPC is in stick mode; (**b**). The residue Val176β with its neighboring non-polar residues

As to the mutation I75βT, the non-polar side residue Ile75β was replaced by the polar residue Thr75β which was located in a loop area in the binding region shown by Fig. (**6a**). The carboxyl group on Thr75β could stabilize the neighboring Asp177β by forming two hydrogen bonds. The C-O group on Asp177β interacted with ND1 on His178β which fixed the amino adipyl moiety in CPC. These structural interactions implied that the mutation from Ile75β to Thr75β was more favorable because the polar side chain contributed to the stability of the binding region by supplying additional hydrogen bonds.

For the mutation I176βV, Val176β had shorter side chain than that of Ile176β, which could avoid the side chain clashes with the neighboring residues, such as Leu163α and Leu175β, as that shown in Fig. (**6b**). And Val176β was located in a loop region near the two important binding residues, i.e., Asp175β and His176β, which interacted with the amino adipyl moiety in CPC, so the effect of reducing spatial clashes was beneficial to stabilize the important interactions for binding.

In addition, the three residues, V121αA, G139αS and S471βC were far from the active region. The mutations V121αA and G139αS were related to expression level as

Table 2. Catalytic Kinetic Parameters for CPC Acylase Acy II and S12

Protein	K_m(mM)	k_{cat}(sec^{-1})	k_{cat}/K_m(sec^{-1}(μM)$^{-1}$)
Acy II	23.71	7.622	0.321
S12	15.26	14.03	0.919

Specific activities for Acy II and S12 are measured to be 2.868 unit/mg proteins and 6.011 units/mg proteins for CPC, respectively (pH8.0). Kinetic parameters were calculated from Lineweaver-Burk plots of the primary velocity of 7-ACA transformed directly from CPC (2.1, 4.2, 6.3, 8.4, 10.5 and 20.9mM) in the presence of Acy II and S12 (1 μ M) at 37℃ for 8 min.

shown in Fig. (**2**) [26]. The mutation S471βC gave S12 the product inhibition as Pollegioni's report [2].

3.4. Experimental Analysis About Mutation of AcyII

From the SDS-PAGE results shown in Fig. (**1**), we observed that both of the CPC acylase AcyII and S12 were expressed with the MW approximately 87 kDa and was composed of two subunits, the 58kDa α-subunit and the 25kDa β-subunit, which was consistent with that reported by Mstsuda *et al.* [27]. The expression levels of AcyII and S12 were measured to be 322 U/L and 291 U/L, respectively, shown in Fig. (**2**). The specific activities of AcyII and S12 were measured and the results were shown in Table **2**. And we could see that the specific activity of S12 was 2-fold higher than that of AcyII and it reaches 6.011 U/mg protein. The catalytic parameters K_m, k_{cat} and k_{cat}/K_m of AcyII and S12 were determined by Lineweaver-Burk plot method. The k_{cat}/K_m of S12 was higher than that of AcyII, which indicated that the mutation of six amino acid residues increased the catalytic efficiency. The result was similar to that of N176 [14]. Combined with the theoretical analysis based on structural modeling presented in Section 3.3, it was implied that the mutations around the active region, i.e., F58βN- I75βT- I176βV, enhanced the binding capability between the enzyme and the transition state of the substrate instead of substrate itself since S12 had larger K_m but higher k_{cat}. By virtue of the transition state theory for enzyme catalysis [28], the enhancement of binding between enzyme and the transition state of substrate reduced the activation energy and led to the increase of turnover number, i.e., k_{cat}. Because the structures between substrate and its transition state was different, the strong binding capability between enzyme and transition state of the substrate increased the dissociation reaction between enzyme and the substrate, which could be certificated by the Michaelis constant, i.e., K_m, shown in Table **2**.

4. CONCLUSION

In this work, Acy II from *Pseudomonas sp.* and the mutation named S12 were structurally modeled and experimentally characterized in order to investigate their catalytic mechanism for further designing highly efficient enzyme to the one-step preparation of 7-ACA from CPC. With the methods of sequence alignment, homology modeling, and molecular docking, the structures of the active region of CPC acylase and the complex were obtained and the mutations around active site were analyzed based on intermolecular binding interaction, including steric hindrance, hydrogen bonding stabilization, solvation, and electrostatic contribution. The further experimentally measured catalytic parame-

ters, i.e., k_{cat} and K_m, for CPC acylase AcyII and S12 confirmed the predicted model and provided strong evidence that the mutations around active region for increased activity would contribute to binding reaction between enzyme and the transition state substrate and decreasing activation energy for reaction. This implied that further designing of highly efficient CPC acylase *in silico* should focus on the amino acid sites for stronger transition state binding capability.

ACKNOWLEDGEMENT

Y.R sincerely appreciates the financial support from the National Science Foundation of China and the National High Technology Research and Development (863) Program of China. Y.R thanks to Dr. Yushan Zhu and Dr. Jianan Zhang for their support and helpful discussion throughout the experimental course of this work.

REFERENCES

[1] Stefanie S, Christian MD, Jeffrey S, *et al.* The cephalosporin antibiotics. Prim Care Update for Ob Gyns 1997; 4: 168-74.
[2] Pollegioni L, Lorenzi S, Rosini E, *et al.* Evolution of an acylase active on cephalosporin C. Protein Sci 2005; 14: 3064-76.
[3] Shibuya Y, Matsumoto K, Fuji T, *et al.* Isolation and properties of 7β-(4-carboxy butanamido) cephalosporanic acid acylase producing bacteria. Agric Biol Chem 1981; 45: 1561-7.
[4] Ichikawa S, Murai Y, Yamamoto S, *et al.* The isolation and properties of *Pseudomonas* mutants with all enhanced productivity of 7β-(4-carboxybutanamido) cephalosporinic acid acylase. Agric Biol Chem 1981; 45: 2225-9.
[5] Zhu SC, Yang YL, Zhao GP, *et al.* A rapid and specific method to screen environmental microorganisms for cephalosporin acylase activity. J Microbiol Methods 2003; 54: 131-5.
[6] Kim DW, Yoon KH. Cloning and high expression of glutaryl 7-aminocephalosp- oranic acid acylase gene from *Pseudomonas diminuta*. Biotechnol Lett 2001; 23: 1067-71.
[7] Aramori I, Fukagawa M, Tsumura M. Cloning and nucleotide sequencing of a novel 7β-(4-carboxybutanamido) cephalosporanic acid acylase gene of *Bacillus laterosporus* and its expression in *Escherichia coli* and *Bacillus subtilis*. J Bacteriol 1991[a]; 173: 7848-55.
[8] Aramori I, Fukagawa M, Tsumura M, *et al.* Isolation of soil strains producing new cephalosporin acylases. J Ferment Bioeng 1991[b]; 72: 227-31.
[9] Aramori I, Fukagawa M, Tsumura M, *et al.* Cloning and nucleotide sequencing of new glutaryl 7-ACA and cephalosporin C acylase genes from *Pseudomonas* strains. J Ferment Bioeng 1991[c]; 72: 232-43.
[10] Zhang QJ, Xu WX. Morphological, physiological and enzymatic characteristics of cephalosporin acylase-producing *Arthrobacter* strain 45-8A. Arch Microbiol 1993; 159: 392-5.

[11]   Deshpands BS, Ambedkar SS, Shewale JG. Cephalosporin C acylase and penicillin V acylase formation by *Aeromonas* sp.ACY95. World J Microbiol Biotechnol 1998; 12: 373-8.

[12]   Oh B, Kim M, Yoon J, *et al.* Deacylation activity of cephalosporin acylase to cephalosporin C is improved by changing the side-chain conformations of active-site residues. Biochem Biophys Res Commun 2003; 310: 19-27.

[13]   Ishii Y, Saito Y, Fujimura T, *et al.* High-level production, chemical modification and site-directed mutagenesis of a cephalosporin C acylase from *Pseudomonas* strain N176. Eur J Biochem 1995; 230: 773-8.

[14]   Saito Y, Ishii Y, Fujimura T, *et al.* Protein engineering of a cephalosporin C acylase from *Pseudomonas* strain N176. Ann NY Acad Sci 1996[b]; 782: 226-40.

[15]   Berman HM, *et al.* The Protein Data Bank. Acta Crystallogr D Biol. Crystallogr 2002; 58: 899–907.

[16]   Brooks BR, Bruccoleri RE, Olafson BD, *et al.* CHARMM: A Program for Macromolecular Energy, Minimization, and Dynamics Calculations. J Comput Chem 1983; 4: 187-217.

[17]   Zhu Y. Mixed-integer linear programming algorithm for a computational protein design problem. Ind Eng Chem Res 2007; 46: 839-45.

[18]   Luo W, Pei J, Zhu Y. A fast protein-ligand docking algorithm based on hydrogen bond matching and surface shape complementarity. J Mol Model 2010; 16: 903-13.

[19]   Matsuda A, Toma K, Komatsu K. Nucleotide sequences of the genes for two distinct cephalosporin acylases from a *Pseudomonas* strain. J Bacteriol 1987[b]; 169: 5821-6.

[20]   Fritz-Wolf K, Koller K-P, Lange G, *et al.* Structure-based prediction of modifications in glutarylamidase to allow single-step enzymatic production of 7-aminocephalosporanic acid from cephalosporin C. Protein Sci 2002; 11: 92-103.

[21]   Kim Y, Yoon K.-H, Khang Y, *et al.* The 2.0Å crystal structure of cephalosporin acylase. Structure 2000; 8: 1059-68.

[22]   Duggleby HJ, Tolley SP, Dodson EJ, *et al.* Penicillin acylase has a single amino-acid catalytic center. Nature 1995; 373: 264-8.

[23]   Ishii Y, Saito Y, Sasaki H, *et al.* Affinity labelling of cephalosporin C acylase from *Pseudomonas* sp. N176 with a substrate analogue, 7β-(6-bromohexanoylamido) cephalosporanic acid. J Ferment Bioeng 1994; 77: 598-603.

[24]   Nobbs TJ, Ishii Y, Fujimura T, *et al.* Chemical modification and site-directed mutagenesis of tyrosine residues in cephalosporin C acylase from *Pseudomonas* strain N176. J Ferment Bioeng 1994; 77:604-9.

[25]   Saito Y, Fujimura T, Ishii Y, *et al.* Oxidative Modification of a Cephalosporin C Acylase from *Pseudomonas* Strain N176 and Site-Directed Mutagenesis of the Gene. Appl Environ Microbiol 1996[a]; 62: 2919–25.

[26]   Saito Y, Ishii Y, Fujimura T, *et al.* Protein engineering of a cephalosporin C acylase from *Pseudomonas* strain N176. Ann N Y Acad Sci 1996; 15:226-40.

[27]   Matsuda A, Matsuyama K, Yamamoto K, *et al.* Cloning and characterization of the genes for two distinct cephalosporin acylases from a *Pseudomonas* strain. J Bacteriol 1987a; 169: 5815-20.

[28]   Fersh F. Structure and mechanism in protein science. New York: W. H. Freeman and Company 1999.

15

Old-field Succession Sequence in Loess Area in Northern Shaanxi of China

Yaojun Bo[a,b], Qingke Zhu[b*] and Weijun Zhao[c]

[a]*College of Life Sciences, Yulin University, Yulin, Shaanxi, 719000, China*

[b]*College of Water and Soil Conservation, Beijing Forestry University, Beijing 100083, China*

[c]*Key Laboratory of Tourism and Resources Environment in Colleges and Universities of Shandong Province, Taishan University, Taian, Shandong 271021, China*

Abstract: Hierarchical clustering method was applied for division of stages of old-field succession of loess area of Northern Shaanxi, and the similarity among different communities in abandoned field of different abandoned year and site condition can be learnt quantitatively, and stages or sequence of old-filed succession can be determined. The result revealed that the stages of old-field succession are: polydominant association of annual weed artemisia scoparia and other perennial herbage→Lespedeza davurica and perennial herbage association→artemisia vestita and perennial herbage association→tuft grass bothriochloa ischaemum and perennial herbage association→undershrub Buddleja alternifolia association. Among them, the appearance of shrub association (undershrub Buddleja alternifolia) contrasts results in former studies that it was impossible to form shrub association in loess area of Northern Shaanxi. Thus, this study is not only a complement of former studies, but also a new discovery and innovation.

Keywords: Community, Ecological restoration, Euclidean distance, Hierarchical clustering, Importance values of species, Succession sequence.

1. INTRODUCTION

Succession is one of the theoretical foundations of properly managing and using natural resources [1]. Old-field succession is an important type of secondary succession of vegetation [2]. Odum deemed that elements of ecological succession are closely associated with the relationship between human and nature, and are the foundation of resolving the contemporary environmental crisis [3]. At present, there are many reports about studies on old-field succession in Loess Plateau and loess area of Northern Shaanxi, for instance, the vegetation succession sequence in old-field succession [4-6], community composition and structure [7], composition of vegetation species [6], alteration of diversity of species [8, 9], features of primary productivity [10], functions of soil and water conservation [11], relationship between characteristics of vegetation and the environment and evolution of physical and chemical properties of soil [12, 13], etc.

The ecological environment of loess area of Northern Shaanxi is rather vulnerable with severe disruption and water and soil loss, which seriously restrains the sustainable development of regional socio-economy. Thus, the study on rules of old-field succession in this area will be of great significance in ecological restoration, development and utilization of natural resources, control of water and soil loss,

improvement of the environment, promotion of regional sustainable development, construction of ecological civilization.

2. MATERIALS AND METHODS

2.1. Overview of the Study Area

The study area Wuqi County is situated in the northwest of Yan'an City, (N36°33′33″-37°24′27″,E107°38′57″-108°32′49″)with an altitude of 1233 to 1890m. Its landform belongs to typical hilly-gully region of Loess Plateau. Its climate model is semi-arid temperate continental monsoon climate, with an annual average temperature of 7.8°C and with an accumulated temperature of ≥10°C reaching 2817.8°C. Its average annual sunlight hour is 2400h. Its frost-free period is 96-146d. Its annual average rainfall is 478.3mm, and the average amount of land evaporation of many years is 400-450mm [14]. Its rainy season is from July to September, during which the precipitation account for 50%-80% of that of the whole year. The precipitation in other times are mainly invalid precipitation, and natural disasters like drought, hail, rainstorm, blustery and frost injury happen frequently. The soil in this area is mainly loessal soil.

Since 1998, Wuqi County closed hillsides to facilitate afforestation, and grazing has been banned in the whole basin, and applied artificial cultivation for vegetation restoration. At present, the vegetation in the basin is mainly herbosa. Among them, the main species include *artemisia scoparia, artemisia vestita, artemisia giraldii pamp, stipa bungeana trin, lespedeza davurica, potentilla chinensi,*

---
*Address correspondence to this author at the College of Water and Soil Conservation, Beijing Forestry University, Beijing 100083, China;
E-mail: zhuqingke2013@126.com

bothriochloa ischaemum and *buddleja alternifolia*. In dissected valleys there are distributions of shrubs, and in lowland of river valley there are fragmentary distributions of arbors and shrubs.

2.2. Layout of the Sampling Plot and Investigation Method

In July of 2012, by replacing time sequence with spatial sequence, different plant communities naturally reserved in different abandoned year were chosen as object of

investigation and study. In total, 29 sample fields were investigated. Typical sampling was applied in setting sample fields with shrub quadrat of 5m×5m and herb quadrat of 1m×1m. And 10 quadrats were measured in each herbal sampling field. In each shrub quadrat, 1 shrub quadrat and 10 herbal quadrats were set. The record mainly includes: abandoned year, gradient, exposure, slope position, microtopography of a certain quadrat, community type, soil type, quantity, cover degree, above-ground biomass, height, frequency, abundance, altitude and number of plant (or number of clusters) etc.(Table **1**).

Table 1.   General status of sampling plot.

Sample Number	Year of Returning	Exposure	Community Type
1	2	Sunny slope	*Salsola collina + Sonchus oleraceus*
2	3	Sunny slope	*Artemisia scoparia*
3	3	semi-shady slope	*Artemisia scoparia*
4	3	Semi-sunny slope	*Artemisia scoparia*
5	4	Shady slope	*Artemisia scoparia*
6	4	Semi-shady slope	*Artemisia scoparia+ Setaria viridis*
7	5	Shady slope	*Artemisia scoparia+ Heteropappus altaicus*
8	5	Shady slope	*Artemisia scoparia+lespedeza davurica*
9	6	Shady slope	*Lespedeza davurica+ Heteropappus altaicus*
10	6	Semi-shady slope	*Lespedeza davurica+ Potentilla chinensis*
11	7	Semi-shady slope	*Lespedeza davurica+ Potentilla chinensis*
12	7	Semi-shady slope	*Lespedeza davurica+ Potentilla chinensis*
13	8	Semi-shady slope	*Lespedeza davurica+Artemisia vestita*
14	8	Semi-sunny slope	*Lespedeza davurica+Artemisia vestita*
15	10	Sunny slope	*Artemisia vestita+ Leymus secalinus*
16	11	Semi-sunny slope	*Artemisia vestita+ Artemisia giraldii Pamp*
17	11	Semi-shady slope	*Artemisia vestita+ Artemisia giraldii Pamp*
18	12	Shady slope	*Artemisia vestita+Artemisia giraldii Pamp*
19	12	Shady slope	*Artemisia vestita+Artemisia giraldii Pamp*
20	13	Semi-shady slope	*Artemisia vestita+Artemisia giraldii Pamp*
21	13	Semi-shady slope	*Artemisia vestita+Artemisia giraldii Pamp*
22	14	Semi-sunny slope	*Artemisia vestita+Artemisia giraldii Pamp*
23	14	Semi-sunny slope	*Artemisia vestita+ Stipa bungeana*
24	17	Shady slope	*Artemisia vestita+ Stipa bungeana*
25	19	Semi-shady slope	*Bothriochlia ischaemum+Artemisia vestita*
26	22	Shady slope	*Bothriochlia ischaemum +Artemisia vestita*
27	29	Semi-shady slope	*Bothriochlia ischaemum + Poa sphondylodes*
28	40	Semi-shady slope	*Buddleja alternifolia*
29	43	Semi-sunny slope	*Buddleja alternifolia*

2.3. Calculation of Importance Values of Species

The important value is a comprehensive quantitative indicator that fully reflects the position and function of a certain plant in a community [15]. The variation tendency of heterogeneity in each stage of the succession of plant community reflects the evolutionary tend of the interaction between the community and the environment, and the characteristics of the community in each stage are presented by the variation of the important values. Thus, in this paper, important value is an important criterion. Its computational formula is:

$$I = \frac{\text{Relative abundance} + \text{Relative height} + \text{Relative coverage}}{3} \quad (1)$$

2.4. Division of Each Succession Stage

In July Population size of the community is indicated by the importance value of species. For division of community, association type and succession stage, hierarchical cluster analysis is applied in classification of sampling plots of different abandoned year. Hierarchical cluster analysis was conducted with the relative importance values of each species. The relative importance values were calculated by combining 10 quadrats in a sampling plot. First, coefficients of Euclidean distance between each two sampling plots were calculated. Then, hierarchical clustering was conducted by UPGMA [16]. Selection of cut-out values in hierarchical clustering method is based on the principle of being divided into the same association.

Coefficient of Euclidean distance (Δ_{jk}):

$$\Delta_{jk} = \sqrt{\sum_{i=1}^{n}(x_{ij}^2 - x_{ik}^2)} \quad (2)$$

3. RESULTS AND DISCUSSIONS

3.1. Changes of Community in Different Abandoned Year

According to the important value of each species in sampling plot, there are in total 14 community types in abandoned land (Table 1). Among these, dominant species are mainly annual weed salsola collina and artemisia scoparia, perennial herb lespedeza davurica and artemisia vestita, bothrioch loa ischaem um, shrub buddleja alternifolia. Fields abandoned in less than 5 years, including Sampling plot 2, 3, 4 and 5, are all artemisia scoparia monodominant community or polydominant community (sampling plot 1, 6, 7 and 8). For fields abandoned in 6 to 8 years, there are mainly 3 community types, namely polydominant community consisting of Heteropappus altaicus, Lespedeza davurica, potrntilla and artemisia vestita. For fields abandoned in 10 to 17 years, there are mainly 3 community types, namely polydominant community consisting of artemisia vestita, leymus, bungeana trin and Artemisia giraldii Pamp. For fields abandoned in 19 to 29 years, there are mainly 2 community types, namely polydominant community consisting of bothriochloa ischaemum, artemisia vestita and poa sphondylodes. For

fields abandoned in 40 to 43 years, there are mainly 1 community types, namely the monodominant community consisting of the shrub buddleja alternifolia.

Regardless of site conditions like terrain factors, according to the abandoned year of each community type (Table 1), the general community succession tendency of loess area of Northern Shaanxi is: in preliminary stage of returning farmland to woodland, the main community is in general monodominant community of artemisia scoparia, or polydominant communityconsisting of salsola collina, artemisia scoparia, sonchus oleraceus, green bristlegrass, heteropappus altaicus, lespedeza davurica, etc. This stage lasts for 2 to 5 years. And other communities according to the abandoned year of the field are respectively: community of lespedeza davurica → community of artemisia vestita → community of bothriochloa ischaemum → community of buddleja alternifolia. In the field having returned the farmland for more than 6 years, lespedeza davurica begin to dominant and its community takes the place of the community of artemisia scoparia. After that, community with artemisia vestita and Bothriochloa ischaemum as dominant species comes into being. The seed of artemisia vestita is relatively big and the vitality of its seedling is very strong, while that of Bothriochloa ischaemum is small and although its diffusion distance is relatively farther, the vitality of its seedling is weak and its survival rate is low. Thus, the distribution range of artemisia vestita is wider (in 10 sampling plots), while that of bothriochloa ischaemum is smaller (in 3 sampling plots). Finally, there is community of buddleja alternifolia, which completes the succession process of the stage of old-field succession of shrub and herbaceous plant.

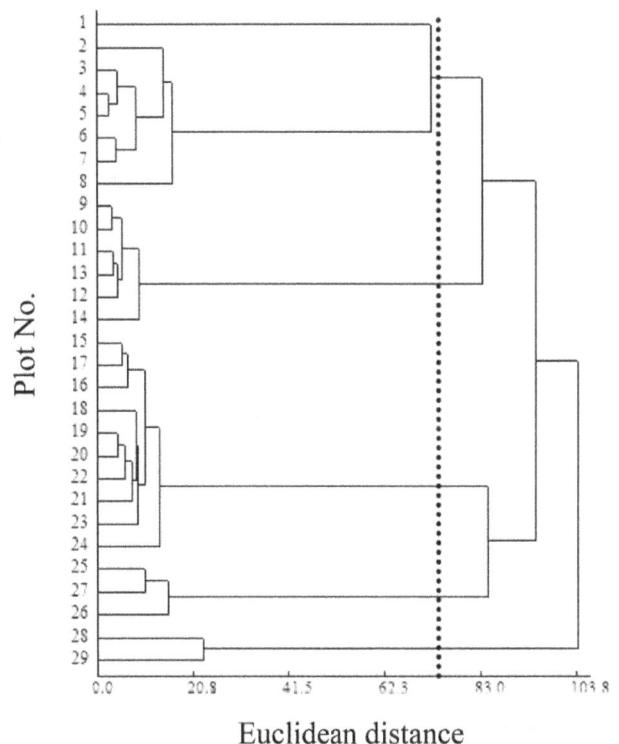

Fig. (1). Tree dendrogram of system clustering of communities in old-land.

3.2. Clustering Analysis of Communities of Different Abandoned Year

The result of hierarchical clustering of serial community in old-field succession reveals that (Fig. (1)) when the cut-out value is 73.0, all the community quadrats cluster into 5 categories: I = {1, 2, 3, 4, 5, 6, 7, 8}, II = {9, 10, 11, 13, 12, 14 }, III = {15, 17, 16, 18, 19, 20, 22, 21, 23, 24}, IV={25, 27, 26}, V ={28, 29}.

The abandoned year of association of annual weed artemisia scoparia andpolydominant association I consisting of annual weed salsola collina, artemisia scoparia and other perennial grass is 2 to 5 years and 3. 6 years in average. The abandoned year of Association II consisting of perennial herb lespedeza davurica and the other perennial herbaceous is 6 to 8 and 7 years in average. The abandoned year of Association III consisting of perennial herb artemisia vestita and perennial herb is 10 to 17 years and 12.7 years in average. The Abandoned year of Association IV consisting of Bothrioch loa ischaem um and other perennial herbs is 19 to 29 year and 23.3 years in average. The abandoned year of association V consisting of buddleja alternifolia is 40 to 43 years and 41.5 years in average. From the perspective of association types and according to the abandoned year, the succession sequence of old-field succession of shrub and herbaceous plant in loess area of Northern Shaanxi is: the polydominant association of annual herb artemisia scoparia and other perennial grass→association of lespedeza davurica and perennial grass→association of artemisia vestita and perennial grass→association of bunch grass bothriochloa ischaemum and perennial grass→association of undershrub Buddleja alternifolia. This succession sequence is different from the former result of the study of vegetation succession sequence of loess area of Northern Shaanxi by Qin Wei, etc. [5]. According to their research, the climax community of vegetation sequence of loess area of Northern Shaanxi is the zonal perennial herbaceous community and shrub community cannot form. Thus, this study is a supplement for the former studies on old-field succession sequence in loess area of Northern Shaanxi. Meanwhile, trees were fund in dissected valley and the bottom of the valley of the study area. However, due to the hard terrain, further investigation was not made, so whether the vegetation of this area can recover to arbor community calls for further study.

In the process of hierarchical clustering, the classification results made by using the original value of importance values of communities of each sampling plot and by using Euclidean distance coefficient tend to be uniform, but the result of division of community type by means of Euclidean distance coefficient is easier for explaining the stages of succession. UPGMA can properly maintain the spatial attribute of quadrats, and different dissimilar coefficient can be chosen for distance between each two quadrats, and there won't be matrix reverse.

3.3. Characteristics of Serial Community

The first stage of succession is association group of annual herb. Its main vegetation includes mainly artemisia scoparia, salsola collina, sonchus oleraceus, green bristle-grass, heteropappus altaicus and lespedeza davurica. And its secondary vegetation includes corispermum hyssopifolium

L, cirsium setosum, gueldenstaedtia stenophylla bunge, potentilla chinensis, calystegia hederacea, euphorbia humifusa, Agriophyllum squarrosum, sonchus arvensis,etc.. Among them, Sonchus arvensis and sonchus oleraceus are weeds with relatively strong root tiller, thus, they have a wide distribution before returning the farmland to the wood and in some districts, they can form dominant species in the preliminary stage of succession. Agriophyllum squarrosum and salsola collina can form dominant community in district with frequent man-made disturbance, like in the field and the roadside. Due to influences of environmental factors like soil moisture and nutrient of the sampling plot, and propagule in nearby grassland and soil seed bank, community types and floristic composition are abundant. Later as old-field succession goes on and the soil environment factors changes, in the 6th year, annual herbs decreases gradually and their vitality decline. Perennial plants begin to take the advantage and community species, composition and biomass all increase. In this stage, the main vegetations include lespedeza davurica, heteropappus altaicus, potentilla chinensis and artemisia vestita. Secondary vegetations include artemisia scoparia, agropyron cristatum, artemisia giraldii pamp, stipa bungeana trin, leymus secalinus and cleistogenes chinensis. And in the 10th to the 43rd year, artemisia vestita and bothriochloa ischaemum of strong competitive and occupation capability and the drought-enduring undershrub buddleja alternifolia begin to take the advantage, and species unable to endure drought and with weaker competitive capacity exit the community, and vegetation species composition reduces. In this stage, secondary vegetations and common species that include potentilla chinensis, poa sphondylodes, lespedeza davurica, artemisia lavandulaefolia, artemisia giraldii pamp, thymus mongolicus, stipa bungeana trin and stipa grandis, etc.

Though old-filed succession begins in secondary bare areas in abandoned fields, the influence of origin vegetation cannot be excluded. The direction and rate of succession will change with discrepancy of site condition and the environment, for example, exposure, gradient, physical and chemical properties of soil, soil moisture, nutrient, etc. Thus, when the regional general climate background and regional species distribution are relatively fixed, the rules of community succession in one region are generally determined by the adaptive capacity fertility, spread capacity and relative competitive ability of species.

CONCLUSION

Hierarchical clustering method was applied in division of stages of old-field sere succession of shrub and herbaceous plant in loess area of Northern Shaanxi, and the result is quite satisfactory. It can analyze the similarity of communities of different site condition and abandoned year. According to the average abandoned year of each association, the sequence of old-field association of loess area of Northern Shaanxi can be approximately determined.

ACKNOWLEDGEMENTS

This research was supported by the Special Fund for Forestry Scientific Research in the Public Interest (201104002-2), Agricultural Attack Project in Shaanxi Province (2014K01-12-03).

REFERENCES

[1] Bazzaz FA. Plant species diversity in old-field successional ecosystems in southern Illinois. Ecology 1975; 56(2): 485-8.

[2] Egler FE. Vegetation science concepts I. Initial floristic composition, a factor in old-field vegetation development. Vegetatio 1954; 4(6): 412-7.

[3] Odum EP, Barrett GW. Fundamentals of ecology. Philadelphia: Saunders Co 1971.

[4] Du F, Shan L, Chen XY, et al. Studies on the vegetation succession of abandoned farmland in the Loess hilly region of northern Shaanxi—succession series after being abandoned. Acta Agrestia Sinica 2005; 13(4): 328-33.

[5] Qin W, Zhu QK, Liu ZQ, et al. Study on natural seres of vegetation and plant species diversity on returning land for farming to forests and grassplots in the hilly-gully regions of the Loess Plateau. Arid Zone Research 2008; 25(4): 507-13.

[6] Cheng, JM, Wan, HE, Hu, XM. Study of Vegetation Restoration and Rebuilding Pattern and the Process of Succession in the Loess Hilly Regions. Acta Agrestia Sinica 2005; 13(4): 324-333.

[7] Du F, Shan L, Liang ZS. Studies on vegetation succession of abandoned arable land in loess hilly regions of northern of Shaanxi Province—analyses of community composition and structure. Acta Agrestia Sinica 2005; 13(2):140-258.

[8] Bai WJ, Jiao JY, Ma XH, et al. Classification and ranking of the forae naturally recuperating on the farming-withdrawn land in the hilly and gully regions of the loess plateau. Acta Botanica Boreali-Occidentalia Sinica 2005; 25(7): 1317-22.

[9] Li YY, Shao MA. The change of plant diversity during natural recovery process of vegetation in zi wu ling area. Acta Agrestia Sinica 2004; 24(2): 252-60.

[10] Ma YS, Li QY, Lang BN, et al. The improvement and utilization on secondary salinized abandoned land in chaidamu basin. Prata Cultural Science 1997; 14(3): 17-20.

[11] Wang BK, Tang KL. The aetivities of waste land reclamation and its effects upon the accelerated soil erosion. Bull Soil Water Conserv 1991; 11(5): 54-60.

[12] Wang GH. Plant traits and soil chemical variables during a secondary vegetation succession in abandoned fields on the Loess Plateau. Acta Botanica Sinica 2002; 44(8): 990-8.

[13] Tang L, Liang ZS, Du F, et al. Vegetation succession of arable old land after being abandoned in Loess Plateau hilly region & ascertaining dominant native herbages in the process, analyzing their chemical nutrient composition. Acta Ecologica Sinica 2006; 26(4): 1165-75.

[14] Bo YJ, Zhu QK, Zhao WJ. Characteristics of soil moisture in relation to microtopography in the Loess region of Northern Shaanxi, China. J Environ Biol 2014; 35(4): 741-9.

[15] Tilman D. Community invasibility, recruitment limitation, and grassland biodiversity. Ecology 1997; 78(1): 81-92.

[16] Tang QY, Feng MG. DPS data processing system for practical statistics. Beijing, China: Science Press 2002; pp. 333-339.

16

Computer Simulation of Stenotic Carotid Bifurcations Hemodynamic and Ultrasonography

Wei Li[*,1] and Qinghua Yang[2]

[1]*Computer School, China West Normal University, Nanchong, Sichuan, 637002 China*

[2]*Medical Imaging Department, North Sichuan Medical College, Nanchong, Sichuan, 637000 China*

Abstract: The carotid sinus of the carotid artery (CA) bifurcation is one of the favored sites for the genesis and development of atherosclerotic lesions. The direct reason is carotid arteries bifurcation and stenoses may lead to great flow pattern change. Aim of this article is to investigate the effect of different eccentric stenosis of internal carotid artery on blood flow. The blood flow in artery is simulated numerically and the simulation is based on convicted reasonable of vascular profile and flow environment *in vivo*. The simulation or the computer flow dynamic calculation reveals the eccentric stenosis can make great effort to the flow pattern. And more serious stenosis proliferation can be generated. Doppler ultrasound (DUS) are widely used in blood flow detection for the diagnosis of cardiovascular diseases (CAD). It is the gold standard of the diagnosis of vessel related diseases and the color sonography display is familiar for clinicians and internist, we simulated the color sonography of stenosis carotid arteries bifurcation mocked the DUS display method. The results are proved to be visual, detailed and accurate that it can more directly use in clinic. All simulation above can alleviate the effects of measurement errors and can be used as the complement of ordinary ultrasound measures.

Keywords: Carotid Bifurcation, Computer Simulation, Sonography, Hemodynamic.

1. INTRODUCTION

Occupying diseases of vascular, such as atherosclerosis, stenosis, abdominal aortic aneurysm are widely and seriously diseases today. For *in vitro* monitoring of vascular anatomy, several medical imaging modalities are used such as MRI, CT and Ultrasound. But by MRI or CT the blood flow cannot measured accuracy, and always need imaging agency and tedious procedure.

Ultrasound or technology called duplex can provides precision velocity measure inexpensive, safely and conventionally. Color flow Doppler imaging (CFDI) can be used for recruitment qualification, pre-treatment classification and post treatment surveillance for remodeling and restenosis.

Due to the ultrasound technology, such as the complication of blood flow, limitation of sample zone and resolution of transducer, signal processing imprecise, we cannot descript the precise flood flow of each voxel in specially appointed time. Recently, scientist and research prefer integrate computer simulation of Doppler ultrasound and computed flow dynamics (CFD) insight into the complex interaction between the acoustic, signal processing and blood flow dynamics [1, 2].

2. METHODS

The medical imaging technology measuring the profile of vascular are seemed to be matured today. That means we can get more reliable anatomy by different imaging modalities. Patent US8, 315, 814, titled" Method and system for patient-specific modeling of blood flow" provide a method to monitor the function of heart and vascular, it consists of the following steps: 1) the obtaining and preprocessing patient-specific anatomical data; 2) creating the three-dimensional model based on obtained anatomical data; 3) preparing the model for analysis and determining boundary conditions; 4) performing the computational analysis and outputting results. It can be used to providing patient-specific treatment planning. On the other side, it can be used to assessing myocardial perfusion and plaque vulnerability [3].

Our procedure is start from the profile of vascular, computing and generating the blood dynamic based on CFD. Comply with ultrasound Doppler imaging method, we change the flow distributes to the color blood exhibition.

2.1. Geometric Model Construction

The aim of medical Ultrasound detection is stenosis, thrombosis and occluding and more, the disturbing of blood including the vascular bifurcation and special twisting. We chose the carotid sinus as our simulation modeling for its typical influence of blood field. It was be study more often by past experiment or instrument measure. Both clinical and post mortem studies indicate that, in humans, the carotid sinus of the carotid artery bifurcation is one of the favored sites for the genesis and development of atherosclerotic lesions [4, 5].

*Address correspondence to these authors at the Computer School, China West Normal University, Nanchong, Sichuan, 637002 China; E-mail: nos036@163.com

Carotid artery supplies blood to brain and neck. It is a Y-shaped artery, which can be divided into three parts. First is common carotid artery (CCA) which is bifurcated into two arteries; one internal carotid artery (ICA) which has a sinus and the other one is of smaller diameter called external carotid artery (ECA).

The carotid artery profile are using the classic model establish by Bharadvaj. As the depict of Fig. (1) Xc means CCA and it is the inlet of blood flow. Xi refers to ICA and Xe refers to ECA, they are outlets. The angle between ICA and ECA is 55 degree. SS1, SS2 and up to SS6 are the diameter of each section of ICA. EC1 to EC4 are the diameter of ECA, and CC1, CC2 means that of ICA. We establish the geometry by AutoCAD® [6].

In some recent research, carotid artery 3D profile can be established by advanced imaging technology such an MRI and X-CT. We segment the profile of vessel in every section, use the profiles pile up and generate a 3D entity through computer graphical method. Due to more convict profile the following up calculate may be more precise. In this paper our research the realize of emphasis on the display and its methodology. A simple model is used instead of.

2.2. Carotid Artery Stenoses

Carotid artery stenoses, always can be found in ICA. There are many reasons that cause the narrow of ICA. Overall the stenoses can be divided into two categories: eccentric and concentric. The former are more common and we discuss there.

In Fig. (1) the distance S1 is fixed and the grade of eccentric stenoses depending on the S2. The stenosis is usually caused by plaques, so we can also express grade of vessel narrowing by the volume of plaques. It can simply define by compare S1 in vascular and diameter. When S1 equal to S2 that means 50% carotid artery stenosis. There we consider the different stenosis of 30%, 45%, 60% and 75%. They refer to from moderate to severe carotid stenosis [7, 8].

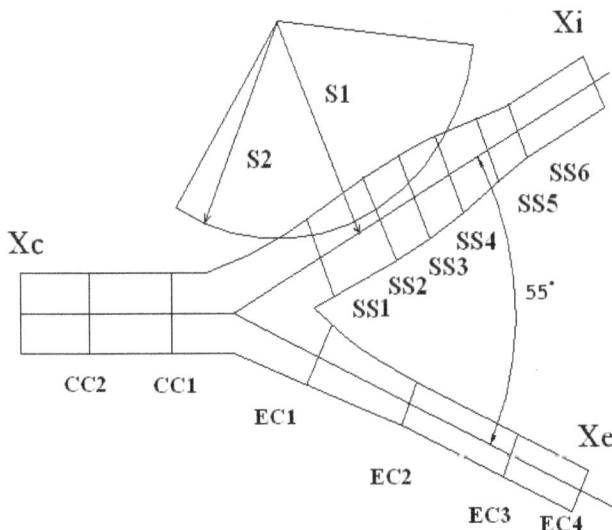

Fig. (1). The carotid artery bifurcation model.

Patent US 7,942,820, titled" Method and system for evaluation of the hemodynamic model in depression for diagnosis and treatment" provides a method for determining the cerebral hemodynamic model for depression using carotid duplex ultrasound to establish percent stenosis of the extracranial right and left internal carotid arteries (RICA, LICA), respectively; and some other site of carotid arteries. It would be of use in a wide range of disease conditions for diagnosis, evaluation of treatment options, effectiveness of surgical measures, and monitoring of progress under certain medications [9].

2.3. Flow Zone grid and Boundary Define

Due to the dedicate profile, it is difficult to calculate by manual programming. There the commercially available computer program FLUENT®6.1 was used in the flow simulations. FLUENT® is a finite volume code capable of efficiently solving the governing equations of fluid mechanics. Before the calculation in software, we must grid the flow zone and define the boundary.

Grid are built by the pre-processing software GAMBIT® 2.1 with much denser grids in the adjacent area to the vessel wall where high variable gradients are expected and detailed flow parameters are important. In our simulation the entity is dividing to 43734 and triangular elements were used to mesh faces, Mesh size was nearly equal in all cases.

We set velocity inlet and pressure outlet. The blood vessel considers being rigid wall and the space in wall, inlet outlets is the computer hydraulic zone. All this setting can be convictable in usually condition.

2.4. Equations Solving

Governing equations of simulation of blood flow are the Navier–Stokes equations for incompressible viscous fluid flow and the pressure equation:

$$\frac{\partial \rho}{\partial t} + \frac{\partial (\rho u_j)}{\partial x_j} = 0 \tag{1}$$

$$\frac{\partial (u_i)}{\partial t} + \frac{\partial (u_i u_j)}{\partial x_j} = F_i - \frac{1}{\rho}\frac{\partial p}{\partial x_i} + \frac{1}{\rho}\frac{\partial \tau_{ij}}{\partial x_j} \tag{2}$$

where ρ and μ are, respectively, the density and viscosity of blood, t is time, u = (ui, uj) is the velocity vector, and p is the pressure. Fi denotes the artificial body force or the feedback signal which is applied at feedback points defined in a feedback domain. The pressure Eq. (2) is derived by substituting the equation of continuity into the divergence of the Navier–Stokes equation, Eq. (1). [10]

2.5. Blood Flow Field Calculation

To imitate the same environment that as *in vivo*, some parameters are list as followings.

Because of ICA is large arteries ,so there blood is treated as an incompressible, Newtonian fluid, viscosity is 4.24×10-3Pa/ms, and a density of 1.0511×103kg/m3. The gauge pressure is 13333Pa (100mm Hg) and the flow was assumed laminar. The more information about the flow, the calculation or the simulation is more reliable. Accuracy, convenient, dynamic are considered to be preferred and important. The Patent US6,814,702, titled "Apparatus for measuring hemodynamic parameters" given a directed toward an ultrasonography apparatus for measuring and/or monitoring hemodynamic activity, such as blood flow. The present invention comprises a doppler ultrasound unit, one or more transducers, and a portable body. maybe it used for special organs (penile), the similar method can be used to other vascular [11].

Due to the heart pump blood with specific rhythm or pulsatile, the velocity of inlet is fluctuation. That means if we must simulation with sequence cycle to get realistic blood field change. Considering the fluctuation of blood *in vivo*, and full precise flow field calculate and deficit can be found in The Patent US8,224,640, titled "Method and system for computational modeling of the aorta and heart", it provide a method for multi-component heart modeling and cardiac disease decision support, comprising: generating a multi-component patient specific 4D geometric model of the heart and aorta estimated from a sequence of volumetric cardiac imaging data of a patient generated using at least one medical imaging modality; generating a patient specific 4D computational model based on one or more of personalized geometry, material properties, fluid boundary conditions, and flow velocity measurements in the 4D geometric model; and estimating patient specific material properties of an aortic wall using the 4D geometrical model and the 4D computational model by determining parameters of a constitutive material model of the aortic wall to minimize a residue between a simulated deformation of the aortic wall determined based on a Fluid Structure Interaction (FSI) simulation and a measured deformation of the aortic wall in the 4D geometric model [12].

There we can depict the blood field in given time spot and with specific velocity. In the measure of blood with ultrasound instrumentation, in clinic doctor is inclined to use freeze technology instead of dynamic imaging [13-15]. In our article the inlet velocity is set to 1 m/s that are a reasonable peak value that measure by other imaging technology such as ultrasound and MRI.

After the input of some parameters and definition of some situation, the software begins the resolve of Navier–Stokes equations with finite volume method. Standard pressure discretization and second-order upwind momentum discretization were used. The pressure-velocity coupling scheme used was the semi-implicit method for pressure-linked equation (SIMPLE) algorithm. For uncomplicated model and not exceed 10000 grid the calculation time is no more than 10 minutes. Grid computer, parallel mainframe is widely used in the sophisticated CFD in recent years. That made 4-D real time calculation possible.

2.6. Ultrasound Display Simulation

Color Doppler flow imaging (CDFI) is a most used way to analysis the blood flow field. The pseudo color often used in pulse Doppler technology. It measure the velocity of little zone that can be called sample zone. For different velocity

direction we use red or blue color. The quantity is converting to the color hue or brightness. And at the same time, we measure the disturbing degree of blood, use green to express. So in ultimate display the two kinds of color are mixed and the color imaging is formed [16].

In the Patent US8,469,887, titled "Method and apparatus for flow parameter imaging" there present invention is directed to a method and an apparatus for performing pulsed-wave spectral Doppler imaging at every color flow range gate location in a two-dimensional (or three-dimensional) region of interest. Spectral processing is necessary to determine the flow parameters. Performing this processing at every color flow range gate location creates the two-dimensional image. The method disclosed herein generates two-dimensional images of flow parameters such as peak velocity, pulsatility index, resistance index, etc. With the two-dimensional image, the user immediately observes where the most critical value of the flow parameter occurs and what that value is [17].

To simulation the velocity field of ultrasound, we must decomposition the velocity to two parts, parallel and vertical to the sound beam as the Fig. (2).

Doppler effect can only get by the parallel part. Using the Doppler frequency of transducer (ft, MHx), the speed of sound through water (c, m/s) and the cosine angle of transducer to the direction of flow (θ ,°), the resulting Doppler shifted frequencies (fd, MHz) were converted to velocities (v, m/s), *via* the Doppler equation:

$$v = \frac{f_d \cdot c}{2 \cdot f_t \cdot \cos\theta} \tag{3}$$

In our simulation the ICA is about $30°$ to horizontal line and equal θ ,it can be used in clinic, so vertical velocity can be discard in Doppler imaging.

As an important medical instrumentation, ultrasound can be considered as golden standard for the diagnosis of soft tissue and vascular. And plenty of new products and patents are emerged these years. The Patent US7,128,713, titled "Doppler ultrasound method and apparatus for monitoring blood flow and hemodynamic"

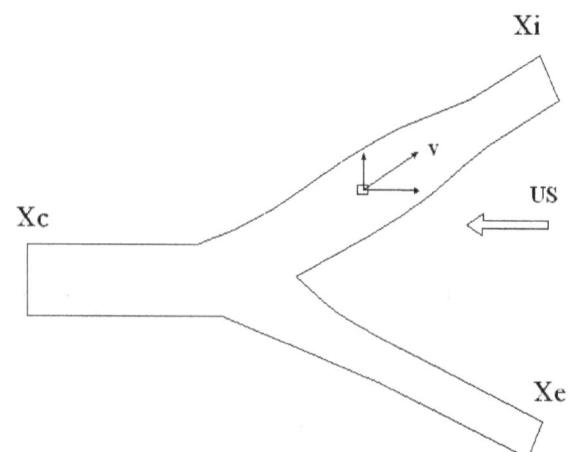

Fig. (2). Velocity decomposition: parallel and vertical component.

Introduced a new Doppler ultrasound system with advantage in solve these two problems: 1) a complicating factor in locating the ultrasound window is determination of the proper depth at which the desired blood flow is located. 2)Once blood flow has been located, it is usually scanned along the course of the vasculature to determine if there are any localized regions in which there are flow abnormalities, which may indicate various diseases [18].

The velocity distribution is calculated by Fluent as above. But in some case the contour velocity display inconvenient for physicians. That is 1) there no turbulence information in velocity map, we often need two map to study the flow; 2) the color bar are different that means we must reading the color with corresponding numerical mean. We use those data to change display mode so that we can get the ultrasound simulation. The method can use the software Tecplot360 to realize.

After all, the key point of operation including:

1) The color on the screen has three attributes: luminosity, hue and saturation. Luminosity is the degree of brightness or shade of the displayed color; hue is the wavelength (i.e., the actual color displayed, from violet through red), and saturation. These three attributes can be used to produce a variety of color scales. It means we can mix or overlay some parameter display in a picture.

2) The horizontal velocity to display in color, if the value is positive it appears in red. If the value is negative it appears in blue. That stands for situation of the flow (blood red cell) toward or away from the transducers. Flow reversal or low velocity zone may cause to opposite color in display.

3) For some US instrumentation, The variance can be displayed along with the mean frequency by using a red and blue scale with increasing amounts of yellow or green introduced as the variance increases, the more brightness of green means more disturbing happens.

4) Stream lines depict the turbulence sometimes can be added. In condition the flow pattern be great disturbed, it is directly perceived through the transducers

3. RESULTS AND DISCUSSION

The procedure as above, after the simulation, formulation we can get the blood flow field in detail. Ultrasound and other imaging technology can only get low resolution imaging with high noise and even with artifacts. This present situation may caused by defect of imaging technology itself. But to get much more reality parameters we can use them [19, 20].

The result of simulation can divide into two sections.

3.1. Blood Velocity Calculates and Display

Blood flow field calculation is time consumption work with more grids divided. Fig. (3) stand for the velocity distribute of different eccentric stenosis. The color bars of display actual an other kind pseudo-color display mode.

From the pictures of Fig. (3) we can see:

At every vessel the velocity in center is much greater than that near wall. The ICA stenosis can cause the blood velocity boosting. And the trend is more obviously when the stenosis is seriously developed. In classic hemodynamic

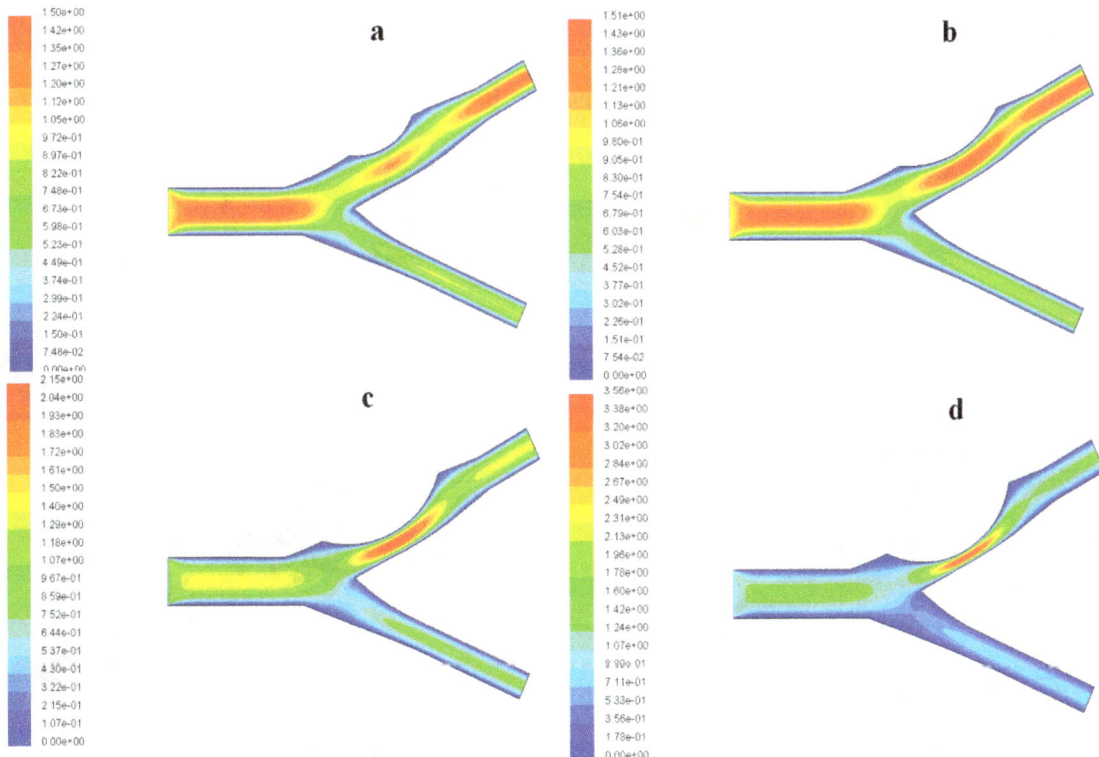

Fig. (3). Contour of velocity in bifurcation a 30% stenosis. b 45% stenosis. c 60% stenosis. d 75% stenosis.

analysis or other dynamics viewpoint that can cause an addition force or pressure in vessel wall. That cause deterioration or cause the stenosis more serious.

The velocity boost may change the distribution of blood to two outlets: ICA and ECA. In 30% stenosis bifurcation the velocity of ECA can maintain, but with the stenosed serious especially that of 75%, the blood in ECA dropped greatly. That means stenosis is not only influence the ICA but also can deteriorate the flow in ECA.

Once stenosis generate, it will create some blood stagnate zone adjacent to the stenosis. This static effect is make the profiling of stenosis or make it worsen. As the simulation result reveal, the more serious stenosed, the more big zone of stagnate exist. That is the pathophysiology of vessel stenosis.

3.2. Flow Display in Color Ultrasound Mode

Calculate and display above can depict the flow pattern detailed. After the processing or flow field calculation in CFD. There result is data with space information and the vector information. To research scientist we can chose to display the flow velocity, kinetic energy or turbulence information. The display is more flexible and quantification by the selecting of special cross or section or in many special mode. But in clinical, the doctors are more cling to color Doppler ultrasonography or called CDFI. That is we must express the flow information more vivid, visual and simply. In other word we lost some quantity information or it can be say these are ignoring in practice clinic use [21].

The overall data is post-processing in Tecplot360. Similar sonography map can establish. We select the 60% stenosis ICA as an example. Fig. (4) is the blue or red contour of horizon velocity component. When the velocity is toward the right it is in red and else is in blue. The black streamlines are drawn to depict the flow tendency more accurate.

Fig. (5) illustrates the distribution of turbulence kinetic energy. The more turbulent of flow the hue of green more dark or deep. It is a simply way to illustrate the change of flow.

Now we get two color flow map, use imaging processing technology, one for velocity direction and magnitude. We mixed the two pictures and color can be joining. The procedure also can be called images fuse. In Fig. (6) the turbulence of flow are consider and turbulence coefficient are indicate as green color to mixed with the velocity. This display method is mocking the ordinary ultrasound instrument display principle.

We can see some place the color mixed greener component or exist the changing of green to cyan, that is the blood stagnant place or highly disturb zone (we can call it dead zone). In Figs. (4 and 6), we can see that kind zones are always close to the bifurcation or plaques. They can proliferation and deteriorate condition of blood flow. If the dead zone is very big that means very dangerous of the patient.

4. CONCLUSIONS

A detailed comparison of effect of different stenosed of ICA on blood flow was done. With the same circumstances, the causing of stagnate zone, blood distribution and the direction of blood field is significant different. Through the simulation we can get more information of the prolife of the stenosis, prediction of vascular complications. Through the

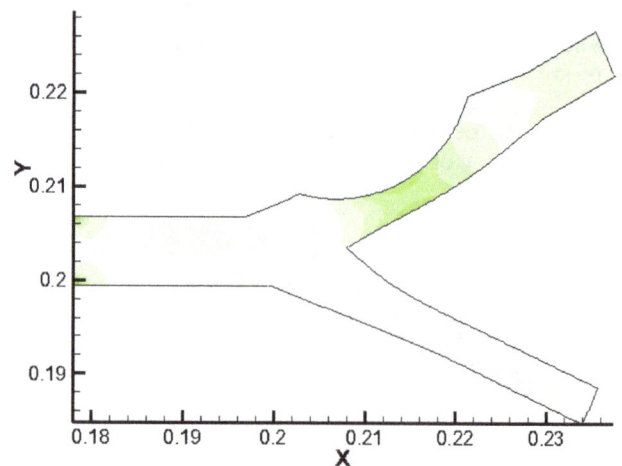

Fig. (5). Filled contour of turbulence kinetic energy.

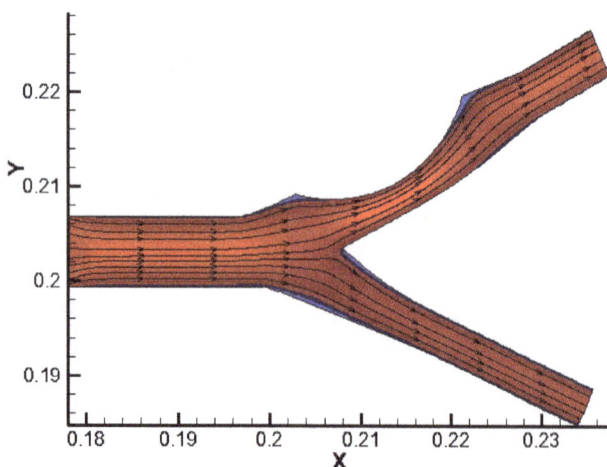

Fig. (4). Filled contour of velocity considering direction and velocity streamline.

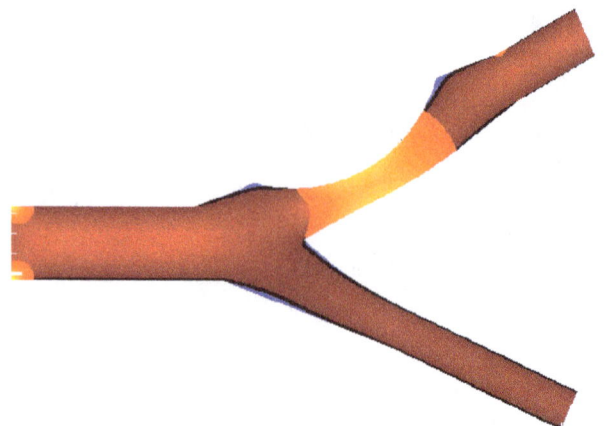

Fig. (6). Simulation of CDFI combines with turbulence information.

post-process, we mocked the display method of ultrasound and get a color Doppler flow field imaging simulated. The picture may more detailed and familiar of clinics or physicians. That can say, traditional blood measure imaging technology can only get overall flow distribution, but they can use the simulation to aid the diagnosis of vascular stenosis.

Quantification of turbulence near carotid stenoses may provide additional diagnostic information, which could enhance the performance of DUS for diagnosis. These blood properties can be managed by lifestyle changes and drug administration, thus guiding the clinician to better manage high risk patients [22].

5. CURRENT & FUTURE DEVELOPMENTS

The Doppler ultrasound system and associated methods are widely used for monitoring blood flow and detecting emboli and stenosis of vascular. Such ultrasound systems are advantageously used both for diagnostic exams (to determine the presence and significance of vascular disease or dysfunction) and during surgical interventions (to indicate surgical manipulations that produce emboli or alter/interrupt blood flow).

The integration of ultrasound technology and CFD can provide explicit hemodynamic information. The intricate imaging emerge in future may including: 1) 4-D, real-time hemodynamic, 2) patient-specific vascular and blood flow imaging and modeling, 3) Pre-surgery evaluating of vascular deforming and flow change.

6. ACKNOWLEDGMENTS

This work was supported by Sichuan Province (P.R.C.) Education Department General Project: (No. 13ZB0012), 2013 teaching reform project of China West Normal University and sponsored research projects of China West Normal University (No.13C002).

REFERENCE

[1] Thrush A, Hartshorne T. Peripheral Vascular Ultrasound: How, Why and When. London: Elsevier 2005.

[2] Fan Y, Jiang W, Deng X, et al. Numerical simulation of pulsatile non-newtonian flow in the carotid artery bifurcation. Acta Mech Sin 2009; 25: 249-55,.

[3] Taylor CA, Fonte TA, Zarins CK. Method and system for patient-specific modeling of blood flow. U.S. Patent 8,315,814, Jan 27, 2011.

[4] Bharadvaj BK, Mabon RF, Giddens DP. Steady Flow in a Model of the Human Carotid Bifurcation. Part I-Flow Visualization. J Biomech 1982; 15: 349-62.

[5] Bharadvaj BK, Mabon RF, Giddens DP. Steady Flow in a Model of the Human Carotid Bifurcation. Part II-Laser Doppler Anemometer Measurements. J Biomech 1982; 15: 363-78.

[6] Swillens A, De Schryver T, Løvstakken L, et al. Two-dimensional flow imaging in the carotid bifurcation using a combined speckle tracking and phase-shift estimator: A Study Based on Ultrasound simulations and in vivo analysis. Ultrasound Med Biol 2010; 36(10): 1722-35.

[7] R.D. Henderson, D.A. Steinman, M. Eliasziw, "Effect of contralateral carotid artery stenosis on carotid ultrasound velocity measurements", stroke, vol. 31, pp. 2636-2640, Nov.2000.

[8] Swillens A, Lovstakken A, Kips L, et al. Ultrasound simulation of complex flow velocity fields based on computational fluid dynamics. IEEE Trans Ultrason, Ferroelectr Freq Control 2009; 56: 546-556.

[9] Njemanze PC. Method and system for evaluation of the hemodynamic model in depression for diagnosis and treatment. U.S. Patent 7,942,820, May 17, 2011

[10] Funamoto K, Hayase T, Saijo Y, et al. Numerical analysis of effects of measurement errors on ultrasonic-measurement-integrated simulation. IEEE Trans Biomed Eng 2011; 58: 653-663.

[11] Redano RT. Apparatus for measuring hemodynamic parameters. U.S. Patent 6,814,702, April 27, 2002.

[12] Sharma P, Georgescu B, Ionasec RI, Comaniciu D. Method and system for computational modeling of the aorta and heart. U.S. Patent 8, 224, 640, July 17, 2012.

[13] Thorne ML, Poepping TL. In vitro Doppler ultrasound investigation of turbulence intensity in pulsatile flow with simulated cardiac variability. Ultrasound Med Biol 2009; 35: 120-8.

[14] Ford MD, Alperin N, Lee SH, et al. "Characterization of volumetric flow rate waveforms in the normal internal carotid and vertebral arteries. Physiol Meas 2005; 26: 477-88.

[15] Szabo TL. Diagnostic ultrasound imaging: inside out. London: Elsevier 2004.

[16] Khoshniat M, Thorne ML, Poepping TL. Real-time numerical simulation of doppler ultrasound in the presence of nonaxial flow. Ultrasound Med Biol 2005; 31: 519-528.

[17] Haider BH. Method and apparatus for flow parameter imaging. U.S. Patent 8,469,887, June 25, 2013.

[18] Moehring MA, Farnsworth JH, Zachariah AP, Baron, Jr. HJ. Doppler ultrasound method and apparatus for monitoring blood flow and hemodynamics. U.S. Patent 7, 128, 713, 2006.

[19] Mo LYL, Cobbold RSC. "Speckle" in continuous wave doppler ultrasound spectra: a simulation study. IEEE Transact Ultrasonics, Ferroelectr Freq Control 1986; 33; 747-53.

[20] Hoskins P, Martin K, Thrush A. Diagnostic ultrasound physics and equipment. Cambridge: Cambridge University Press 2010.

[21] Younis HF, Kaazempur-Mofrad MR, Chan RC. Hemodynamics and wall mechanics in human carotid bifurcation and its consequences for atherogenesis: investigation of inter-individual variation. Biomech Model Mechanbiol 2004; 3(1): 17-32.

[22] Jabbar AU, Ali RU, Parvez K, et al. Three-Dimensional numerical analysis of pulsatile blood flow around different plaque shapes in human carotid artery. Int J Biosci Biochem Bioinforma 2012; 2(5): 305-08.

17

Response Surface Optimized Ultrasonic Assisted Extraction of Total Flavonoids from QingLi Cao and *In Vitro* Antioxidant Activities

Xudong Jiang[1], YaoLing Liao[1], GuiXi Lu[1] and Zhike Xiao[2]

[1]*School of Medicine, Guangxi University of Science and Technology, Liuzhou, 545005, China*

[2]*Liuzhou WanYou Pest Control Research Institute, Liuzhou, 545616, China*

Abstract: An ultrasound-assisted extraction technique was used to extract the total flavonoids from QingLi Cao. The optimal conditions were ethanol concentration 59.20%, liquid-to-solid ratio 31.15 mL/g, extraction time 57.42 min and extraction temperature 58.57°C, which were determined using response surface methodology. The antioxidant activities including reducing power, ABTS+, DPPH, superoxide anion and hydroxyl radical were evaluated, which suggested significant antioxidant activities.

Keywords: Antioxidant, flavonoids, Qing Li Cao, response surface methodology, Ultrasound-assisted extraction.

1. INTRODUCTION

QingLi Cao, a traditional herb has been used for eczema and itching in the minority of Guangxi province, China. It is distributed in Guangxi province and Vietnam. The previous studies show that the extracts are rich in flavonoids, which have been associated with their antioxidant activities. However, insufficient studies have been conducted on flavonoids from QingLi Cao and its antioxidant activity [1].

Ultrasound-assisted extraction (UAE) is an efficient and simple extraction technique. Response surface methodology (RSM) is an effective statistical method for optimizing experimental conditions and investigation of critical processes [2].

The objective of this study was to use RSM to optimize the UAE of total flavonoids from QingLi Cao and evaluate its antioxidant activities. The information obtained will be helpful to further utilization of QingLi Cao.

2. EXPERIMENTAL

2.1. Chemicals and Reagents

2,2'-Azino-bis(3-ethylbenzothiazoline-6-sulfonic acid) diammonium salt (ABTS) (Ruibio, Germany), 1,1,1-Tris (hydroxymethyl) ethane (Tris) (Amresco, USA), 1,1-Diphenyl-1 -picrylhydrazyl (DPPH) (TCI, Japan). The others used purchased from Sinopharm Chemical Reagent Co., Ltd (SCRC, China) and Xiya Reagent Co., Ltd (Xiya, China) used without further purification.

2.2. Sample Preparation

The QingLi Cao was collected from the Qinzhou City of Guangxi Province in November 2013 and authenticated by Prof. Guangwei Huang. It was dried under shade and ground to powder (40 meshes) in a grinding mill. The powder was kept in refrigerator at 0~5°C until use.

2.3. Ultrasound-Assisted Extraction of Flavonoids

The dried powder of QingLi Cao was mixed with ethanol, and the extraction process used an ultrasonic device according to the method described in references [3]. The sample was centrifuged at 3500 rpm for 10 min to collect the supernatant, UV-Vis analyzed the diluted solution. The UAE device was an ultrasonic device (B2200S, Branson Ultrasonics (Shanghai) Company) with 40 kHz and 120 W.

2.4. Experimental Design

RSM was employed to establish the optimum conditions for extraction parameters. A Box-Behnken experiment was employed and a four independent variable at three levels was used, including ethanol concentration (50-70%), liquid-solid ratio (25:1-35:1), extraction time (40-80min) and extraction temperature (50-70°C) Table **1**.

2.5. Determination of Total Flavonoids

The amount of total flavonoids was measured following a previously reported method [4].

2.6. Evaluation of Antioxidant Activity

2.6.1. Reducing Power

The ability of sample to reduce ferric was determined by the method as is described [5].

*Address correspondence to this author at the School of Medicine, Guangxi University of Science and Technology, Liuzhou, 545005, China; E-mail: okhbz@yahoo.com

Table 1. Factors and levels in response surface design.

Levels	Independent Variables			
	A Liquid-Solid Ratio / mL/g	B Concentration / %	C Temperature / °C	D Time / min
1	25	50	50	40
0	30	60	60	60
-1	35	70	70	80

2.6.2. ABTS Radical Scavenging Activity

Determination of the scavenging activity of ABTS radical was based on the procedure described in the study [6].

2.6.3. DPPH Radical Scavenging Activity

The DPPH free radical scavenging activity was determined according to the method [7].

2.6.4. Superoxide Radical Scavenging Activity

The scavenging ability of superoxide radical was measured by the previously described [8].

2.6.5. Hydroxyl Radical Scavenging Activity

The hydroxyl radical assay was according to the previously described [9].

2.7. Statistical Analysis

Data for antioxidant activity are expressed as mean ± SD for analysis performed in triplicate. The mean values and standard deviation were calculated with the Excel program from Microsoft Office 2003 package.

3. RESULTS AND DISCUSSION

3.1. Extraction Parameters for Flavonoids

3.1.1. Fitting the RSM

The extraction yield from QingLi Cao was further optimized through the RSM approach. The experimental points were designed as shown in Table 2. The response value in designed was the average of triplicates.

After fitting to the experimental findings, the response extraction yield of total flavonoids and test variables are related by the following second-order polynomial equation:

$Y = +1.38 + 0.068A + 0.015B - 0.015C - 0.05D - 0.05AB + 0.04AC + 0.05AD + 0.16BC - 0.03BD + 0.07CD - 0.13A^2 - 0.076B^2 - 0.11C^2 - 0.16D^2.$

Table 3 indicates that the coefficient of determination R-Squared is variability in the data explained. The R-Squared

was 0.8658, suggesting that a high correlation was achieved [10]. The F-value of 6.45 and Values of "Prob>F" less than 0.05 indicated the model were significant.

The effects of ethanol concentration, liquid-to-solid ratio, extraction time and extraction temperature on total flavonoids extraction yield of QingLi Cao, as well as their interactions, are shown in Fig. (1).

The plots showed interaction effects of two factors on the response while other factors were kept at constant level. When the contour plots are oval, it means the interaction of two independent variables is significant [11]. According to Table 3 and Fig. (1), the interaction between extraction time and extraction temperature was significant.

3.1.2. Verification of Predictive Model

The optimal extraction conditions as follows: ethanol concentration 59.20%, liquid-to-solid ratio 31.15 mL/g, extraction time 57.42 min and extraction temperature 58.57°C. The maximum predicted yield of total flavonoids was 13.90 mg/g. The mean value (13.80±0.04 mg/g) obtained from experiment which was close to the predicted result.

3.2. Evaluation of Antioxidant Activity

3.2.1. Reducing Power

The reducing power was evaluated based on the reduction of ferric to divalent iron in which the yellow color of the test solution changes to green or blue, depending on the different reducing power of each sample. Rising absorbance at 700nm indicate an increase in reducing power [12].

As shown in Fig. (2), the absorbance of the concentration of 1μg/mL sample and ascorbic acid were 0.106 and 0.397 but sharply increased to 0.425, 0.801 at the concentration of 10 μg/mL, respectively. On reducing power, the extraction had significant effects with increasing concentration in the range of 1-10μg /mL, but compared with the contrast, the effect of sample was slight.

3.2.2. ABTS Radical Scavenging Activity

The abilities of extracts assayed to be scavenging the ABTS radical in comparison with ascorbic acid [13], are

Table 2. Box-Behnken design matrix and the experimental observed responses.

Run	X₁/Liquid-Solid Ratio (mL/g)	X₂/Concentration (%, v/v)	X₃/ Extraction Time (min)	X₄/Temperature (°C)	Total Flavonoids Yield (mg/g)
1	1	0	0	-1	1.19
2	0	1	1	0	1.35
3	0	0	0	0	1.46
4	0	1	0	-1	1.15
5	0	0	1	-1	1.11
6	-1	0	0	1	0.91
7	0	0	0	0	1.21
8	0	0	0	0	1.43
9	-1	1	0	0	1.19
10	0	0	-1	-1	1.22
11	1	1	0	0	1.22
12	0	-1	0	1	1.15
13	-1	0	1	0	0.94
14	0	-1	0	-1	1.09
15	0	0	0	0	1.39
16	0	-1	-1	0	1.37
17	0	1	-1	0	1.09
18	0	-1	1	0	1.01
19	0	1	0	1	1.09
20	-1	0	0	-1	1.21
21	0	0	1	1	1.15
22	1	0	-1	0	1.21
23	1	0	0	1	1.09
24	0	0	0	0	1.39
25	-1	-1	0	0	1.03
26	1	0	1	0	1.22
27	0	0	-1	1	0.97
28	1	-1	0	0	1.24
29	-1	0	-1	0	1.09

Table 3. Variance for response surface quadratic model.

Source	Sum of Squares	Df	Mean Square	F value	P value	Significance
Model	0.50	14	0.036000	6.45	0.0006	significant
A	0.057	1	0.057000	10.33	0.0063	

Table 3. contd...

Source	Sum of Squares	Df	Mean Square	*F* value	*P* value	Significance
B	0.002408	1	0.002408	0.43	0.5211	
C	0.0027	1	0.002700	0.49	0.4973	
D	0.030	1	0.030000	5.40	0.0358	
AB	0.011	1	0.011000	1.98	0.1809	
AC	0.0064	1	0.006400	1.15	0.3014	
AD	0.001	1	0.001000	1.80	0.2012	
BC	0.096	1	0.096000	17.29	0.0010	
BD	0.0036	1	0.003600	0.65	0.4344	
CD	0.020	1	0.020000	3.53	0.0814	
A^2	0.11	1	0.110000	19.87	0.0005	
B^2	0.037	1	0.037000	6.65	0.0219	
C^2	0.077	1	0.077000	13.93	0.0022	
D^2	0.16	1	0.160000	29.59	<0.0001	
Residual	0.078	14	0.005559			
Lack of Fit	0.040	10	0.003991	0.42	0.8782	Not significant
Pure Error	0.038	4	0.009480			
Total	0.58	28				

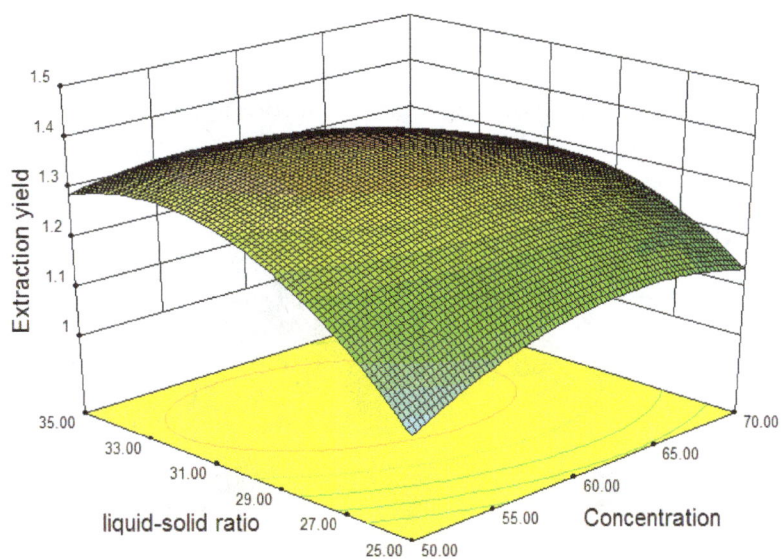

(**A**) Liquid-to-solid ratio and concentration

Fig. (**1**). Contd...

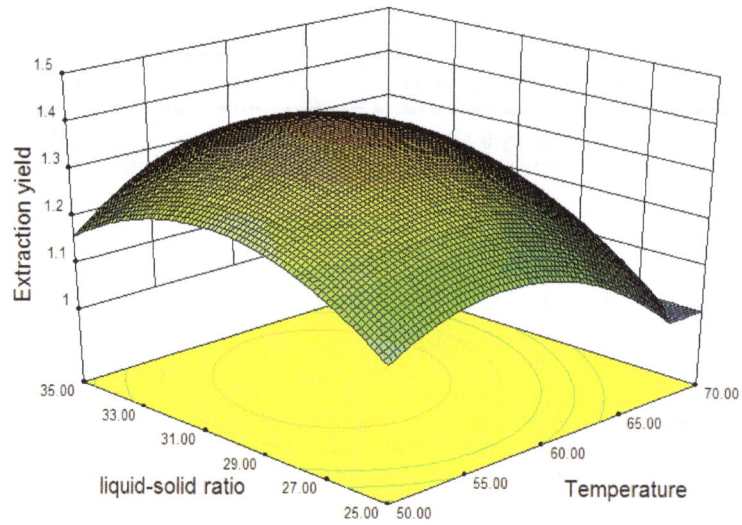

(**B**) Liquid-to-solid ratio and temperature

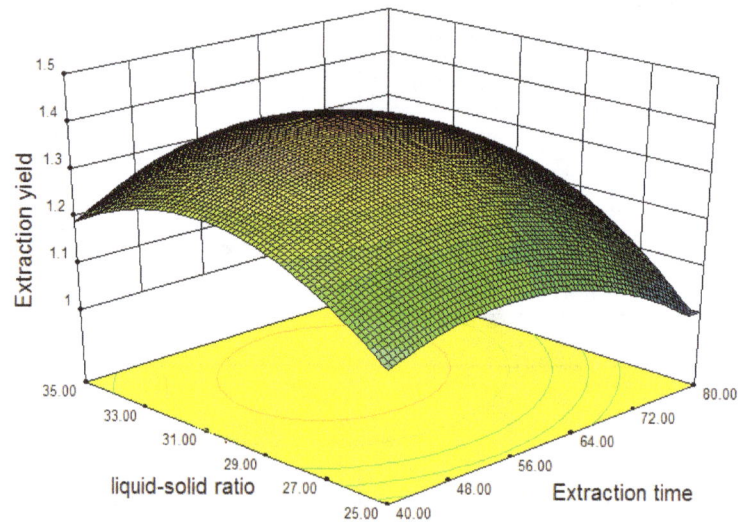

(**C**) Liquid-to-solid ratio and Extraction time

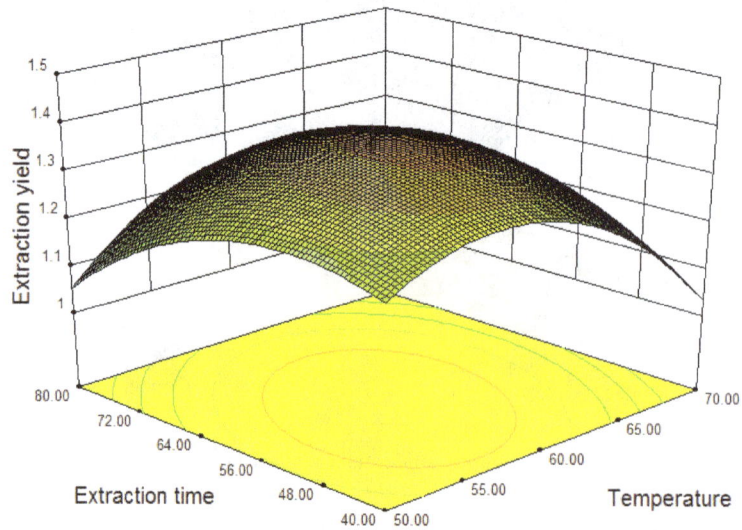

(**D**) Extraction time and temperature

Fig. (**1**). Contd…

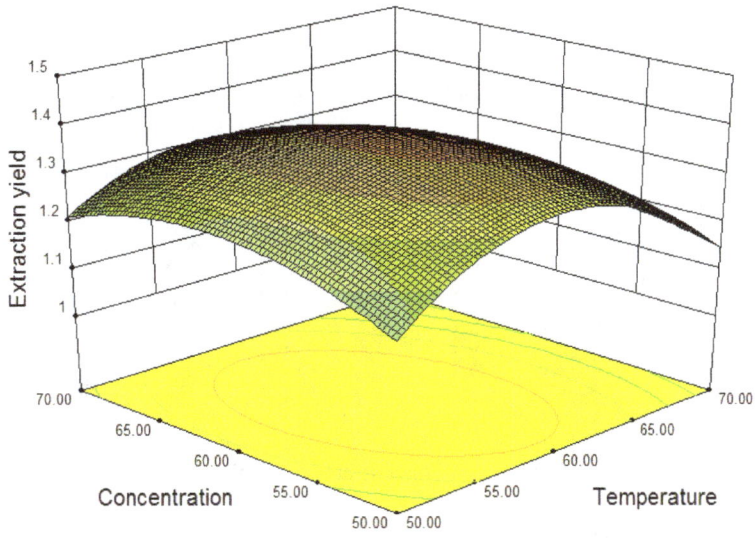

(**E**) Concentration and temperature

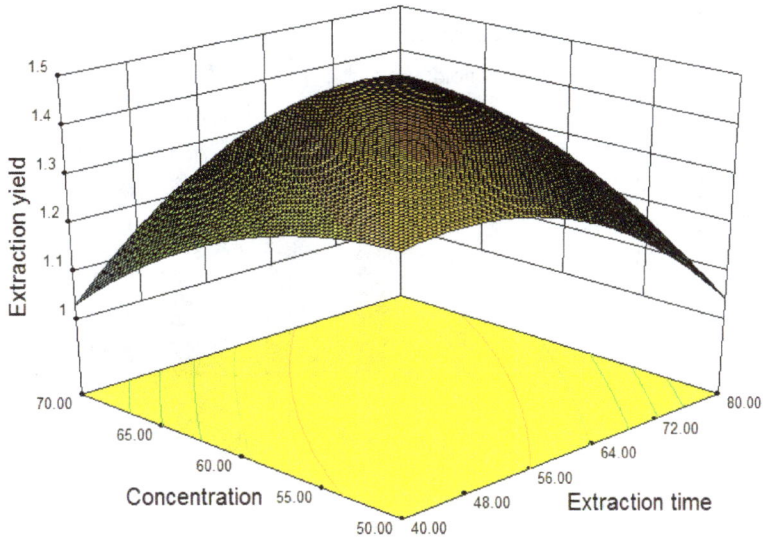

(**F**) Concentration and Extraction time

Fig. (1). Response surface graphs for the effects of concentration, liquid-to-solid ratio, extraction time and temperature on total flavonoids extraction yield

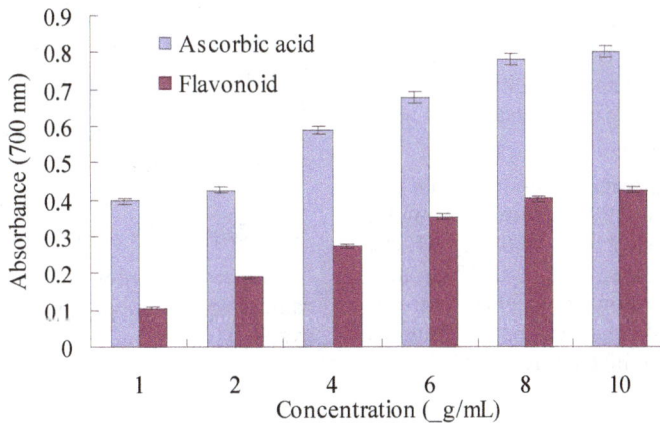

Fig. (2). Reducing power of sample and ascorbic acid.

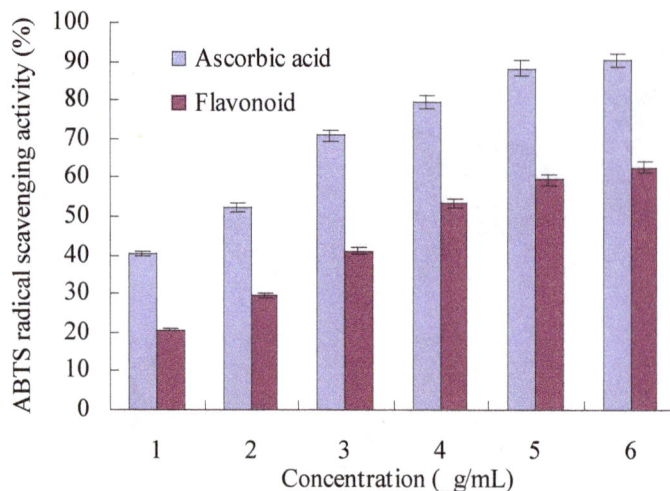

Fig. (3). ABTS radical scavenging activity of sample and ascorbic acid.

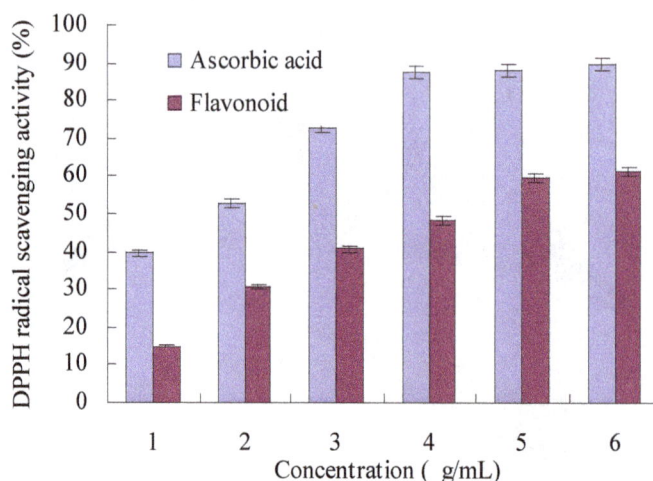

Fig. (4). DPPH radical scavenging activity of sampleand ascorbic acid.

shown in Fig. (**3**). The extracts had effective ABTS in a concentration-dependent manner (1 to10µg/mL), sharply increased from 20.6% to 62.7%. The scavenging effect of the contrast (90.5%) was observed to be higher obviously.

3.2.3. DPPH Radical Scavenging Activity

The absorbance is decreased and the solution changes from purple to light yellow when DPPH concentration is reduced. The mechanism of scavenging DPPH radical is caused by the fact that natural compounds can transfer an electron or a hydrogen atom to DPPH [14].

In the present investigation, a comparison of sample and ascorbic acid is shown in Fig. (**4**). DPPH radical scavenging abilities of sample sharply increased from 15.0% to 61.3%, when the concentration was increased from 1 to 10 µg/mL. The results show that sample showed excellent percent inhibition of DPPH activity at the concentration of 10µg/mL but significantly lower than that of the contrast (89.6%).

3.2.4. Hydroxyl Radical Scavenging Activity

Hydroxyl radical exhibits the strongest oxidative activity in terms of its very high redox potential and extremely fast kinetics [15]. Thus, it is an important parameter for evaluating the antioxidant activity of sample extracts.

As shown in Fig. (**5**), hydroxyl radical scavenging effect of sample increased with concentrations. At concentration of 10µg/mL, it was 60.7% and 89.7% respectively for sample and ascorbic acid. Results indicated that sample had strong capability of scavenging hydroxyl radical but significantly lower than contrast.

3.2.5. Superoxide Anion Radical Scavenging Activity

Superoxide anion radical is considered as an initial free radical and formed from mitochondrial electron transport system [16].

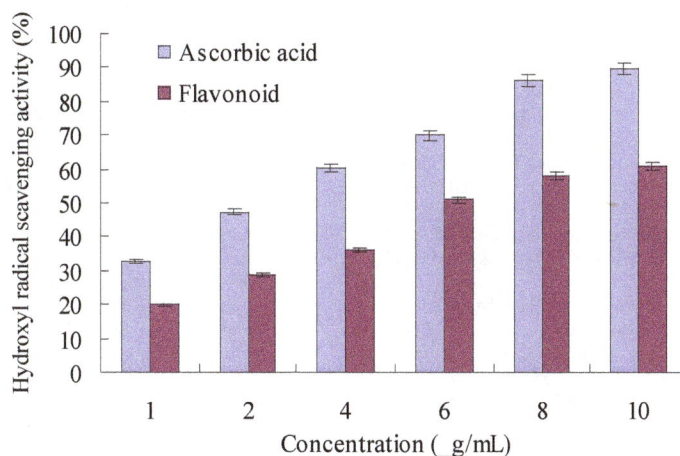

Fig. (5). Hydroxyl radical scavenging activity of sample and ascorbic acid.

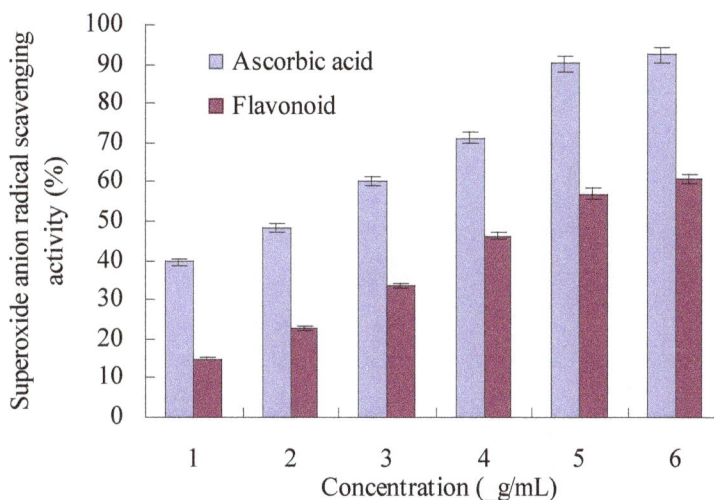

Fig. (6). Superoxide anion radical scavenging activity of sample and ascorbic acid.

Fig. (**6**) shows the percentage inhibition of superoxide radical generation from 1 to 10 μg/mL concentration of sample and ascorbic acid. In this study, superoxide anion radicals were scavenged by sample and ascorbic acid in a concentration dependent manner. The inhibition of superoxide anion radical scavenging at 10μg/mL of sample was 60.7%, but ascorbic acid showed stronger that is 92.4%.

CONCLUSION

In this study, the RSM was successfully employed to optimize the ultrasonic-assistant extraction of total flavonoids from QingLi Cao and evaluated. The results showed that all the factors had significant effects on the extraction rate of total flavonoids. The optimum extraction conditions were obtained at ethanol concentration 59.20%; liquid-to-solid ratio 31.15mL/g; extraction time 57.42 min and extraction temperature 58.57°C. The extraction rate of total flavonoids was in agreement with the predicted ones.

The antioxidant activities including reducing power, ABTS+ radical scavenging activity, DPPH radical scavenging activity, superoxide anion scavenging activity and hydroxyl radical scavenging activity were evaluated, which suggested significant antioxidant activities.

ACKNOWLEDGEMENTS

This study was financially supported by Liu Dong New Area science and technology development plan (20120009) and Liuzhou WanYou pest control research Institute.

REFERENCES

[1]   Zsuzsanna H, Ana M, Orbn-Gyapa O. Xanthine oxidase-inhibitory activity and antioxidant properties of the methanol extract and flavonoids of *Artemisia asiatica*. Rec Nat Prod 2014; 8: 3: 299-302.

[2]   Wang XS, Wu QN, Chen GY. Response surface optimized ultrasonic-assisted extraction of flavonoids from Sparganii rhizoma and evaluation of their *in vitro* antioxidant activities molecules. 2012; 17: 6769-83.

Recent Progress in Biotechnology

[3]   Zhang C, Wu S, Xu N. Optimization of ultrasonic-assisted extraction of polysaccharides from the stem of *Actinidia arguta*. Sci Technol Food Ind 2012; 33: 297-9.

[4]   Mustafa O, Mustafa B, Kubilay G. Antioxidant /antiradical properties of microwave-assisted extracts of three wild edible mushrooms. Food Chem 2014; 157: 323-31.

[5]   Tian F, Li B, Ji BP. Antioxidant and antimicrobial activities of consecutive extracts from *Galla chinensis:* The polarity affects the bioactivities. Food Chem 2009; 113: 173-9.

[6]   Heo BG, Park YJ, Park YS. Anticancer and antioxidant effects of extracts from different parts of indigo plan. Ind Crops Prod 2014; 56: 9-16.

[7]   Tukappa A, Londonkar RL. Evaluation of antibacterial and antioxidant activities of different methanol extract of *Rumex vesicarius*. L. Am J Drug Discov Dev 2013; 3: 72-83.

[8]   Kannadhasan1 R, Venkataraman S. *In vitro* capacity and *in vivo* antioxidant potency of sedimental extract of Tin*ospora cordifolia* in streptozotocin induced type 2 diabetes. Avicenna J Phytomed 2013; 3: 7-24.

[9]   Chien PJ, Li CM, Lee CH. Influence of micronized chitosan on antioxidative activities in grape juice. Food Nutr Sci 2013; 4: 224-8.

[10] Zheng XL, Liu BG, Li LM. Microwave-assisted extraction and antioxidant activity of total phenolic compounds from pomegranate peel. J Med Plants Res 2011; 5: 1004-11.

[11] Wang YX, Lu FX, Lu ZX. Optimization of cultivation conditions for exopolysaccharide and mycelial biomass by *Clitocybe* sp. using Box-Behnken design. Sci Agric Sin 2005; 1:145-50.

[12] Meir S, Kanner J, Akiri B. Total phenolic content and antioxidant activity of seed extract of *Lagerstroemia speciosa*. L. Chem Sci Trans 2013; 2: 75-80.

[13] Shon MY, Kim TH, Sung NJ. Antioxidants and free radical scavenging activity of *Phellinus baumii* (Phellinus of Hymenochaetaceae) extracts. Food Chem 2003, 82: 593-7.

[14] Sasikumar JM, Patharaj J, Adithya ES. Antioxidant capacity and phenolic content of Elaeagnus kologa schlecht. An underexploited fruit from India. Free Radicals Antioxidants 2012; 3: 28-35.

[15] Tang EL, Rajarajeswaran J, Fung SY. Antioxidant activity of *Coriandrum sativum* and protection against DNA damage and cancer cell migration. BMC Complement Altern Med 2013; 13: 347-54.

[16] Roginsky V, Lissi EA. Review of methods to determine chain breaking antioxidant activity in food. Food Chem 2005; 92: 235-54.

18

Optimization for Fermentation of Cattle Manure to Produce Bio-Fertilizer by Inoculating Complex Microbial Agents

Ruimin Fu[1,2] Fang Lin[3], Hong Zhang[1], Wenhui Xing[1], Huiping Chang[1] and Wuling Chen[*,2]

[1]*Department of Life Science, Henan Normal of Education, Zhengzhou 450046, China*

[2]*College of Life Science, Northwest University, Xi'an 710069, China*

[3]*Department of Life Science and Technology, Xinxiang University, Xinxiang, 453003, China*

Abstract: In this study, in order to determine the optimization technology of bio-fertilizer, the aerobic fermentation test was conducted by using fresh manure and rice husk powder as materials. The inoculation, moisture content, C/N ratio and turning frequency analysed by single-factor test. Following this, the optimal conditions in the first fermentation were obtained through orthogonal experiments. After the first fermentation, the second fermentation was carried out by adding complex microbes with the ability to release potassium, dissolve phosphate and fix nitrogen. The termination time of second fermentation was determined by testing the growth condition of beneficial bacteria in the fermentation process. The results show, that the optimal parameter in the first fermentation contained moisture content of 70%, C/N ratio of 20:1, inoculation of 3‰, and turning frequency of once every three days. After a period of fifteen days of fermentation, the germination index reached 91.3%. Following this, second fermentation was conducted at the sixth day and the germination index was increased to 98.8%. Under this optimum condition, the material could heat up fast and the top temperature could be high with a long duration.

Keywords: Bio-organic fertilizer, cattle manure, complex microbial agents, fermentation.

1. INTRODUCTION

With the establishment of the large amount of intensive pasture, more manure has been produced [1, 2]. The manure has become a major source of ecological environmental pollution [3], which is not only harmful for the environment and human health, but also influences the sustainable development of livestock industry. Moreover, manure contains a large amount of organic matter and other nutrients composition such as N, P, S, K, *etc*. After processing, it can become a good agricultural fertilizer [4-6]. Aerobic fermentation is the most effective way of treating manure harmlessly and resourcefully [7, 8]. In the fermentation process, a large number of nitrogen, phosphorus and potassium compounds useful for plant growth are produced by microbial degradation. Moreover, the synthesis of new organic polymer - humus could be an important active substance for soil fertility [9]. However, due to limited indigenous microorganisms and fermentation technology, the traditional fermentation process has not been widely used before. Currently, some reports have indicated that the fermentation efficiency could be improved by inoculating microbial agents [10-14]. However, there is little research on the optimization of process parameters. In this study, the main parameter of first fermentation was optimized by single factor test and orthogonal experiments. In the second fermentation, the additional phosphate-dissolving bacteria released soluble phosphorus to plants by dissolving

insoluble tricalcium phosphate. The silicate bacteria degraded silicate minerals and released the potassium for plant growth. Meanwhile, the nitrogen fixation bacteria fixed the nitrogen and supplied it to the plant. Therefore, these beneficial microbes can be added in the soil to provide crop nutrients or produce hormones to stimulate plants.

In this study, aiming at producing biological organic fertilizer, secondary fermentation process was conducted and different composite microbial agents were added at different stages of fermentation. The main process parameter was optimized and the optimal fermentation conditions were explored to improve the fermentation efficiency.

2. MATERIAL AND METHODS

2.1. Fermentation Materials

Fresh cattle manure and rice husk powder were used as fermentation materials. The main physical and chemical properties are shown in Table 1.

2.2. Preparation of Complex Microbial Agents in the First Fermentation

Trichoderma viride, Aspergillus oryzae, Bacillus subtilis and *Pseudomonas* sp., which were isolated and saved in the laboratory, were respectively increased for expanding culture step by step to prepare the solid microbial agents. *Following this, Trichoderma viride, Aspergillus oryzae, Bacillus subtilis* and *Pseudomonas* sp., were mixed in the proportion of 2:1:1:2 to produce mixed microbial agents.

*Address correspondence to this author at the College of life Science, Northwest University, Xi'an, China;

E-mails: wulingchen@yeah.net; angelaminmin@163.com

Table 1.   Main physical and chemical characteristics of compost material.

Material	Moisture Content (%)	pH Value	Organic Matter (%)	C / N Ratio	Total Carbon (%)	Total Nitrogen (%)
cattle manure	90	8.0	66.2	20.6	39.8	1.93
rice husk powder	10	7.0	71.6	70.6	45.6	6.46

2.4. Preparation of Complex Microbial Agents in the Second Fermentation

Azotobacter choococcum which could fix nitrogen, *Bacillus megaterium* which could dissolve phosphorus and *Bacillus mucilaginosus* which could dissolve potassium, were respectively cultured in liquid under the condition of 180rpm, 30°C. When the cell concentration reached 10^9CFU/mL, they were respectively mixed with sterilized bran in the proportion of 1:1 to prepare solid agents. The solid agents of *Azotobacter choococcum*, *Bacillus megaterium* and *Bacillus mucilaginosus* were mixed together with the same ratio to produce mixed microbial agents.

2.5. The First Fermentation Design

Static composting experiment was carried out by piling up the material to 2 meters long, 1 meters wide, and 0.8 meters high. Inoculation, moisture content, C/N ratio and turning frequency were analyzed by single factor test. In this study, temperature and germination indexes were considered as evaluation index to identify the optimal range of four factors. After testing four factors by orthogonal test, the optimum fermentation parameters were determined With the ambient temperature being about 30 °C.

2.6. The Second Fermentation Design

When the first fermentation temperature dropped to 45 °C, adding *Azotobacter choococcum*, *Bacillus megaterium* and *Bacillus mucilaginosus* released potassium, dissolved phosphate and fixed nitrogen to conduct a second fermentation,with the turning frequency being once a day and keeping the temperature suitable for bacterial growth and reproduction. Viable count of bacteria was tested to determine the terminal time of the secondary fermentation process.

2.7. Determination of the Indicators

The temperature was detected by a glass thermometer. The moisture content and C/N ratio were tested according to the organic research method [15] and potassium dichromate - sulfuric acid digestion method [16]. The seed germination index was tested as follows: 30 rape seeds were put into a petri dish lined with filter paper. After adding 5 mL of fermentation product extracts into the petri dish and taking distilled water as control, all the petri dishes were kept into the seed germination boxes at 25 °C for 96 h. Following this, the germination rate and the root length were measured and calculated according to the following formula:

Germination index % = (seed germination rate in test group × the length of seed root in the test group)/(seed germination rate in control group× the length of seed root in the control group) ×100%.

3. RESULTS

3.1. Influence of Moisture Content on Cattle Manure Fermentation

The effect of moisture content on germination index and on fermentation temperature is respectively shown in Table **2** and Fig. (**1**). When the moisture content reached 60%, the material warming increased with the maximum temperature (Fig. 1) of 68 °C and the high temperature could hold 8.5d. All the treatment was in line with the Chinese Ministry of Agriculture industry standards (NY / T 394-2000) hygiene standards. When the moisture continued to increase, the temperature gradually declined, due to excessive moisture limiting the aerobic microorganisms' contact with oxygen. Meanwhile, the germination index (Table **2**) was maintained above 80% when the moisture content was between 60%-80%.

3.2. Influence of C/N Ratio on Cattle Manure Fermentation

The effect of C/N ratio on germination index and on fermentation temperature is respectively shown in Table **3** and Fig. (**2**). When the C/N ratio reached 30:1, the material warming increased with the maximum temperature (Fig. 2) of 65 °C and the high temperature could hold 8.5d. All the treatment was in line with the Chinese Ministry of Agriculture industry standards (NY/T 394-2000) hygiene standards. Meanwhile, the germination index (Table **3**) was maintained above 86.6% when the C/N ratio was 30:1.

3.3. Influence of Turning Times on Cattle Manure Fermentation

The effect of turning frequency on germination index and on fermentation temperature is respectively shown in Table **4** and Fig. (**3**). Due to different turning frequencies, the temperature trends in the fermentation were basically the same, which experienced heating period, high temperature period and cooling period. When the turning frequency was once every 3 days, the material warming increased with the maximum temperature (Fig. 3) being 65°C. This manifested because the turning frequency increased the ventilation rate in the fermentation. However, extensive turning could cause the heat loss. The germination index (Table **4**) could be maintained above 80% when the turning frequency was once every 3 days or once a day.

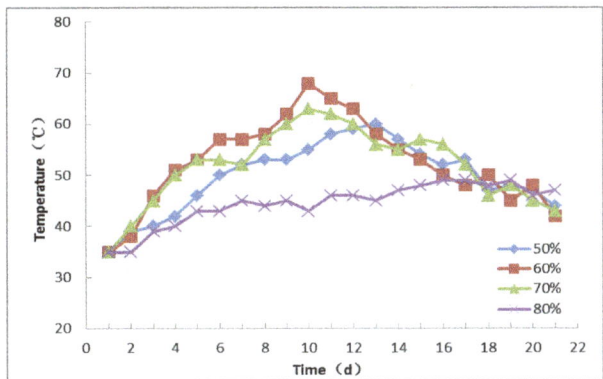

Fig. (1). Effect of moisture content on fermentation temperature.

Table 2.   Effect of moisture content on germination index.

Moisture content (%)	50	60	70	80
Germination index (%)	80.8	88.6	86.6	<50

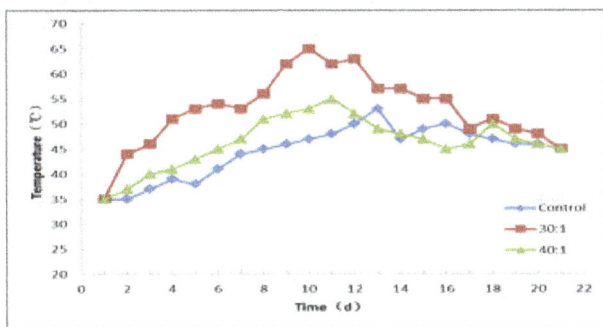

Fig. (2). Effect of C/N ratio on fermentation temperature.

Table 3.   Effect of C/N ratio on germination index.

C/N	Control (20:1)	30:1	40:1
Germination index (%)	70	86.6	72

3.4. Influence of Inoculation on Cattle Manure Fermentation

The effect of inoculation on germination index and on fermentation temperature is respectively shown in Table **5** and Fig. (**4**). From Fig. (**4**), it was observed that the temperature increased rapidly with an increase in the rate of inoculation. Meanwhile, the germination index (Table **5**) could be maintained above 80% when the inoculation amount was between 2‰-3‰.

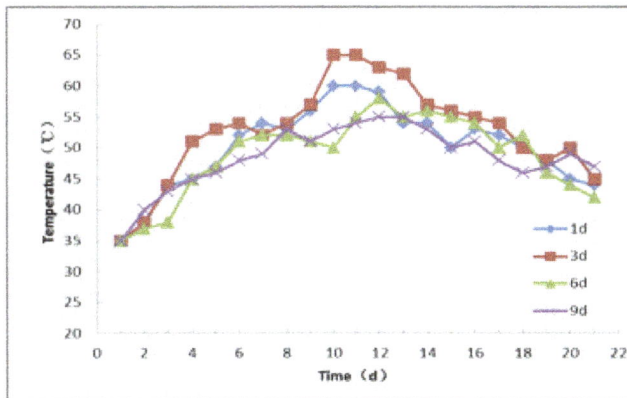

Fig. (3). Effect of turning frequency on fermentation temperature.

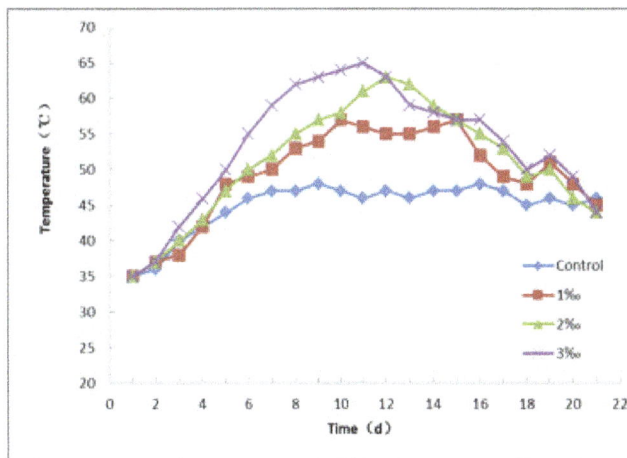

Fig. (4). Effect of inoculation on fermentation temperature.

3.5. The Determination of Optimum Conditions in the Fermentation

In order to optimize the fermentation process, four factors were analyzed and three levels orthogonal experiment were conducted according to the optimal range of factors resulting from single-factor test. The result is shown in Table 6. By analyzing the range, the relationship between the primary and secondary factors was observed to be $D>B>A>C$, which suggests that the complex microbial inoculants were the main source of influence on the fermentation. The optimum fermentation conditions were combined into $A_2B_1C_2D_3$, which inculded moisture content of 70%, C/N ratio of 20:1, turning frequency of once every three days, and inoculation of 3‰. Under this condition, the germination index could reach 91.3%.

Table 4.   Effect of turning frequency on germination index.

Turning Frequency	Once a Day	Once Every Three Days	Once Every Six Days	Once Every Nine Days
Germination index (%)	70	86.6	72	68

Table 5.   Effect of inoculation on germination index.

Inoculation	Control (without Microbe Addition)	1‰	2‰	3‰
Germination index (%)	<50	71.6	89.6	91.8

Table 6.   Results of orthogonal experiment.

No.	A Moisture Content %	B C/N Ratio	C Turning Frequency	D Inoculation	Germination Index (%)
1	60	20	2	1	82.3
2	60	30	3	2	84.2
3	60	40	4	3	86.6
4	70	20	3	3	91.3
5	70	30	4	1	88.5
6	70	40	2	2	81.3
7	80	20	4	2	82.2
8	80	30	2	3	89.2
9	80	40	3	1	83.6
K1	84.367	85.267	84.267	84.800	--
K2	87.033	87.300	86.367	82.567	--
K3	85.000	83.833	85.767	89.03	--
R	2.666	3.467	2.100	6.466	--

CONCLUSION

In this study, single factor test and orthogonal test were used to optimize bio-organic fertilizer fermentation. The optimum parameters involved; moisture content of 70%, C/N ratio of 20:1, turning frequency of once every three days, and inoculation of 3‰. Under this condition, the germination index could reach 91.3% after the first fermentation. Subsequently, the second fermentation was conducted by adding complex microbial agents to improve soil fertility. After the second fermentation, the germination index was increased to 98.8%. Under the optimal conditions, the cattle manure heated up fast and the maximum temperature was achieved with a long duration. Moreover, fermentation cycles did not reduce with high maturity. This study provides necessary parameters and theoretical basis for the production of bio-organic fertilizer.

ACKNOWLEDGEMENTS

This work was financially supported by Agricultural Science and Technology Achievements Transformation Fund Project (2012GB2G000451) and Youth Scientific Research Projects in Henan Institute of Education (20100103).

REFERENCES

[1]   Montes F, Meinen R, Dell C, et al. SPECIAL TOPICS—mitigation of methane and nitrous oxide emissions from animal operations: II. A review of manure management mitigation options. J Anim Sci 2013; 91(11): 5070-94.

[2]   Guo D-J, Wu H-S, Ma Y, Chang Z-Z. Study on the Amount of Manure and Urine Excreted by Sheep and Rabbits in Intensive Pasture. J Ecol Rural Environ 2011; 1: 010.

[3]   Heuer H, Schmitt H, Smalla K. Antibiotic resistance gene spread due to manure application on agricultural fields. Curr Opin Microbiol 2011; 14(3): 236-43.

[4]   Bouwman L, Goldewijk KK, Van Der Hoek KW, et al. Exploring global changes in nitrogen and phosphorus cycles in agriculture induced by livestock production over the 1900–2050 period. Proc Natl Acad Sci USA 2013; 110(52): 20882-7.

[5]   Holden S, Lunduka R. Do fertilizer subsidies crowd out organic manures? The case of Malawi. Agric Econ 2012; 43(3): 303-14.

[6]   Chakraborty A, Chakrabarti K, Chakraborty A, Ghosh S. Effect of long-term fertilizers and manure application on microbial biomass and microbial activity of a tropical agricultural soil. Biol Fertil Soils 2011; 47(2): 227-33.

[7]   Müller C, Johansson M, Salomonsson AC, Albihn A. Effect of anaerobic digestion vs livestock manure and inorganic fertilizer on the hygienic quality of silage and haylage in bales. Grass Forage Sci 2014; 69(1): 74-89.

[8]   Elsaidy N, Abouelenien F, Kirrella GA. Impact of using raw or fermented manure as fish feed on microbial quality of water and fish. Egypt J Aquatic Res 2015; 41(1): 93-100.

[9]   Erhart E, Siegl T, Bonell M, et al. Fertilization with liquid digestate in organic farming-effects on humus balance, soil potassium contents and soil physical properties. EGU General Assembly Conference Abstracts 2014.

[10]   López I, López R, Santamaría P, Torres C, Ruiz-Larrea F. Performance of malolactic fermentation by inoculation of selected Lactobacillus plantarum and Oenococcus oeni strains isolated from Rioja red wines. Vitis 2015; 47(2): 123.

[11]   Jiang J, Liu X, Huang Y, Huang H. Inoculation with nitrogen turnover bacterial agent appropriately increasing nitrogen and promoting maturity in pig manure composting. Waste Manag 2015; 39: 78-85.

[12]   Liu L, Li T, Wei X, Jiang B, Fang P. Effects of a nutrient additive on the density of functional bacteria and the microbial community structure of bioorganic fertilizer. Bioresource Technol 2014; 172: 328-34.

[13]   Xiong S, Xu W, Yang Y, et al. Effects of microbes and cellulase in pig manure fermentation at different temperature. Acta Scientiae Circumstantiae 2014; 12: 028.

19

Correlation between Selenium and Heavy Metal Content in *Camellia oleifera* in Hainan, China

Jun Yuan[1,2], Deyi Yuan[1,2], Xiaofeng Tan[1,2,*], Feng Zou[1] and Shixin Xiao[2]

[1]*Hunan Provincial Cooperative Innovation Center of Non-wood Forest Cultivation and Utilization, Central South University of Forestry and Technology, Changsha, Hunan, 410004, China*
[2]*Key Laboratory of Cultivation and Protection for Non-Wood Forest Trees (Central South University of Forestry and Technology), Ministry of Education, Changsha, Hunan, 410004, China*

Abstract: Eighteen forests of *Camellia oleifera* in Hainan province, China were selected to measure the contents of selenium (Se), zinc (Zn), mercury (Hg), cadmium (Cd), chromium (Cr), lead (Pd) and arsenic (As) in soils, roots, leaves, and kernels, and analyze the correlation between the soil content and the corresponding content in the plant. It was found that the content of Se in soils ranged between 0.489 and 2.110 mg/kg, and was higher than 1.0 mg/kg in 10 out of 18 forests. The average content of six heavy metals was low and fulfilled the requirements of soil environment for green-food production, except that the contents of Cr, Pb, and Cd exceeded standards in few regions. The content of Se in *Camellia oleifera* was relatively low. Only five regions showed detectable levels of Se in kernels, with the highest in Fuwen Town of Dingan county of 0.085 mg/kg. The contents of Zn, Cr, Pb, Cd, As, and Hg in roots of *Camellia oleifera* were higher than in the leaves and kernels. The contents of Zn, Cr, and Pb were higher than that of Cd, As, and Hg. The As content in soil was in significant correlation with that in the roots of *Camellia oleifera*. The Pd content in the soil was positively correlated with that of the roots of *Camellia oleifera*. The Cr, Pd and Cd contents in roots were in positive correlation with that in the leaves significantly. However, the Hg content of the roots was positively correlated with that of the kernels. Furthermore, the Se content in roots of *Camellia oleifera* was negatively associated with that of the other six heavy metals, having a significantly negative correlation between Se and As.

Keywords: Correlation, Hainan province, Heavy metal, Oiltea, Selenium.

1. INTRODUCTION

Soil is an integral part of human ecological environment, and the foundation of agriculture and forestry production. Heavy metal pollution adversely affects the growth and development of plants and the quality of agricultural products [1]. Studies on rice [2] and persimmon [3] suggested that Se not only reduced the uptake of heavy metals such as cadmium (Cd) and lead (Pd), but also exerted anti-cancer effect on the human body [4, 5]. Studies focused on selenium (Se) and the other heavy metals in the soils cultivating rubber and pepper found abundant Se in soil in the Hainan province [6-8]. *Camellia oleifera* is an important and major source of edible oil in South China [9, 10]. It has a long history of planting and utilization in Hainan province. Also known as 'pomelo on the hill', Camellia oil from Hainan province is of high quality due to excellent production environment. Studies have been published on *Camellia oleifera* germplasm resources [11], varieties

selection [12], ecological adaptability [13], and high yield cultivation techniques [14]. However, the correlation between Se and heavy metal contents in soil and their corresponding contents in *Camellia oleifera* have yet to be reported. In the current study, 18 forests of *Camellia oleifera* from 11 major production counties in Hainan province were selected to measure the contents of selenium (Se), zinc (Zn), mercury (Hg), Cd, chromium (Cr), Pd and arsenic (As) in soils, roots, leaves, and kernels. The correlation between soil levels of Se and heavy metals and their corresponding contents in *Camellia oleifera* was analyzed in order to provide a scientific basis for pollution-free and Se-rich oil tea production in Hainan province, China.

2. MATERIAL AND METHODS

2.1. Materials

The forests of *Camellia oleifera* covering an area over one acre were selected from 11 counties including Chengmai, Dingan, Qionghai, Wenchang, Qiongzhong, Danzhou, Tunchang, Wuzhishan, Baoting, Baisha, and Haikou. Detailed information was collected regarding the location, latitude, longitude, cultivation method and age of trees (Table **1**).

*Address correspondence to this author at the Hunan Provincial Cooperative Innovation Center of Non-wood Forest Cultivation and Utilization, Central South University of Forestry and Technology, Changsha, Hunan, 410004, China;
E-mail: tanxiaofengcn@126.com

Table 1. General characteristics of sampling sites.

No.	County (City)	Town	Village	Longitude	Latitude	Altitude (m)	Type of Forest	Age of Forest (a)
1	Wenchang	Penglai	Hongqi farm	110°31'53.8"	19°31'43.5"	160	rubber mixed forest	45
2	Haikou	Dongshan	Dongshan forest farm	110°30'05.6"	19°37'48.4"	41	artificial pure forest	56
3	Chengmai	Jiale	Potang village	110°01'36"	19°35'02"	100	artificial pure forest	80
4	Chengmai	Heling farm	Qunchang village	109°47'05"	19°33'25"	157	artificial pure forest	50
5	Chengmai	Zhongxing	Liwan village	110°54'38"	19°37'57"	122	rubber mixed forest	82
6	Dingan	Fuwen	Baihe village	110°12'21.3"	19°26'47.7"	110	artificial pure forest	100
7	Dingan	Long-Hu	Gaolin village	110°29'0.2"	19°35'04.6"	60	artificial pure forest	10
8	Danzhou	Dacheng	Qiangtu village	109°24'38.1"	19°30'40.8"	155	artificial pure forest	210
9	Tunchang	Nankun	Shanyoupo village	109°58'39.4"	19°09'49.3"	217	artificial pure forest	45
10	Tunchang	Poxin	Xinpo village	110°30'5"	19°37'48.4"	40	artificial pure forest	46
11	Baisha	Xishui	Shihan village	109°32'19.2"	19°12'33.4"	210	artificial pure forest	32
12	Qiongzhong	Changzheng	Xinzhai village	109°51'46"	18°58'33"	352	rubber mixed forest	43
13	Qiongzhong	Wanling	Shiwan village	109°57'13.8"	19°10'26.7"	276	artificial pure forest	65
14	Qionghai	Huishan	Zhongjiu village	110°18' 20.2"	19°05' 17.6"	60	artificial pure forest	300
15	Qionghai	Jiaji	Hongxing committee	110°27'36"	19°15'02"	40	areca-nut mixed forest	54
16	Wuzhishan	Changhao	Shiha village	109°27'55.7"	18°40'47.6"	728	artificial pure forest	62
17	Wuzhishan	Tongshi	Nanding village	109°34'41.5"	18°49'50.3"	748	artificial pure forest	58
18	Baoting	Xiangshui	Maorui forest farm	109°31'40"	18°38'55.2"	738	artificial pure forest	46

Five mid-sized trees from each forest were selected as the samples. The samples of soil were collected by wooden spade one meter away from the sampling trees, at a depth of 0 to 20 cm. The soil samples were then mixed and stored up to 1 kg by quartation method. The slim roots of 5 g from each tree sample were taken after removing the surface soil. Forty mature leaves were collected from east, south, west, and north of each tree; four mid-sized fruits were collected as well.

Soil samples were naturally air-dried indoors and filtered through 0.147 mm nylon sieve. The plant samples were repeatedly cleaned by water and deionized water thrice to remove dirt and mud. The kernels were removed from the fruits. The leaves, roots and kernels were fixed at 105°C for 30 minutes and crushed after drying at 65°C to obtain a constant weight.

2.2. Determination of Se and Heavy Metals in Soil

The soil contents of Se, Hg, and As were measured according to the standard industrial methods in NYT1104-2006 (Chinese Standard, the same below), NY/T 1121.10-2006 and NY/T 1121.11-2006 using atomic fluorescence

spectrophotometry (AFS-9120, Beijing Jitian Instruments Co., Ltd, China). The soil contents of Zn, Cr, Pb and Cd were measured according to the processing methods described in the national standard (GB/T 17141-1997) by inductively coupled plasma mass spectrometry (ICP-MS) (ELAN, DRC-e, PerkinElmer, US).

2.3. Determination of Se and Heavy Metals in Plant

The content of Se in roots, leaves, and kernels of *Camellia oleifera* was measured according to the processing method described in the Chinese national standard GB/T 21729-2008. Zn, As, Cr, Hg, Pd and Cd were measured according to the processing method described in the national standards GB/T 5009.14-2003, GB/T 5009.11-2003, GB/T 5009.123-2003, GB/T 5009.17-2003, GB 5009.12-2010, and GB/T 5009.15-2003, respectively.

2.4. Statistical Analysis

All measurements were performed in triplicate. Excel 2010 and SPSS 17.0 software were utilized for statistical analysis. Certain data and analysis were absent due to undetectable low content of metal elements.

Table 2. Se and six heavy metals in the forest soil of *Camellia oleifera* in Hainan (mg·kg^{-1}).

Location	pH	Zn	Cr	Pb	Cd	As	Hg	Se
Wenchang-Penglai	4.5±0.2	110±0.6	267±0.4	5.51±0.2	0.1500±0.030	1.34±0.05	0.050±0.012	1.540±0.135
Haikou-Dongshan	4.4±0.2	98.8±0.5	145±0.3	10.2±0.3	0.1130±0.016	2.92±0.20	0.062±0.003	0.944±0.032
Chengmai-Jiale	4.8±0.3	36.0±0.3	10.4±0.2	25.7±0.1	0.0569±0.003	1.81±0.06	0.030±0.001	1.100±0.062
Chengmai-Heling farm	5.1±0.2	24.3±0.5	24.0±0.2	14.5±0.2	0.0508±0.001	1.64±0.05	0.075±0.003	1.200±0.017
Chengmai-Zhongxing	4.5±0.2	8.26±0.2	5.64±0.3	7.93±0.2	0.0195±0.002	1.48±0.07	0.017±0.001	0.489±0.032
Dingan-Fuwen	4.6±0.3	42.7±0.3	39.9±0.3	27.9±0.4	0.0491±0.004	6.40±0.36	0.059±0.004	0.996±0.099
Dingan-Longhu	4.2±0.1	98.5±0.4	174±0.4	37.4±0.4	0.1420±0.007	1.88±0.05	0.072±0.003	1.610±0.035
Danzhou-Dancheng	4.7±0.2	65.0±0.5	81.0±0.5	25.8±0.4	0.0521±0.007	2.61±0.15	0.118±0.007	0.999±0.095
Tunchang-Nankun	4.6±0.1	68.4±0.4	43.1±0.7	62.0±0.3	0.0726±0.001	0.61±0.01	0.048±0.003	0.857±0.006
Tunchang-Poxing	5.0±0.3	78.0±0.3	7.54±0.4	40.2±0.4	0.0494±0.010	1.24±0.18	0.042±0.009	1.230±0.192
Baisha-Xishui	5.6±0.2	30.3±0.6	16.4±0.6	19.4±0.6	0.0569±0.003	2.46±0.07	0.031±0.002	0.665±0.055
Qiongzhong-Changzheng	5.3±0.2	65.2±0.4	42.9±0.5	38.3±0.4	0.0569±0.003	1.72±0.19	0.053±0.003	1.150±0.178
Qiongzhong-Wanling	5.0±0.3	26.0±0.4	54.2±0.3	28.2±0.3	0.0504±0.002	1.18±0.17	0.065±0.004	1.700±0.142
Qionghai-Huishan	4.0±0.2	20.3±0.7	17.1±0.4	10.1±0.9	0.0570±0.008	1.21±0.10	0.040±0.014	0.520±0.087
Qionghai-Jiaji	4.6±0.3	96.4±0.4	43.6±0.4	37.6±0.4	0.3070±0.022	4.28±0.30	0.110±0.021	2.110±0.173
Wuzhishan-Changhao	5.4±0.1	55.0±0.5	9.32±0.4	30.8±0.4	0.1260±0.024	2.87±0.33	0.060±0.003	0.804±0.031
Wuzhishang- Tongshi	4.9±0.1	68.4±0.4	21.5±0.4	32.9±0.3	0.0397±0.005	0.73±0.14	0.097±0.016	1.070±0.253
Baoting-Xiangshui	4.5±0.2	39.5±0.4	6.56±0.2	41.1±0.6	0.0813±0.002	2.18±0.15	0.080±0.003	1.590±0.030

3. RESLUTS AND ANALYSIS

3.1. Soil Contents of Se and Heavy Metals

It was found from 18 forests that Se content ranged between 0.480 and 2.110 mg/kg, which was high according to the classification method [3]. The Se content exceeded 1.0 mg/kg in 10 forests such as in Penglai of Wenchang county. All the soil samples were acidic, and the heavy metal content was relatively low; the 'Soil Environmental Quality Standard' (GB, 15618 - 1995) was used as a reference. The As and Hg contents of all the tested regions were lower than first-grade standard level. The Zn content was lower than first-grade standard level except of Penglai in Wenchang. The Cr content of all the tested regions was lower than the first-grade standard level except for Penglai in Wenchang county, Dongshan in Haikou county, and Longhu in Dingan county. Though Pb content of six counties such as Longhu in Dingan county exceeded first-grade standard level, all tested samples were lower than second-grade standard. The Cd content in all tested regions was lower than natural background, except in Jiaji of Qionghai county, which was higher than second-grade. All tested soil samples fulfilled the requirements of soil environmental quality for green food production using 'green food producing area environment quality' (NY/T 319 - 2013) as reference, except soil Cr in Penglai of Wenchang county, Dongshan of Haikou county, and Longhu of Dingan county, soil Pb in Nankun of Tunchang county, and soil Cd in Jiaji of Qionghai county (Table 2).

3.2. Se and Heavy Metals Contents in Plant

As shown in Table 3, there was difference in all root samples obtained from 18 forests in terms of Se and heavy metals contents. The Se content in roots was relatively low, with the maximum of only 0.104 mg/kg being the highest in Heling farm and Jiale in Chengmai county, followed by Wuzhishan county. The Zn, Cr and Cd, Pb, As and Hg contents in roots were the highest in Nankun of Tunchang county, Jiale in Chengmai county, Changhao of Wuzhishan county, Dacheng of Danzhou county and Heling farm of Chengmai county, respectively.

As shown in Table 4, Se content in leaves of *Camellia oleifera* was the highest in Heling farm of Chengmai county, followed by Jiale of Chengmai county; and it was the lowest

Table 3. Se and six heavy metals in the roots of *Camellia oleifera* in Hainan (mg·kg^{-1}).

Location	Zn	Cr	Pb	Cd	As	Hg	Se
Wenchang-Penglai	47.1±0.2	26.10±0.27	2.90±0.19	0.52±0.10	0.13±0.01	<0.01	<0.02
Haikou-Dongshan	38.6±0.8	13.26±1.24	6.18±0.16	0.33±0.02	0.31±0.01	0.042±0.001	<0.02
Chengmai-Jiale	58.7±0.2	53.16±0.13	15.4±0.32	1.60±0.15	0.29±0.02	0.110±0.024	0.092±0.004
Chengmai-Heling farm	17.9±0.4	12.79±0.26	6.30±0.28	0.54±0.17	0.16±0.02	0.220±0.012	0.104±0.004
Chengmai-Zhongxing	31.1±1.9	29.08±0.25	9.10±0.31	1.30±0.14	0.22±0.02	0.036±0.002	<0.02
Dingan-Fuwen	13.2±0.5	31.78±0.25	25.1±0.30	0.83±0.08	0.35±0.01	0.045±0.003	0.030±0.001
Dingan-Longhu	15.4±0.3	30.71±0.14	6.30±0.17	0.43±0.06	0.17±0.01	0.032±0.002	<0.02
Danzhou-Dancheng	12.4±0.2	48.18±0.15	23.9±0.20	1.30±0.14	0.72±0.02	0.041±0.002	0.034±0.002
Tunchang-Nankun	61.2±0.2	31.54±0.29	23.2±0.41	0.95±0.14	0.20±0.02	0.072±0.002	<0.02
Tunchang-Poxing	31.1±0.2	16.86±0.24	13.8±0.19	0.58±0.07	0.12±0.01	<0.01	<0.02
Baisha-Xishui	21.2±0.3	7.42±0.16	22.3±0.26	1.40±0.13	0.50±0.04	0.048±0.002	0.032±0.001
Qiongzhong-Changzheng	9.5±0.1	11.68±0.15	23.3±0.10	0.88±0.06	0.10±0.03	0.048±0.001	0.042±0.006
Qiongzhong-Wanling	8.7±0.2	14.60±0.28	17.6±0.10	0.47±0.07	<0.1	0.030±0.006	0.016±0.002
Qionghai-Huishan	6.4±0.1	14.63±0.46	3.50±0.06	0.41±0.01	<0.1	0.022±0.001	0.035±0.007
Qionghai-Jiaji	6.0±0.1	7.85±0.16	3.10±0.17	0.05±0.01	<0.1	0.046±0.001	0.041±0.007
Wuzhishan-Changhao	25.2±0.2	26.57±0.37	49.1±0.44	0.61±0.13	0.65±0.01	0.095±0.001	0.074±0.004
Wuzhishang-Tongshi	18.7±0.5	12.16±0.17	8.55±0.43	0.40±0.13	0.10±0.02	0.088±0.001	0.080±0.006
Baoting-Xiangshui	14.4±0.1	10.26±0.19	6.57±0.08	0.26±0.03	0.09±0.01	0.040±0.002	0.088±0.005

Table 4. Slenium and six heavy metals in the leaves of *Camellia oleifera* in Hainan (mg·kg^{-1}).

Location	Zn	Cr	Pb	Cd	As	Hg	Se
Wenchang-Penglai	<0.4	0.82±0.05	0.42±0.05	0.045±0.006	<0.1	<0.01	0.025±0.004
Haikou-Dongshan	<0.4	1.35±0.12	0.60±0.04	0.028±0.004	<0.1	0.022±0.003	0.057±0.002
Chengmai-Jiale	<0.4	2.28±0.05	0.54±0.05	<0.01	<0.1	0.032±0.004	0.077±0.005
Chengmai-Heling farm	<0.4	0.59±0.04	1.50±0.07	<0.01	<0.1	<0.01	0.120±0.003
Chengmai-Zhongxing	<0.4	1.56±0.05	1.00±0.04	<0.01	<0.1	0.095±0.005	0.056±0.010
Dingan-Fuwen	<0.4	2.45±0.05	1.30±0.07	0.012±0.002	<0.1	0.014±0.005	0.046±0.005
Dingan-Longhu	<0.4	3.03±0.07	0.66±0.05	0.031±0.006	<0.1	0.035±0.005	0.045±0.005
Danzhou-Dancheng	<0.4	0.89±0.04	0.51±0.05	0.024±0.005	<0.1	0.086±0.005	<0.02
Tunchang-Nankun	<0.4	1.38±0.07	0.30±0.03	0.022±0.005	<0.1	<0.01	0.020±0.002
Tunchang-Poxing	<0.4	1.99±0.11	0.30±0.02	0.018±0.004	<0.1	0.046±0.005	0.064±0.005

Table **4**. contd...

Location	Zn	Cr	Pb	Cd	As	Hg	Se
Baisha-Xishui	<0.4	1.44±0.11	0.50±0.08	0.089±0.007	<0.1	0.030±0.002	0.039±0.006
Qiongzhong-Changzheng	<0.4	1.25±0.06	0.20±0.04	<0.01	<0.1	0.032±0.006	0.051±0.006
Qiongzhong-Wanling	<0.4	0.06±0.01	0.15±0.01	0.013±0.003	<0.1	0.045±0.003	0.059±0.007
Qionghai-Huishan	<0.4	0.35±0.06	0.89±0.05	0.038±0.003	<0.1	0.033±0.004	<0.02
Qionghai-Jiaji	<0.4	0.48±0.01	0.09±0.02	0.034±0.006	<0.1	0.021±0.005	0.025±0.005
Wuzhishan-Changhao	<0.4	1.52±0.05	0.34±0.05	0.015±0.005	<0.1	0.062±0.007	0.047±0.006
Wuzhishang- Tongshi	<0.4	1.55±0.08	0.09±0.01	0.012±0.006	<0.1	0.073±0.003	0.025±0.002
Baoting-Xiangshui	<0.4	0.28±0.02	0.25±0.01	0.014±0.001	<0.1	0.038±0.006	0.035±0.004

Table 5. Se and six heavy metals in *Camellia oleifera* Kernels in Hainan (mg·kg^{-1}).

Location	Zn	Cr	Pb	Cd	As	Hg	Se
Wenchang-Penglai	<0.4	0.59±0.03	0.04±0.01	0.024±0.002	<0.1	<0.01	0.038±0.004
Haikou-Dongshan	<0.4	1.35±0.04	0.59±0.02	0.020±0.002	<0.1	0.011±0.001	<0.02
Chengmai-Jiale	<0.4	4.54±0.06	0.06±0.01	0.031±0.004	<0.1	0.010±0.004	<0.02
Chengmai-Heling farm	<0.4	4.25±0.05	0.21±0.04	0.025±0.005	<0.1	<0.01	<0.02
Chengmai-Zhongxing	<0.4	0.63±0.05	0.94±0.01	0.039±0.004	<0.1	0.012±0.001	<0.02
Dingan-Fuwen	<0.4	0.16±0.01	<0.05	0.019±0.004	<0.1	0.039±0.004	0.085±0.006
Dingan-Longhu	4.80±0.1	0.18±0.01	0.20±0.01	0.016±0.004	<0.1	0.037±0.005	0.060±0.006
Danzhou-Dancheng	3.80±0.1	0.22±0.02	0.14±0.02	0.030±0.002	<0.1	0.032±0.004	0.041±0.001
Tunchang-Nankun	11.1±0.1	0.22±0.03	0.08±0.01	0.023±0.004	<0.1	0.015±0.004	<0.02
Tunchang-Poxing	11.7±0.2	3.92±0.06	0.08±0.01	0.024±0.004	<0.1	<0.01	<0.02
Baisha-Xishui	11.9±0.1	0.60±0.01	0.15±0.03	0.025±0.004	<0.1	0.010±0.002	<0.02
Qiongzhong-Changzheng	9.50±0.1	0.29±0.04	0.20±0.02	0.048±0.003	<0.1	0.026±0.003	<0.02
Qiongzhong-Wanling	1.40±0.1	0.40±0.05	<0.05	0.018±0.003	<0.1	0.018±0.006	<0.02
Qionghai-Huishan	10.8±0.1	4.13±0.04	0.14±0.02	0.016±0.005	<0.1	0.011±0.005	<0.02
Qionghai-Jiaji	6.00±0.1	2.74±0.05	0.31±0.01	0.037±0.022	<0.1	0.013±0.004	<0.02
Wuzhishan-Changhao	16.8±0.1	1.55±0.04	0.19±0.02	0.013±0.004	<0.1	<0.01	<0.02
Wuzhishang- Tongshi	24.3±0.1	0.65±0.04	0.33±0.03	0.036±0.006	<0.1	0.0135±0.004	<0.02
Baoting-Xiangshui	17.7±0.1	0.12±0.02	0.13±0.01	0.014±0.002	<0.1	0.0120±0.001	0.022±0.005

in Dacheng of Danzhou county and Huishan of Qionghai county. Cr content in leaves was the highest in Longhu of Dingan county, followed by Fuwen of Dingan county. Pd, Cd, and Hg contents in the leaves were the highest in Heling farm of Chengmai county, Xishui of Baisha county, and Zhongxing of Chengmai county, respectively. The Cd content was undetectable in leaves in four counties including

the Heling farm of Chengmai county; the content of Hg was undetectable in leaves in three counties which include Penglai of Wenchang county. The Zn and As contents in leaves from all 18 counties were undetectable.

As shown in Table **5**, Se content in kernels of *Camellia oleifera* was very low, only detectable in five regions and

Table 6. Coefficients of Se and heavy metal elements in soils and different parts of *Camellia oleifera* in Hainan.

	Zn	Cr	Pb	Cd	As	Hg	Se
Root	0.191	0.134	0.308*	-0.497**	0.436**	0.075	-0.214
Leaf	-	0.088	-0.176	-0.127	-	0.004	-0.139
Kernel	-0.107	-0.218	0.015	0.809**	-	0.164	-

Note: **indicates correlated at P<0.01 level; *indicates correlated at P<0.05 level Certain data are absent due to undetactable content of Zn and Se in leaves and As and Se in kernels in some areas. Similar conditions apply in the following tables.

Table 7. Coefficients of Se and heavy metal levels in roots, leaves, and seeds of *Camellia oleifera* in Hainan.

	Zn	Cr	Pb	Cd	As	Hg	Se
Leaf	-	0.469**	0.151	0.576**	-	-0.157	-0.076
Kernel	0.244	-0.038	0.075	-0.305*	-	0.605**	-

Table 8. Coefficients of Se and heavy metal concentrations in roots and leaves of *Camellia oleifera* in Hainan.

	Zn	Cr	Pb	Cd	As	Hg
Root	-0.141	-0.079	-0.147	-0.095	-0.368*	-0.119
Leaf	-	0.239	-0.057	-0.025	-	0.159

was highest in Fuwen of Dingan county. Zinc content in kernels was highest in Tongshi of Wuzhishan county, and undetectable in six regions including Penglai of Wenchang county. Cr content in kernels exceeded 1.0 mg/kg in seven regions such as Dongshan of Haikou county. Pd content in kernels was over 0.5 mg/kg in Dongshan of Haikou county and Zhongxin of Chengmai county. The contents of Cd, As and Hg were very low, undetectable in kernels of all the 18 forests.

3.3. Correlation of Se and Heavy Metals Between Soil and Roots, Leaves, and Kernels of *Camellia oleifera*

Significant positive correlation in As and Pb contents was observed between soil and roots of *Camellia oleifera*. However, significant negative correlation of Cd content between soil and roots of *Camellia oleifera* was observed. Significant positive correlation between soil and kernels was only found for Cd content; suggesting that only soil As and Pb content showed high biological availability for *Camellia oleifera* (Table 6).

To investigate the transferability of metal elements from roots to leaves and kernels, the correlation study was used to compare the contents of each element in different parts of *Camellia oleifera* (Table 7). It was found that there was significant positive correlation between roots and leaves in terms of Cr and Cd content; and there was positive correlation between roots and leaves in terms of Pd content. However, significant negative correlation was found between roots and leaves in terms of Hg and Se contents. In addition, there was significant positive correlation between roots and kernels with regard to Hg content; whereas, significant negative correlation was found between roots and

kernels in terms of Cd content, implying that Cr and Cd are easily translocated from roots to leaves, and Hg from roots to kernels.

3.4. Correlation Between Se and Six Metal Elements in Roots and Leaves

The relationship between the roots and leaves regarding the Se content and 6 metal elements is shown in Table **8**. There was negative correlation of content between Se and the other six metal elements in the roots of *Camellia oleifera*, and significant negative correlation was found between Se and As in roots. A positive correlation between Se and Cr content as well as Hg in leaves, and negative correlation between Se and Pd as well as Cd, suggested that Se reduced the absorption of heavy metals by roots, especially the absorption of As.

4. DICUSSION

Se-enriched soil is widely distributed in Hainan Island of China. Se content was high in the soils of Haikou and Chengmai counties [15]. It has been confirmed in the current study that Se content ranged between 0.489 and 2.110 mg/kg in 18 forests of *Camellia oleifera*, which is higher than that in the soil of rice paddy fields [15]. The soil Se content was higher than 1.0 mg/kg in 10 forests out of 18. However, there was no significant correlation between the soil and roots of *Camellia oleifera* with respect to Se content. The Se content in leaves and kernels was also low, probably due to acidic soil conditions and impaired biological availability of Se [16]. Our findings suggest that blindly supplementing Se fertilizer in the nutrient management of

Camellia oleifera should be avoided. Instead, alkaline fertilizer to promote the activation of Se and improve its bioavailability in soil [17], or screen for suitable *Camellia oleifera* varieties should be recommended to produce effective and economical Se-enriched camellia oil [2, 18]. The average content of heavy metals in soil was low in *Camellia oleifera* forests of Hainan province and fulfilled the requirements for green-food production. However, the heavy metal content in soil and Camellia trees greatly varied geographically with different cultivation methods, which might be attributed to diverse soil characteristics and use of chemical fertilizers [19]. Therefore, instead of chemical fertilizers containing heavy metals, ecofriendly cultivation using manures [20] and forest intercropping [21] should be adopted.

The content of heavy metals in roots was higher than that in leaves and kernels, suggesting that Camellia restricts the translocation of heavy metals from roots to leaves and kernels [22], which is the key to successful utilization of seeds of *Camellia oleifera*. The mechanism of heavy metal distribution in the plant following absorption from soils should be further elucidated. Also, production processes that limit heavy metal transfer from roots to kernels should be adopted. Variations in heavy metal accumulations by different organs of plants are a consequence of species and climate diversity [23], which is also known as heavy metal enrichment coefficient.

CONCLUSION

In the current study, the enrichment coefficient was relatively higher for Zn, Cr and Pb by the roots of *Camellia oleifera*, which was consistent with the study based on tropical trees such as *Casuarina equisetifolia* [24]. Significant positive and negative correlations were found between soil and roots with regard to As and Pb contents, respectively, implying that As and Pb are easily activated and absorbed by *Camellia oleifera*, since they are associated with high utilization efficiency of phosphate [25]. Therefore, it is necessary to further analyze the relationship between essential elements and heavy metal activation in Camellia [26]. A significant positive correlation between the roots and leaves in terms of Cr and Cd contents was observed. A significant positive correlation between roots and kernels in terms of Hg content suggested that the transportation of Hg into fruits was relatively easier. It is thus essential to control the Hg content in soil to reduce the risk of Hg absorption. Se and six heavy metals were in negative correlation in roots, especially Se and As, implying that strengthing of Se enrichment effectively reduced the accumulation of heavy metals [27]. Therefore, in addition to increasing Se bioactivity in soil, Se supplementation through leaves is recommended to regulate heavy metal distribution in fruits [28], after carrying out additional testing and analysis.

ACKNOWLEDGEMENTS

This work is supported by Special Fund for Forestry Scientific Research in the Public Interest (201504705).

REFERENCES

[1] Li B, Zhao C. Current situation of heavy metals pollution in soil at farmland and detection technologies analysis in China. J Agric Resour Environ 2013; 30: 1-7.

[2] Geng JM, Wu LL, Yu A, Tang SM. Screening enriched-selenium hybrid rice cultivars of Hainan province. Chin Agric Sci Bull 2010; 26: 376-80.

[3] Yang Y, Liu X, Ning C, et al. Effects of foliar feeding of selenium on fruit quality and accumulation of cadmium, lead and mercury in sweet persimmon. Acta Hort Sin 2013; 40: 523-30.

[4] Chen S, Sun G, Chen Z, Chen F, Zhu YG. Progresses on selenium metabolism and interaction with heavy metals in higher plants. Plant Physiol J 2014; 50: 612-24.

[5] Xiong SJ, Liu JY, Xu WH, Xie W. Progresses on selenium metabolism and interaction with heavy metals in higher plants. Environ Sci 2015; 36: 286-94.

[6] Lin D, Wang L, Zhang Y. Present situation of soil heavy metals in banana gardens of Hainan and its changing tendency. J Saf Environ 2006; 6: 54-8.

[7] Guo Y, Fu Y, Bai X, Yang Y, Zhang G. Source analysis and evaluation of soil heavy metals in pepper fields in Hainan Province. Chin J Soil Sci 2012; 43: 54-8.

[8] Tan Y, Wei J, Chen Z, Gao W. Distribution and evaluation of heavy metal contents in the soil of *Areca catechu* plantations in Hainan. Chin Environ Sci 2011; 31: 815-9.

[9] Zhuang R. Oiltea in China. Beijing: Forestry press in China, 2012, pp. 3-5.

[10] State Forestry Bureau of China, National Development Planning of Oiltea(2009 ~ 2020). Beijing: Forestry press in China, 2012, pp. 15-8.

[11] Yuan J, Han Z, He S, Huang L, Zhou N. Investigation and cluster analysis of main morphological and economical characters for oiltea resource in Hainan. J Plant Genet Resour 2014; 15: 1380-84.

[12] Yang W, Chen L, Wang X. Preliminary report on introduction and trial planting of subtropical excellent varieties in Hainan area. Acta Agricul Jiangxi 2012; 24: 63-5.

[13] Yang W, Chen L, Wang X. Analysis on ecological adaptability of *Camellia oleifera* in central area of Hainan. Acta Agricul Jiangxi 2010; 22: 93-5.

[14] Fu D, Chen L, Yang W. High yielding cultivation technique for oil Camellia in Hainan. Chin J Trop Agric 2012; 32: 23-7.

[15] Geng J, Wang W, Wen C, Yi Z, Tang S. Concentrations and distributions of selenium and heavy metals in Hainan paddy soil and assessment of ecological security. Acta Ecol Sin 2012; 32: 3477-86.

[16] Punz WF, Sieghardt H. The response of roots of herbaceous plant species to heavy metals. Environ Exp Bot 1993; 33: 85-98.

[17] Du Q, Chen C, Zeng B. Ways to increase effective selenium in laterite soil of Hainan. Chin. J Trop Agric 2013; 33: 9-12.

[18] Du Q, Zhang Y, Tang S, Long K, Zeng B, Li G. The absorption and distribution mechanisms of selenium from Hainan selenium – rich soil by different rice varieties. Soil Fertil Sci Chin 2009; 6: 37-40.

[19] He T, Dong L, Liu Y, Shu Y, Luo H, Liu F. Change of physical-chemical properties and heavy mental element in soil from different parent material /rock. J Soil Water Conserv 2006; 6: 157-62.

[20] Li X, Peng X, Wu S, Li Z, Feng H, Jiang Z. Effect of arbuscular mycorrhizae on growth, heavy metal uptake and accumulation of Zenia insignis chun seedlings. Environ Sci 2014; 35: 3142-8.

[21] Wei Z, Guo X, Wu Q, Long X. Continuous remediation of heavy metal contaminated soil by co-cropping system enhanced with chelator. Environ Sci 2014; 35: 4305-12.

[22] Dahmani-Muller H, Gelie B, Balabane M, Van F. Strategies of heavy metal uptake by three plant species growing near a metal smelter. Environ Pollut 2000; 109: 231-8.

[23] Weis JS, Weis PH. Metal uptake, transport and release by wetland plants: implications for phytoremediation and restoration. Environ Int 2004; 30: 685-700.

[24] Jin M, Ding Z, Zhou H, Ye G. Absorption and enrichment of heavy metals by Casuarina equisetifolia of different stand ages in a coastal zone. Chin J Ecol 2014; 33: 2183-7.

[25]   Bolan N, Mahimairaja S, Kunhikrishnan A. Phosphorus–arsenic interactions in variable-charge soils in relation to arsenic mobility and bioavailability. Sci Total Environ 2013; 463: 1154-62.

[26]   Zhu W, Yang Y, Bi H, Liu Q. Research on hte total bioavailble concentrations and biaoabailabity of Zn, Pb, Cu and Cd in soils in Hainan province. Acta Mineral Sin 2004; 24: 239-44.

[27]   Hu Y, Duan G, Huang Y, Liu Y, Sun G. Interactive effects of different inorganic As and Se species on their uptake and translocation by rice (*Oryza sativa L.*) seedlings. Environ Sci Pollut Res 2014; 21: 3955-62.

[28]   Zhao J, Gao Y, Li Y, *et al.* Selenium inhibits the phytotoxicity of mercury in garlic (*Allium sativum*). Environ Res 2013; 125: 75-81.

20

Microorganism Quantity and Enzyme Activities in Wheat Field Subjected to Different Nitrogen Fertilizer Rate

Sun Jingjing, Ma Jiaheng, Wu Xiaoying, Wang Cheng and Yao Jun[*]

School of Civil & Environmental Engineering and National "International Cooperation Based on Environment and Energy" and Key Laboratory of "Metal and Mine Efficiently Exploiting and Safety" Ministry of Education, University of Science and Technology Beijing, Beijing 100083, P.R. China

Abstract: Field experiments are described involving nitrogen fertilizer in the wheat field to analyze its effect on microbial quantity and enzyme activity. The results showed that 250 kg N hm^{-2} applied in the wheat field (1) can increase bacteria, fungi and the quantity of actinomycetes by 39.9%, 56.7% and 70.5% compared to 0 kg N hm^{-2}. (2) Similarly, the activities of soil urease, catalase and FDA exhibited 60.8%, 18.3% and 49.1% improved variation compared to 0 kg N hm^{-2}, respectively. (3) Moreover, Shannon-Wiener diversity index (H) and Evenness index (E) reached the peak (H=0.37, E=0.33). Correlation analysis of microbial quantity with enzyme activity indicated that they were related to each other. These findings suggested that soil microbial quantity and enzyme activity were significantly influenced by nitrogen fertilizer and the application of 250 kg N hm^{-2} in the wheat field was observed to be the best.

Keyword: Enzyme activity, microorganism quantità, nitrogen fertilizer, wheat field.

1. INTRODUCTION

Nitrogen (N) fertilizer applied to agricultural soil influences the crop and soil system, which has made the current nitrogen fertilizer management practices lead to negative environmental effects [1]. The North China Plain as the major wheat production area in China plays an important role in securing national food security [2]. It is well known that in the last three decades, in this region, farmers increased the use of nitrogen fertilizer in the soil which caused an abundance of microorganisms [3] (bacterial, fungal and actinomycetes) and enzyme activities (urease, phosphatase, catalase, dehydrogenase activities) [4] which impaired the nutrient (especially nitrogen, carton and phosphorus) cycles [5, 6] of soil.

Microorganisms perform a critical role in nutrient transformation, cycling, and in many soil biochemical processes. Microbial quantity, biomass and enzyme activity express the functional relationship among microbial compositions [7] when different nitrogen fertilizer rates are applied to the soil [8]. The quantity of microorganisms, especially of bacteria is directly and indirectly related to the wheat field in the soil organic matter decomposition [9] and soil respiration [10], which significantly influence the absorption of nutrients of wheat. The difference in the population of soil fungal was due to nitrogen fertilizer [11]. A number of bacteria, fungi and actinomycetes are associated with the amount of nitrogen fertilizer, and the optimum nitrogen fertilizer in the soil results in larger populations. In addition, their diversity is an important soil microbial parameter. Shannon-Wiener diversity index is sensitive to changes in the microbial quantity, the more the number of species, the higher is the value of microbial diversity [12]. Soil enzyme activities can be associated with soil properties, active cells and climate which are used as the indicators of soil fertility [13, 14]. Urease plays an important role in nitrogen cycling, which catalyzes urea to carbon dioxide and ammonia [15]. Catalase is used to catalyze hydrogen peroxide (H_2O_2) to water (H_2O) and molecular oxygen (O_2) in the soil, and prevent the toxicity of H_2O_2 in the biological body [16]. Fluorescein diacetate (FDA) activity was observed to be directly related to microbial activity. Previous studies suggested that nitrogen addition increases enzyme activities [17, 18], but other studies have shown that excessive nitrogen fertilizer reduces the enzyme activity [19] and microbial diversity [20]. As a consequence, the quantity of microorganisms and enzyme activity in the soil need to be better understood.

We selected a wheat field in the North China Plain to study the relationships between microbial quantity and enzyme activity. Our aims were to analyze the changes on the soil under different nitrogen fertilizers from biological perspective and investigate the microbial diversity of the soil to determine its correlation with the nitrogen fertilizer.

2. MATERIALS AND METHODS

2.1. Experiment Sites

The experiment site is located in Qingyuan county, Hebei province, China (38°5'N, 115°30'E). The area is a temperate monsoon climatic region, with annual average temperature of 12°C and annual precipitation of 550mm. The experiment field was established with a wheat-maize rotation. Before the experiment, the physicochemical properties of initial soil

*Address correspondence to this author at the School of Civil & Environmental Engineering and National "International Cooperation Based on Environment and Energy" and Key Laboratory of "Metal and Mine Efficiently Exploiting and Safety" Ministry of Education, University of Science and Technology Beijing, Beijing 100083, P.R. China;
E-mail: sunjing0314@126.com

samples were collected from the surface layer (0-20cm) in May 2010, while the content of soil organic matter, total N, Olson-P and Olson-K was 16.8g kg^{-1}, 0.9g kg^{-1}, 16.6 mg kg^{-1} and 99.3 mg kg^{-1}, respectively.

2.2. Experiment Design and Soil Sampling

A randomized block design was used with three replicates of each treatment, with the area of each plot being 40 m^2 (8m ! 5m). The experiment comprised of five treatments: 0 (N0), 100 (N100), 180 (N180), 250 (N250), and 300 (N300) kg N hm^{-2}. Fertilizers used were urea (46%), phosphorus pentoxide (12%) and potassium sulfate (60%). Urea was applied three times during the whole stage; 40% of urea was applied as the basal fertilizer, 40% of urea was applied at the reviving stage and20% was applied at the blossoming stage. Total P and K fertilizers were applied as basal fertilizers, in the quantity of 120 kg P$_2$O$_5$ hm^{-2} and 120 kg K$_2$O hm^{-2}.

Dates used in this study were collected from the surface layer (0-30cm) in May, 2013. The fresh soil samples were sieved 2mm and preserved for further experiment and analysis.

2.3. Microbial Population

The microbial populations were determined by the dilution plate method [21]; bacteria were determined using beef extract peptone medium; fungi, using Martin medium, and actinomycetes using Gause's I medium.

2.4. Enzyme Activity

Soil urease activity was based on the colorimetric determination [22], where 2.5g soil with 0.5 mL toluene was added in 50mL volumetric flask for 15 min; then 2.5 mL of 10%urea and 5mL of citrate buffer (pH 6.7) were added in the constant temperature incubator (38°C) for 24h. Following this, the soil sample with 38°C distilled water was diluted to 25 ml (toluene should float in the scale above) mixture, containing 1ml filtrate in 50ml volumetric flask with 10 ml distilled water. In this mixture, 4 ml sodium phenate and 3ml sodium hypochlorite were added in constant volume, and determined at 578nm 20 min later and expressed as mg NH3-N g^{-1} 24h^{-1}.

Soil catalase activity was based on permanganate titration method [23]. In this method, 2 g of soil with 40 mL of distilled water and 5 mL of 0.3% H$_2$O$_2$ were added into 100 mL triangular flask vibrated for 20 min. Following this, 5 mL of 3 mol L^{-1} H$_2$SO$_4$ terminated the reaction, leaving 25 mL of filtrate with 0.1 mol L^{-1} potassium permanganate titration. The catalase was expressed as mL 0.02 mol L^{-1} KMnO$_4$ g^{-1} 20min^{-1}.

Soil fluorescein diacetate (FDA) hydrolase activity was based on colorimetric determination [24], where 5 g of fresh soil with 15 mL of 60 mmol L-1 phosphate buffer (pH 7.6) and 0.2 mL of 1000μg mL^{-1} fluorescent diacetate (FDA) reserves were added into the 50 mL triangular flask vibrated for 20 min at 30°C. Following this, 15 mL of chloroform/methanol solution terminated the reaction, with the absorbance of the released fluorescein at 490 nm.

2.5. Data Analysis

Shannon-Wiener index (H), H = -∑P$_i$! [lnP$_i$], where P$_i$ is the proportion of each taxon in the total quantity. Evenness index (E), E= -H/ lnS, where S is the total number of species [25]. Correlation analysis was analyzed with SPSS 18.0 and other data analysis was performed with origin 9.0.

3. RESULTS AND DISCUSSION

3.1. Microbial Population

Soil microbial quantity varied greatly under the different N fertilizer treatments used in this study. The quantity of soil bacteria was increased by 20.4%, 31.95, 39.9% and 35.6% compared to N0 (Fig. 1A). Application of N fertilizer brought significant increase in the soil fertility, as well as caused changes in the microbial quantity. The quantity of soil fungi was increased by 22.0%, 53.3%, 56.7% and 49.4% compared to N0, (Fig. 1B). Similarly, the quantity of soil actinomycetes was increased by 12.9%, 54.25, 70.5% and 59.1% compared to N0, (Fig. 1C). The increase in the total microbial quantity (bacteria, fungi and actinomycetes) was greater in N250. However, the microbial quantity in N300 showed a slight decrease compared to N 250 (Fig. 1D). N0 had the lowest quantity in all treatments, which can be attributed to less N nutrition suggesting that there was not enough demand for soil microorganism [26]. Some researchers have already reported that microbial quantity had no significant change at low fertilizer application in the black soil [27]. In addition, the response of fungi number to N fertilizer was highly variable [28]. After 13 years' fertilizer experiment [29], the bacteria, actinomycetes and fungi in the red soil changed greatly and indicated that protecting the diversity of microorganisms sustained the soil development in the agroecological system.

3.2. Enzyme Activity

The enzyme activity was significantly influenced by N fertilizer [30], as the activity of urease increased with increase in the N fertilizer rate and N300 had the highest urease activity in five treatments; being significantly higher than other treatments (Table 1). This result was similar to a maize-wheat experiments in India [31]. Compared to the N0, N100, N180, N250 and N300 increased the catalase activity by 15.7%, 18.3%, 16.5% and 11.3%, respectively. But there was no significant effect on catalase activity, which was different from other studies [32]. FDA activity increased by 10.9%, 45.5%, 49.1% and 40.0% in N100, N180, N250 and N300 compared to N0, respectively, and closely related to bacteria, fungi and actinomycetes quantity (Table 3). An equation from Nayak indicated that FDA hydrolysis activity was one of the important factors in the soil biochemical possesses [30].

Higher enzyme activities in N250 probably resulted from optimum N fertilizer rate, which increased the soil fertility. On the other hand, crops growing better not only directly benefited the proper fertilizer [33], but alsocreated more root exudation (carbohydrate, amino acids and organic acids) [34] and the turnover indirectly stimulated the soil microbiological metabolism [35].

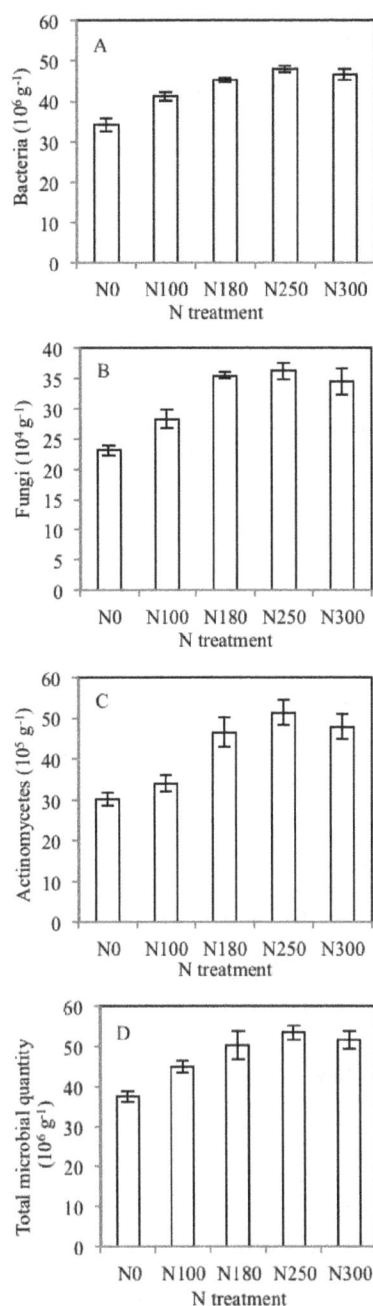

Fig. (1). Effects of different nitrogen fertilizer rate on soil microbial quantity (CFU/g soil).

3.3. Microbial Diversity Index

Fertilizer in the agricultural soil is a major impact factor that influences the diversity of microorganisms [36]. In the results of this study, H and E (Table **2**) did not show a trend of increase with increased N fertilizer rate. They both reached the peak (H=0.37, E=0.33) at N250, and decreased at N300, which indicated that applying low and high rate was not good for microbial diversity. Sarathchandra reported that N fertilizer affected soil microbial functional diversity and with increased N fertilizer, H was significantly reduced [8]. This trend also appeared in the enzyme activity. Because NH_4 and NO_3 as the nutrients of soil microorganism were obtained from N fertilizer, thus, different rate resulted in different microbial diversity [8]. Based on our study, the application of 250 kg N hm^{-2} can promote microbial diversity in the wheat field.

3.4. Relationships between Microbial Quantity and Enzyme Activity

According to the correlational analysis (Table **3**), microbial quantity was positively correlated with the soil enzyme activity. All microbial quantity was significantly correlated with FDA activity. In few of the analyses, the quantity of bacteria was significantly positively correlated with FDA (r=0.982, P<0.05), along with the quantity of fungi and actinomycetes, which was significantly positively related to FDA (r=0.985, r=0.975, P<0.01). These positive correlations were observed through biochemical processes [37]. In addition, this result showed that increasing the total quantity of microorganisms increased the FDA activity of the soil [38]. No significant relationship was found between microbial quantity, urease and CAT activity.

CONCLUSION

In our study, we tested differences in microorganism quantity, enzyme activities and soil microbial diversity. All these parameters were related to each other and more importantly, were directly influenced by the addition of nitrogen fertilizer. Applying 250 kg N hm^{-2} in this region not only improved soil enzyme activity but also the soil microbial quantity as well as the soil microbial diversity.

Table 1.   Effects of different nitrogen fertilizer rate on soil enzyme activity.

Treatment	Urease Activity mg/kg·24h	CAT Activity 0.002 mol/L KMnO₄ mL/g	FDA Activity μg/ g·20min
N0	7.39ab	1.15a	0.55a
N100	8.04ab	1.33a	0.61a
N180	7.63ab	1.34a	0.80a
N250	11.88a	1.36a	0.82a
N300	11.96a	1.28a	0.77a

Table 2.   Effect of nitrogen fertilizer on Shannon-Wiener diversity and Evenness index.

Diversity Index	N0	N100	N180	N250	N300
H	0.32	0.31	0.34	0.37	0.35
E	0.29	0.28	0.31	0.33	0.32

H:Shannon-Wiener diversity index, *E*: Evenness index.

Table 3.   Correlation analysis microbial quantity and enzyme activity.

	Bacteria	Fungi	Actinomycetes	Urease	CAT	FDA
Bacteria	1					
Fungi	0.960**	1				
Actinomycetes	0.884*	0.974**	1			
Urease	0.500	0.642	0.751	1		
CAT	0.846	0.794	0.654	0.256	1	
FDA	0.928*	0.985**	0.975**	0.589	0.727	1

* Correlation is significant at the 0.05 level (2-tailed).
** Correlation is significant at the 0.01 level (2-tailed).

ACKNOWLEDGEMENTS

This work was supported in part by grants from the International Joint Key Project from Chinese Ministry of Science and Technology (2010DFB23160), Key project from the National Science Foundation of China (41430106), National Natural Science Foundation of China (41273092), Public welfare project of Chinese Ministry of Environmental Protection (201409042), Overseas, Hong Kong and Macau Young Scholars Collaborative Research Fund (41328005).

REFERENCES

[1]   Zhao S, Qiu S, Cao C, Zheng C, Zhou W, He P. Responses of soil properties, microbial community and crop yields to various rates of nitrogen fertilization in a wheat–maize cropping system in north-central China. Agric Ecosyst Environ 2014; 194: 29-37.

[2]   Lu C, Fan L. Winter wheat yield potentials and yield gaps in the North China. Plain Field Crops Res 2013; 143: 98-105.

[3]   Yu C, Hu XM, Deng W, et al. Changes in soil microbial community structure and functional diversity in the rhizosphere surrounding mulberry subjected to long-term fertilization. Appl Soil Ecol 2015; 86: 30-40.

[4]   Saha S, Prakash V, Kundu S, Kumar N, Mina BL. Soil enzymatic activity as affected by long term application of farm yard manure and mineral fertilizer under a rainfed soybean–wheat system in NW Himalaya. Eur J Soil Biol 2008; 44: 309-15.

[5]   Melero S, López-Bellido RJ, López-Bellido L, Muñoz-Romero V, Moreno F, Murillo JM. Long-term effect of tillage, rotation and nitrogen fertiliser on soil quality in a Mediterranean Vertisol. Soil Tillage Res 2011; 114: 97-107.

[6]   Chirinda N, Olesen JE, Porter JR, Schjønning P. Soil properties, crop production and greenhouse gas emissions from organic and inorganic fertilizer-based arable cropping systems. Agric Ecosyst Environ 2010; 139: 584-94.

[7]   Meena VS, Maurya BR, Verma JP. Does a rhizospheric microorganism enhance K(+) availability in agricultural soils? Microbiol Res 2014; 169: 337-47.

[8]   Sarathchandra SU, Ghani A, Yeates GW, Burch G, Cox NR. Effect of nitrogen and phosphate fertilisers on microbial and nematode diversity in pasture soils. Soil Biol Biochem 2001; 33: 953-64.

[9]   Xu Z, Yu G, Zhang X, et al. The variations in soil microbial communities, enzyme activities and their relationships with soil organic matter decomposition along the northern slope of Changbai Mountain. Appl Soil Ecol 2015; 86: 19-29.

[10]   Sapp M, Harrison M, Hany U, Charlton A, Thwaites R. Comparing the effect of digestate and chemical fertiliser on soil bacteria. Appl Soil Ecol 2015; 86: 1-9.

[11]   Marschner P. Structure and function of the soil microbial community in a long-term fertilizer experiment. Soil Biol Biochem 2003; 35: 453-61.

[12]   Li P, Dong J, Yang S, et al. Impact of β-carotene transgenic rice with four synthetic genes on rhizosphere enzyme activities and bacterial communities at different growth stages. Eur J Soil Biol 2014; 65: 40-6.

[13]   Marcote I, Hernández T, García C, Polo A. Influence of one or two successive annual applications of organic fertilisers on the enzyme activity of a soil under barley cultivation. Biores Technol 2001; 79: 147-154.

[14]   Li J, Zhao B, Li X, Jiang R, Bing SH. Effects of long-term combined application of organic and mineral fertilizers on microbial biomass, Soil Enzyme Activities and Soil Fertility. Agric Sci China 2008; 7: 336-343.

[15]   Tan B, Wu FZ, Yang W, He X. Snow removal alters soil microbial biomass and enzyme activity in a Tibetan alpine forest. Appl Soil Ecol 2014; 76: 34-41.

[16]   Orikasa Y, Nodasaka Y, Ohyama T, et al. Enhancement of the nitrogen fixation efficiency of genetically-engineered Rhizobium with high catalase activity. J Biosci Bioeng 2010; 110: 397-402.

[17]   Wang QK, Wang SL, Liu YX. Responses to N and P fertilization in a young Eucalyptus dunnii plantation: Microbial properties, enzyme activities and dissolved organic matter. Appl Soil Ecol 2008; 40: 484-90.

[18]   Liu YR, Li X, Shen QR, Xu YC. Enzyme activity in water-stable soil aggregates as affected by long-term application of organic manure and chemical fertiliser. Pedosphere 2013; 23: 111-9.

[19]   Piotrowska A, Wilczewski E. Effects of catch crops cultivated for green manure and mineral nitrogen fertilization on soil enzyme activities and chemical properties. Geoderma 2012; 189: 72-80.

[20]   Li F, Liu M, Li Z, Jiang C, Han F, Che Y. Changes in soil microbial biomass and functional diversity with a nitrogen gradient in soil columns. Appl Soil Ecol 2013; 64: 1-6.

[21]   Yang SS, Fan HY, Yang CK, Lin IC. Microbial population of spruce soil in Tatachia mountain of Taiwan. Chemosphere 2003; 52: 1489-98.

[22]   Guo H, Yao J, Cai M, et al. Effects of petroleum contamination on soil microbial numbers, metabolic activity and urease activity. Chemosphere 2012; 87: 1273-80.

[23]   Chang-sheng Y. Research Method of Soil Fertility. Chinese Agriculture Press: Beijing, China (in Chinese) 1988: pp. 277-9.

[24]   Grosso F, Temussi F, De Nicola F. Water-extractable organic matter and enzyme activity in three forest soils of the Mediterranean area. Eur J Soil Biol. 2014; 64: 15-22.

[25]   Ruan W, Ren T, Chen Q, Zhu X, Wang JG. Effects of conventional and reduced N inputs on nematode communities and plant yield under intensive vegetable production. Appl Soil Ecol 2013; 66: 4855.

[26]   Yun-fu G, Xiang Y, Xiao-ping Z, Shi-hua T, Xi-fa S, Lindström K. Effect of different fertilizer treatments on quantity of soil microbes and structure of ammonium oxidizing bacterial community in a calcareous purple paddy soil. Agric Sci China 2008; 7: 1481-9.

[27]   Hou S, Xin M, Wang L, Jiang H, Li N, Wang Z. The effects of erosion on the microbial populations and enzyme activity in black soil of northeastern China. Acta Ecol Sin 2014; 34: 295-301.

[28]   Qin H, Lu K, Strong PJ, et al. Long-term fertilizer application effects on the soil, root arbuscular mycorrhizal fungi and community composition in rotation agriculture. Appl Soil Ecol 2015: 89: 35-43.

[29]   Wenhui Z, Zucong C, Lichu Y, He Z. Effects of the long-term application of inorganic fertilizers on microbial community diversity in rice-planting red soil as studied by using PCR-DGGE. Acta Ecol Sin 2007; 27: 4011-8.

[30]   Nayak DR, Babu YJ, Adhya TK. Long-term application of compost influences microbial biomass and enzyme activities in a tropical Aeric Endoaquept planted to rice under flooded condition. Soil Biol Biochem 2007; 39: 1897-906.

[31]   Saha S, Gopinath KA, Mina BL, Gupta HS. Influence of continuous application of inorganic nutrients to a Maize–Wheat rotation on soil enzyme activity and grain quality in a rainfed Indian soil. Eur J Soil Biol 2008; 44: 521-31.

[32]   Yang L, Li T, Li F, Lemcoff JH, Cohen S. Fertilization regulates soil enzymatic activity and fertility dynamics in a cucumber field. Sci Hortic 2008; 116: 21-6.

[33]   Chu H, Lin X, Fujii T, et al. Soil microbial biomass, dehydrogenase activity, bacterial community structure in response to long-term fertilizer management. Soil Biol Biochem 2007; 39: 2971-6.

[34]   Macdonald LM, Paterson E, Dawson LA, McDonald AJS. Defoliation and fertiliser influences on the soil microbial community associated with two contrasting Lolium perenne cultivars. Soil Biol Biochem 2006; 38: 674-82.

[35]   Marinari S, Masciandaro G, Ceccanti B, Grego S. Influence of organic and mineral fertilisers on soil biological and physical properties. Biores Technol 2000; 72: 9-17.

[36]   Murase J, Hida A, Ogawa K, Nonoyama T, Yoshikawa N, Imai K. Impact of long-term fertilizer treatment on the microeukaryotic community structure of a rice field soil. Soil Biol Biochem 2015; 80: 237-43.

[37]   Li F, Yu J, Nong M, Kang S, Zhang J. Partial root-zone irrigation enhanced soil enzyme activities and water use of maize under different ratios of inorganic to organic nitrogen fertilizers. Agric Water Manag 2010; 97: 231-9.

[38]   Chen X, Song B, Yao Y, Wu H, Hu J, Zhao L. Aromatic plants play an important role in promoting soil biological activity related to nitrogen cycling in an orchard ecosystem. Sci Total Environ 2014; 472: 939-46.

21

Free Triplet Conjecture and Equivalence Classes Derived using Group Theory

Zhang Dakun[*], Song Guozhi and Huang Cui

School of Computer Science and Software Engineering, Tianjin Polytechnic University, Tianjin300387, China

Abstract: All proteins are made up of 20 different amino acids which contain 4 kinds of nucleotides . Three consecutive nucleotides on the gene, called triplet codons, are used to code an amino acid, and 64 triplet codons comprise the genetic code table. Central dogma (DNA-RNA-protein) has been acknowledged, but the process and mechanism of mRNA passing through the nuclear membrane still require further investigation. For these two problems mentioned above, this paper proposed a conjecture of nucleotide free triplet and obtained 20 equivalence classes of mapping from free triplet vertex set to nucleotide set using group theory. Whether the four numbers 3, 4, 20 and 64 have relevance are taken into consideration here. Subsequently, the numbers 3, 4, 20 and 64 were connected together which was important for the analysis of triplet code and protein composition.

Keywords: DNA/mRNA/protein, equivalence class, free triplet, group theory, nucleotide.

1. INTRODUCTION

The deciphering of the genetic code is one of the most outstanding achievements in the history of science. It not only provides a theoretical basis for the study of protein synthesis, but also confirms the precision of the central dogma. The presence of the genetic code has been further validated with the help of rapid development in molecular biology techniques such as DNA, RNA sequence determination and amino acid sequencing technology advances since the 1970s [1].

Though research on the decoding of genetic code has already had a history of more than forty years, interpretation of codon distribution and analysis of the nature and amount of amino acid code are still a challenge [2]. So the study of the genetic code continues to attract many scientists [3-7] and so many interdisciplinary research topics like life sciences, and molecular biology are to be further explored. Newton once said: "bold conjecture can be a great discovery." The emergence of any scientific theory in the history has bold conjecture and assumptions involved in the gestation process. For example, Einstein's general relativity theory, DNA double helix structure, triplet genetic code, *etc*, all have their origin aided by conjecture. The study on challenging interdisciplinary subjects like; life science and molecular biology, still needs bold conjecture and hypothesis. Galileo said. "The nature is a book written in Mathematical language", also it is necessary to use more mathematical methods in the study of life sciences and molecular biology [8, 9].

2. THE DERIVATION OF CONJECTURE OF NUCLEOTIDE FREE TRIPLET

2.1. Genetic Code Composition

The genetic code is a set of rules by which information encoded within the genetic material (DNA or mRNA) is translated into proteins. Each amino acid in the protein is expressed by codon composed of three consecutive nucleotides in the gene [10]. DNA has four different nucleotides, each containing four different bases: adenine (A), guanine(G), thymine(T) and cytosine(C). The order of these bases in the gene determines the sequence of amino acids in the protein which forms the primary structure of protein. When the nucleotide sequence of DNA transcripts to mRNA, uracil(U) replaces thymine(T). The 64 possible triplet codes are mentioned in the 9th reference.

The composition of triplet code was first proposed by astrophysicist Gamow in 1954 [11]. Though it was questioned by many biologists and geneticists, it was confirmed by the Institute of Science in 1961 (Crick et al first confirmed the correctness of the composition of triplet code from the perspective of genetics) [12].

2.2. Central Dogma and mRNA

1. Central Dogma of molecular biology

In 1958, Crick proposed to name the genetic pathway as central dogma, suggesting that chromosomal and DNA are the templates for the RNA molecule. Subsequently, It was transferred to the cytoplasm after synthesis to determine the amino acid sequence of the protein [12]. The DNA is the template used for self-replication and also for the synthesis of RNA (transcription), and RNA is the template for the synthesis of protein (translation). The latter two processes can only be executed in one direction.

*Address correspondence to this author at School of Computer Science and Software Engineering, Tianjin Polytechnic University, Tianjin 300387, China; E-mail: zhangdakun2013@163.com

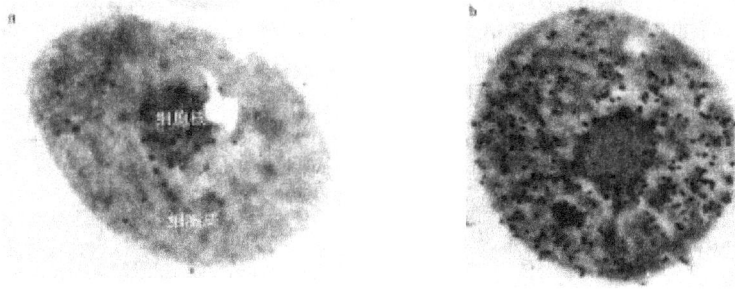

Fig. (1). Process of mRNA through nuclear membrane.

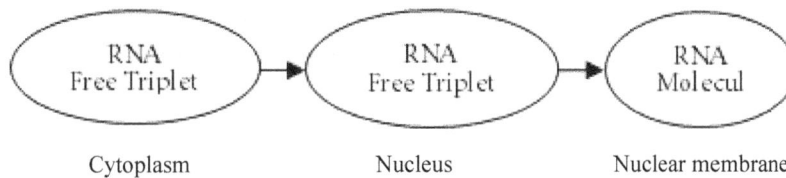

Cytoplasm   Nucleus   Nuclear membrane

Fig. (2). Schematic diagram of the process of triplet through nuclear membrane.

2. The discovery of mRNA

Through indepth study it was discovered that RNA can be divided into three kinds, i.e. mRNA(messenger RNA), tRNA(transfer RNA) and rRNA(ribosomal RNA), with mRNA only accounting for a few percent in the cells. To code different proteins in the cells, the length of mRNA and composition of nucleotide vary widely. Within a given time, only a small section of mRNA binds to the ribosome, therefore, an mRNA can be examined simultaneously by multiple ribosomes.

3. mRNA moved to cytoplasm after synthesis in the nucleus

Until now, there is sufficient evidence to prove that RNA indeed moves from the DNA-containing nucleus to the cytoplasm containing ribosome [13]. Following this, the cells are briefly exposed to the radioactive labeling precursor, and an excess of unlabeled amino acids (pulse-labeled experimental) are added. mRNA synthesis manifests in a short period of time. The study shows that the mRNA synthesized in the nucleus. After about one hour, it was observed that most of the mRNAs left the nucleus and moved into the cytoplasm (Fig. **1**) [12]. mRNA moves through the nuclear membrane in the protein complex called nuclear pore after being modified, which allows it to be expressed [13]. It is needed to be further explored in detail that how mRNA travels from the nucleus through the nuclear membrane to the cytoplasm.

2.3. The Derivation of Conjecture of Nucleotide Free Triplet

For the derivation of conjecture of nucleotide free triplet three consecutive nucleotides in the gene must have a triplet code and mRNA must move from the nucleus into cytoplasm through nuclear membrane. In this paper, the derivation of conjecture of nucleotide free triplet was obtained at some points, and three consecutive nucleotides were bound as separate clusters. These clusters were made up of tiny triangles, called nucleotide free triplet composed of three nucleotides. For example, when mRNA moved through the nuclear membrane, three nucleotides formed a group and mRNA free triplet, and after free triplet moved through the nuclear membrane into cytoplasm, it restored the linear state and again synthesized mRNA molecules . The process of free triplet through nuclear membrane is shown in Fig. (**2**).

Cytoplasm Nucleus Nuclear membrane

3. TRIPLET CODE 20 EQUIVALENCE CLASSES DERIVED USING GROUP THEORY

3.1. Group Theory and Permutation Groups

1. Brief Introduction of Group Theory

Group theory is an important part of modern algebra. The concept of "group" was first put forward by the young French mathematician Galois (E. Galas, 1811-1832) and only defined the permutation group. Abstract group can be perceived as a collection of a class of objects, which have binary operator relations similar to multiplication; such operations satisfy the following properties:

⊖ Closure. If a and b are in the group, then a·b is also in the group;

⊖ Associativity. If a, b and c are in the group, then (a·b)·c= a·(b·c).

⊖ There is an element I of the group such that for any element a of the group a·I=I·a=a;

⊖ For any element a of the group there is an element a-1 such that a·a-1=a-1·a=I.

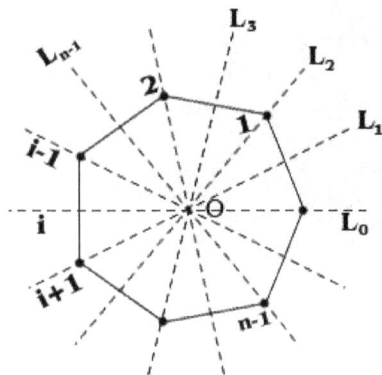

Fig. (3). Regular n quadrate.

In general, group theory is a universal tool to describe the symmetry ??. In the late 19th century, group theory was applied to study crystal structure, which had a wide range of applications in many areas. In modern physics, group theory has become a powerful tool for the study of elementary particles, and quantum mechanics. Algebra also experienced a renaissance due to the introduction and development of the concept of group theory [14].

2. Permutation Groups

In theoretical research and experimental application, permutation groups hold immense importance . On one hand, any finite group can be expressed by them; On the other hand, permutation groups are used to solve the problem of algebraic equations using square root. Permutation groups play an important role in the certification and application of Burnside lemma and Pólya theorem,. Permutation groups have a wealth of practical background, for example, these groups are used for the observation of proofs and application of polygon symmetry by observing all possible rotation and flipping that let polygon back to the original state, to obtain the permutation group called dihedral group [15].

3.2. Dihedral Group

Suppose X={0,1,2,...n-1}(without loss of generality, 0~n-1 was taken as sequence number) is the vertex set of the regular n(n>=3) quadrate and arranged counter-clockwise, as shown in Fig. (3).

When the regular n quadrate was rotated according to 2 π/n counter-clockwise, vertex i moved to the position originally occupied by vertex i+1(mod n), sothis rotation resulted in the conversion on X, marked as

R_1:

$$R_1 = \begin{pmatrix} 0 & 1 & 2 & \cdots & n-1 \\ 1 & 2 & 3 & \cdots & \cdots & 0 \end{pmatrix} \quad (1)$$

The conversion according to 2kπ/n marked R_k :

$$R_k = \begin{pmatrix} 0 & \cdots & n-1 \\ k & \cdots & \cdots & k+n-1 \end{pmatrix} \quad (2)$$

Where the addition and subtraction are the modulo n operation (as the same for the entire paper); R_1 is the identity; and R_k can be shown as:

$$R_k(i) = k + 1, i = 0,1, \ldots , n - 1 \quad (3)$$

Another conversion is reflection in the symmetric axis according to π, named reflectivity conversion. Because of n symmetric axis, the axis was marked through vertex 0 as L_0, and the axis through the vertex of the midpoint of edge marked [0,1] as L_1 ,..., until L_{n-1} . The corresponding reflectivity conversion was marked as $M_0, M_1, \ldots , M_{n-1}$. For instance,

$$M_n = \begin{pmatrix} 0 & 1 & \cdots & n-1 \\ 0 & \cdots & n-1 & \cdots & \cdots & 1 \end{pmatrix} \quad (4)$$

We can prove that

$$M_k = k + n - i, k = 0,1, \ldots n - 1 \quad (5)$$

Let

$$D_n = \{R_k, M_k | k = 0,1, \ldots n - 1\} \quad (6)$$

Following this, the D_n was closed under the composite operation of the conversion with the identity R_0 and each element was inversed , as a result, D_n formed the dihedral group [16].

3.3. Role of Group on Collection and Necklace Problem

The concept and theory of roles of groups in the collection provide theoretical basis for their application, and also provide powerful tools for the analysis of finite group. Let X={1, 2, ..., n}, G was a permutation group on X, for any g $\in$G and x $\in$X, call g(x) was the group element with g acting on x, and group G acting on collection X. X is called the target set. Here, g(x) represents a reversible transformed g corresponding to X. The notion of permutation group on the role of the target set can be popularized to pop groups, and both Burnside lemma and Pólya theorem can solve the problem of track number (equivalence class) of collection acted on the group [15].

Necklace problem is described as making an n bead necklace using beads of m colors, so how many different types of necklace can be made?

When solving the necklace problem using Pólya theorem, initially the n-bead necklace is abstracted towards regular polygon of n vertices, and regular polygon exports the dihedral group, the equivalence necklace program number can be derived using Pólya theorem.

3.4. Nucleotide Free Triplet Vertex Set and Free Triplet Coding Complete Works

Nucleotide free triplet was abstracted to equilateral triangle (Fig. 4), whose vertex set can be defined as X={1,2,3}. The set composed of four nucleotides is defined as B={U,C,A,G} Each mapping f: X→B represents a free triplet coding (it represents a label necklace in necklace problem). Let Ω={f|f:X→B}=BX i.e. the coding completes the function of all free triplets, apparently Ω=|B||x|=|4|3=64,

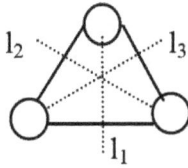

Fig. (4). Abstract model of free triplet.

Table 1. Elements of D3 group and the rotation numbers.

Elements of D₃ Group	Group Elements Rotation Resolving Expression	Rotation Numbers in Group Elements λ(gₖ)
g_0	(1)(2)(3)	3
g_1	(1 2 3)	1
g_2	(1 3 2)	1
g_3	(1)(2 3)	2
g_4	(2)(1 3)	2
g_5	(3)(1 2)	2

they are fit with the triplet genetic code set (see in the 9th reference) exactly when listed one by one.

3.5. 20 Equivalence Classes of Triplet Code Derived Using Group Theory

The dihedral group (D₃ group) was calculated corresponding to the triplet code, and the group elements R_k, M_k(k=0,1,2) as g_k(k=0,1,...,5) were represented as shown in Table 1 [17]. The equivalence class number of triplet code can be solved directly using Pólya theorem.

Pólya Theorem: Let G be a permutation group of the set of n objects, and n objects be dyed using m colors, then the number of nonequivalent colorings is given by

$$L_G = \frac{1}{|G|}\sum_{q_k \in G} m^{\lambda(g_k)} \tag{7}$$

Where G={$g_1,g_2,...,g_p$}, $\lambda(g_k)$ is the number of rotations in the permutation g[K][15].

Solving the triplet code equivalence class is equivalent to the necklace problem of three beads dyed by four colors, where m=4, |G|=6, $\lambda(g_k)$ (k=0,1, ..., 5) as shown in Table **2**, these values are substituted in equation (7)

$$L_{D_1} = \frac{1}{6}(4^3 + 2 * 4^1 + 3 * 4^2) = 20 \tag{8}$$

Therefore, the triplet code set (64 code) can be divided into 20 equivalence classes using group theory.

Table 2. The triplet code table divided by 20 equivalence classes.

Equivalence Classes Number	Representative Elements	Triplet Code in Equivalence Classes	Code Number in Equivalence Classes	Subtotal
1	UUU	UUU	1	
2	CCC	CCC	1	
3	AAA	AAA	1	4
4	GGG	GGG	1	
5	UCC	UCC,CCU,CUC	3	
6	UAA	UAA,AAU,AUA	3	9
7	UGG	CGG,GGU,GUG	3	
8	CUU	CUU,UUC,UCU	3	
9	CAA	CAA,AAC,ACA	3	9
10	CGG	CGG,GGC,GCG	3	
11	AUU	AUU,UUA,UAU	3	
12	ACC	ACC,CCA,CAC	3	9
13	AGG	AGG,GGA,GAG	3	
14	GUU	GUU,UUG,UGU	3	
15	GCC	GCC,CCG,CGC	3	9
16	GAA	GAA,AAG,AGA	3	
17	UCA	UCA,CAU,AUC,ACU,UAC,CUA	6	
18	UCG	UCG,CGU,GUC,GCU,UGC,CUG	6	
19	UAG	UAG,AGU,GUA,GAU,UGA,AUG	6	24
20	CAG	CAG,AGC,GCA,GAC,CGA,ACG	6	

3.6. Representative Elements of 20 Equivalence Classes and Division of Equivalence Classes

64 triplet codes were divided into 20 equivalent classes and the results are shown in Table **2**.

Initiating, equivalence , any triplet code in every equivalence class can be used as representative element, As shown in Table **2**, triplet codes in every equivalence class had the same nucleotide composition. It is important for the analysis of triplet code and protein composition.

CONCLUSION AND FUTURE WORK

In this paper, the nucleotide free triplet conjecture of nucleotide is proposed and the research based on the abstract model of nucleotide free triplet using group theory method was conducted. It has been found that the total number of mapping from free triplet set of vertices to nucleotide collection was 64, which matches exactly with the element number of the triplet code set (64). 64 triplet codes were divided into 20 equivalence classes using group theory (Pólya Theorem), which were different from the triplet code set for 20 amino acids, but the inner link needs to be further studied in depth.

Studies have shown that there are 20 kinds of amino acids in nature and each amino acid can be represented by more than one triplet code. The triplet codes representing the same amino acid are generally different in the third base, which is called "silent codon". Chinese writer Lu Xun said that "To explode in silence, or to die in it". The secret role of "silent codon" may one day be found like recessive gene.

ACKNOWLEDGEMENTS

This work was supported by the National Natural Science Foundation of China (NSFC) (61272006).

REFERENCES

[1]   Zhu Y, Li Y, Zheng Z. Modern Molecular Biology, 6th ed, Beijing, China: Higher Education Press 2011; pp.108-9.

[2]   Baranov PV. Codon size reduction as the origin of the triplet genetic code. Plos One 2009; 4: 1-9.

[3]   Petra S, Lucila OM. A robust feature selection method for novel pre-microRNA identification using a combination of nucleotide-structure triplets. IEEE 2nd Conference on Healthcare Informatics, Imaging and Systems Biology: California, 2012; p. 61.

[4]   Wang K, Schmied WH, Chin JW. Reprogramming the Genetic Code: From Triplet to Quadruplet Codes. Angew Chem Int Ed Engl 2012; 51(10): 2288-97.

[5]   Fernando A, Michael F. Symmetry breaking in the genetic code: finite groups. Math Comp Model 2011; 53: 1469-88.

[6]   Tsvi T. A colorful origin for the genetic code: Information theory, statistical mechanics and the emergence of molecular codes. Phys Life 2010; 7: 362-76.

[7]   Wu B. Evolution of the genetic triplet code *via* two types of doublet codons. J Mole Evol 2006; 63: 54-64.

[8]   Sherwood GB. Functional dependence and equivalence class factors in combinatorial test designs. IEEE 7th International Conference on Software Testing, Verification and Validation Workshops: United Kingdom 2014; pp. 108-17.

[9]   Lu Y, Cha J. A fast algorithm for identifying minimum size instances of the equivalence classes of the pallet loading problem. Eur J Oper Res 2014; 237: 794-801.

[10]   Petsko GA, Ringe D, Xiaochun GT. Proteins structure and function. Beijing: Science Press 2009, pp. 6-7.

[11]   Zheng T, Yao F. Science and the mysteries of life. Knowledge Beijing, Press 2003, pp. 28-29.

[12]   Waston JD. Molecular biology of the gene. 6th ed. Beijing: Science Press 2010, pp. 40-60.

[13]   Sylvain WL, Wang Y. Introduction to molecular biology. Beijing: Chemical Industry Press 2008, pp.34-6.

[14]   Li W. Introduction to the history of mathematics. 3rd ed. Beijing: Higher Education Press 2005, pp. 211-2.

[15]   Vanlint JH. Combinatorial mathematics tutorial. Beijing: Mechanical Industry Press 1993, pp. 189-207.

[16]   Hu G. Application of modern algebra. 2nd ed. Beijing: Tsinghua University Press 1999; pp. 47-9, 103-8.

[17]   Zhang D. Study of key technology of three-dimensional for LEO/MEO satellite networking based on virtual reality. MS thesis, Shenyang: Northeastern University 2004, pp. 26-7.

22

Study of Intelligent Agricultural Cultivation Management Plan Model based on Geographic Information System

Chuang Lu, Bo Wang*, Xiu-Yuan Peng, Xiao-Lei Hou, Bing Bai and Chun-Meng Wang

Liaoning Academy of Agricultural Sciences, Shenyang, Liaoning, China

Abstract: Management plan model of agricultural planting information technology research and application of design system for agricultural production and digital has important theoretical and practical significance of agricultural planting. The study concluded, extracted the relevant agricultural planting design theory and technology research based on the show, applying the system analysis principle and mathematical modeling technique, the construction and perfection of the cropping system, ecological regionalization, precision farming and productivity of quantitative analysis of the agricultural planting management knowledge model, by using the technology of software component, with GIS as spatial information management platform, the establishment of the digital system design based on GIS and model plant. The system has realized the design of cropping system of regional cropping information standardization management and different levels of for the realization of crop planting design, quantitative and digital laid the foundation.

Keywords: Internet of things, agriculture, agricultural planting management plan, geographic information system.

1. INTRODUCTION

Agricultural cultivation management plan model reflects the management of agricultural production, processing, sales throughout the chain, the realization of all the resources of the whole chain and process management [1]. Embodies the lean production, management and agile production idea, by drawing on the concurrent engineering in industrial production and the concept of agile manufacturing, the macro implementation of synchronization of the management processes in the management of agriculture, and can quickly obtain market demand and respond to guarantee the agricultural product, high quality, implementation of lean production. Plan and control the affair beforehand thinking, prenatal, control to the agricultural industry in the production process, postpartum [2]. Reflect the thought of process management, the scientific definition and Realization of each link of planting industry management, scientific and rational organizational processes, improve efficiency, and ensure the quality of products [3].

Single narrow agricultural information, its content includes agriculture economy information, agricultural science and technology information, the information of agricultural resources, agricultural policy and agricultural related information, with the generalized social information [4]. Agricultural information covers the following aspects: agricultural production and management, the agricultural information acquisition and processing of agricultural expert system,

simulation of agricultural system, decision support system for agriculture, agricultural network. Including the application of agricultural information technology in information storage and processing: computer, communication, network, multimedia, artificial intelligence. In general has network the whole process, comprehensive and other features [5, 6].

According to the characteristics of agriculture information space and dynamics, the GIS technology will be applied to agriculture information management, attribute and spatial database were established, and the attribute database and spatial database of interrelated; taking Visual Basic as the development platform, the independence of the component based software development environment, the spirit of the upgrade, convenience development of integration, explore GIS and model coupling technology, the further development of digital agriculture cropping design system based on GIS and model. System consisted with file management, map operation, information query, planting design, ecological regionalization, productivity analysis, precise management, expert consultation, system maintenance, system help and other major function.

2. RELATED THEORY AND RESEARCH STATUS

2.1. The Existing Problems in Our Country's Agricultural Production

The current production and management of agriculture in China is relatively backward, agricultural information resource sharing is poor, agricultural production process controllability, low quality, lack of agricultural products,

*Address correspondence to this author at the Liaoning Academy of Agricultural Sciences, Shenyang, Liaoning, China;
E-mail: aidilunzi@hotmail.com

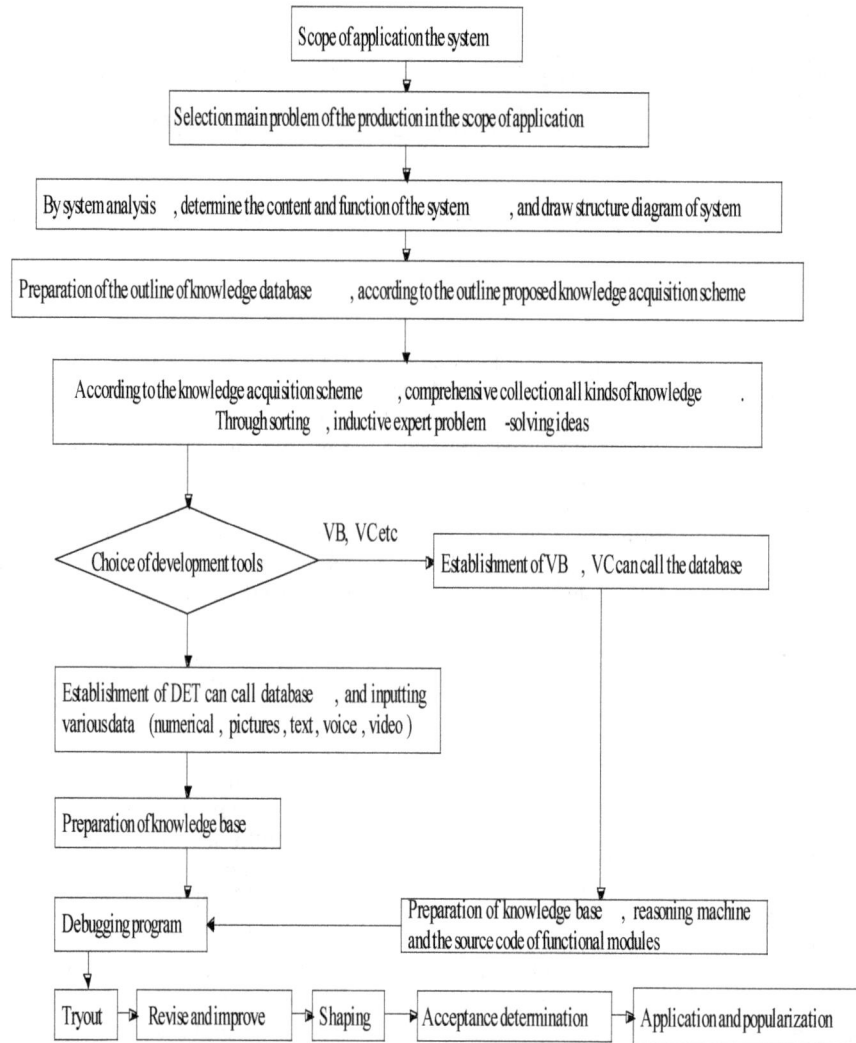

Fig. (1). The procedures and knowledge acquisition model.

agricultural products market demand peasants. Because the agricultural production and natural environment are closely linked, the fundamental way to improve agricultural productivity is the scientific production and scientific management. Because the standard production namely to agricultural product quality standards as the goal, to organize agricultural production [7].

It is an important means of organization, implementation, evaluation of the modernization of production, is the scientific management of the important content [8-10]. So the agricultural development is the inevitable trend of standardization. By drawing on the standardization management experience in industrial production, according to the production characteristics of agriculture, exploring an Internet of things technology as the core of the new planting management and mode of production - agricultural planting management plan (ACMP) model, based on information technology, network technology and agricultural production and management needs, with systematic management thought, provide a decision tool management platform for agricultural production decision layer and implementation layer, realize safety production, information management [11].

2.2. Agricultural Information Technology

Artificial intelligence research is an important aspect of information construction, and also is one of the hotspot research. Artificial intelligence is regarded as one of the century three major scientific and technological achievements, expert system and the middle of the 60's began to appear, as an application field of artificial intelligence, the research of artificial intelligence from the solid face room into the real world, the application of artificial intelligence technology is currently recognized as the most widely used and the most successful in the field. Foot l expert system is also called intelligent system, is a computer system program design method based on knowledge built up. It integrates the major achievements and experience of experts in a particular field, and can use these knowledge such as a human expert, make decisions through the simulation of human experts to solve complex problems. The basic structure of expert system is a rule based system, including knowledge base, inference engine, database, man-machine interface, to explain the procedures and knowledge acquisition (Fig. 1).

3. AGRICULTURAL CULTIVATION MANAGEMENT PLAN MODEL

3.1. Climate Adaptability Evaluation

a) Light evaluation model

Solar radiation is the ability to source of agricultural ecological system, the amount of income and population transfer efficiency determines the level of crop colony productivity, this research according to the relationship between energy and crop growth and yield formation, choose daily radiation and percentage of sunshine as light evaluation index.

(1) Evaluation model of daily radiation

See evaluation equation of daily radiation model:

$$LSI_i = \begin{cases} 0 & (ASH_i \leq CSH_i) \\ \frac{ASH_i - CSH_i}{OSH_i - CSH_i} & (CSH_i < ASH_i < OSH_i) \\ 1 & (ASH_i \geq OSH_i) \end{cases} \quad (1)$$

In the formula, LSIi is the daily radiation membership values; CSHi crop critical daily radiation (MJ/m^2); the amount of solar radiation ASHi practical (MJ/m^2); OSHi is the most suitable daily radiation (MJ/m^2); i for crop growth stages, i=1,2, ..., 5.

(2) The percentage of sunshine evaluation model

The percentage of sunshine evaluation model was seen equation (2):

$$RSI_i = \begin{cases} 0 & (ASR_i \leq CSR_i) \\ \frac{ASR_i - CSR_i}{OSR_i - CSR_i} & (CSR_i < ASR_i < OSR_i) \\ 1 & (ASR_i \geq OSR_i) \end{cases} \quad (2)$$

In the formula, RSIi is the percentage of sunshine membership values; CSRi crop critical percentage of sunshine; ASRi for the actual percentage of sunshine; OSRi is the most suitable percentage of sunshine.

b) Temperature evaluation model

Three basis points in crop life activity process (optimum temperature, maximum temperature and minimum temperature) and temperature index make crops suffer death or together known as the five basic points of temperature on crop, according to the ecological characteristics of crops and the principle of fuzzy mathematics, can build the temperature of the membership function, see equation (3)

$$TSI_i = \begin{cases} 0 & (AT_i \leq LTT_i, AT_i \geq HTT_i) \\ \frac{0.8 \times (AT_i - LTT_i)}{OTLL_i - LTT_i} & (LTT_i < AT_i < OTLL_i) \\ \frac{AT_i - OTLL_i}{OTT_i - OTLL_i} + \frac{0.8 \times (AT_i - OTT_i)}{OTLL_i - OTT_i} & (OTLL_i < AT_i < OTT_i) \\ 1 & (AT_i - OTT_i) \\ \frac{AT_i - OTUL_i}{OTT_i - OTUL_i} + \frac{0.8 \times (AT_i - OTT_i)}{OTUL_i - OTT_i} & (OTT_i < AT_i < OTUL_i) \\ \frac{0.8 \times (AT_i - HTT_i)}{OTUL_i - HTT_i} & (OTUL_i < AT_i < HTT_i) \end{cases} \quad (3)$$

c) Evaluation of precipitation model

Seasonal distribution of precipitation amount and the years of the satisfaction degree determines the length of the growing season of crops and crop water requirement, which determines the crop production and distribution. In addition, precipitation suitability, depends not only on the precipitation depends on how much and how much water at the crop, and the effective degree of precipitation and the relative rate of change. For the early crop, excessive rainfall will cause waterlogging. This research chose the characterization of total precipitation and precipitation rhythm of daily precipitation and the ten day precipitation days two factors to establish the evaluation model of precipitation.

See evaluation equation of daily precipitation model (4):

$$PSI_i = \begin{cases} 1 & (AR_i \leq CR_i) \\ \frac{AR_i - ER_i}{CR_i - ER_i} & (CR_i < AR_i < ER_i) \\ 0 & (AR_i \geq ER_i) \end{cases} \quad (4)$$

In the formula, CRi is the most suitable for daily precipitation, ARi is the actual daily precipitation (mm); ERi for wet injury lethal critical daily precipitation (mm).

d) Climate adaptability evaluation model

Indices of climate adaptability of crops is made of light, temperature, precipitation and other factors common decision, in a certain environment or some kind of creature, not all factors have the same importance. In addition, according to the law of the minimum factor, the growth of the organism is determined by the number of the missing factor. Therefore, when the membership function of a one factor is less than or equal to 0, and the comprehensive effects of other factors, also established the following climate adaptability index model (equation 5) accordingly.

$$CSI = \begin{cases} 0 & (\forall K_j \leq 0, j = 1,2,...,n) \\ \sum_{j=1}^{n} K_j W_j & (\forall K_j > 0, j = 1,2,...,n) \end{cases} \quad (5)$$

Among them, CSI is the climate adaptability index; Kj index the adaptability of J evaluation indexes; Wj is the weight of the first j evaluation indexes.

3.2. Put Forward the Concept of Agricultural Planting Management Plan Model

According to the integrated management of agriculture antenatal, production, post natal needed in agricultural resources, namely agricultural production, processing, sales of integrated management system, based on computer information technology, networking technology, agricultural planting process management, follow the standardization of agricultural production requirements, a plan of agricultural planting management model, the realization of prenatal, agricultural production in producing, postpartum information, technology, materials, management of whole process management, the full realization of the rational allocation of agricultural resources.

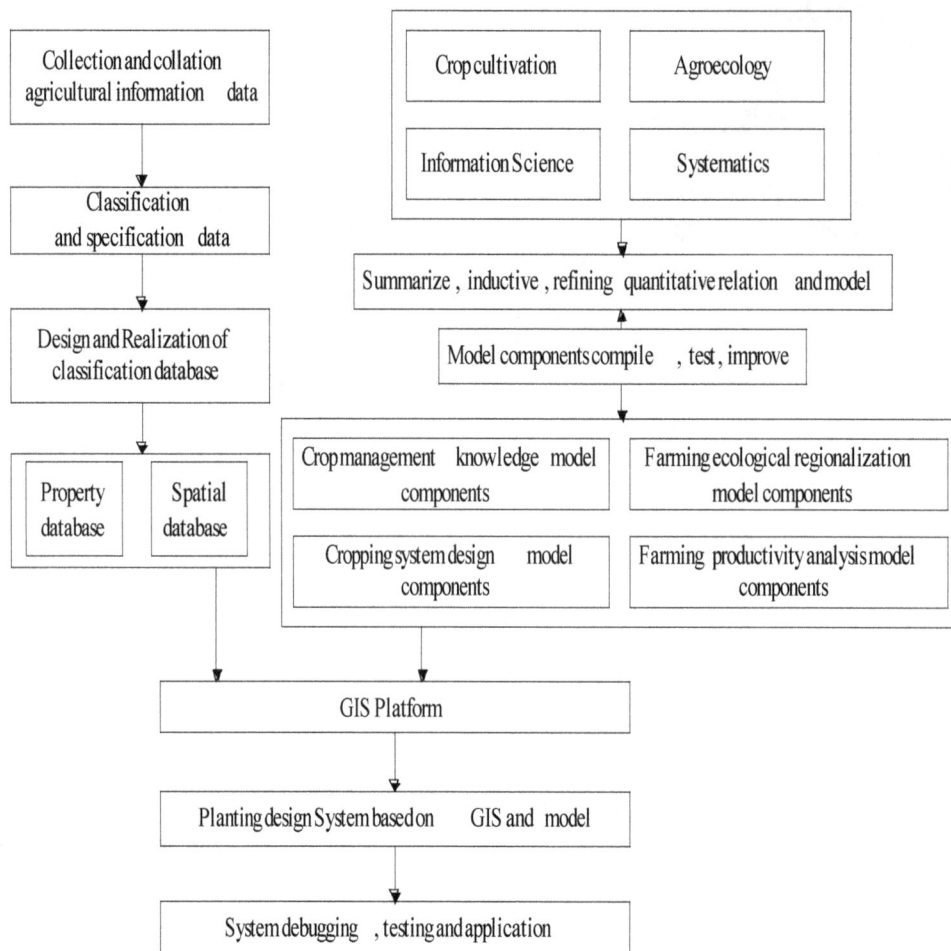

Fig. (2). The technical route.

a)   Pre production

On the basis of this model based on Internet of things technology of real time acquisition of planting production data, to obtain the current has cultivated crop information, including the planting of species and have the planting area, planting area and other information, according to the mining market information analysis, planting area of soil moisture, soil testing with Fang suggestions such as the information, with the expert consultation, reasonable arrangement of planting plan, and follow the material standardization requirements, prepare the required data of agricultural production, ready for production.

b)   In the process of production

On the process of agricultural production management, to ensure that the crops planted in accordance with the standard requirements, realize the whole course standard of crop production, standardized operation. The production of information can also be used to support agricultural resources survey, land suitability investigation, land utilization, agricultural regional planning, agricultural output estimation. Manage production management based mainly on planting standards on production environment factor of the acquisition, storage, monitoring, analysis, early warning and control.

c)   After production

To realize information management of production obtained in the crops, transport, processing, storage and sale process. The analysis results of the crop production data, the planting area and yield of crop varieties and obtain information through the technology of the Internet of things, the macro-control of the market of agricultural products to provide data support, to provide efficient and transparent information guarantee for the trade of agricultural products, production and marketing to achieve docking, reduce agricultural risks, ensure agricultural reproduction. Data mining analysis of agricultural product market information, provide information assurance determined for agriculture antenatal work, and the formation of antenatal, production, post natal in information circulation.

3.3. The Idea of Management of Agricultural Planting Management Plan Model

According to the basic ideas of the system the research content and the research work setting, established the technical route of this study (Fig. **2**).

Table 1. Meteorological data table structure.

Number	Field Name	Unit	Data Type	Range	Default Value	Remarks
1	Country Names	—	nvarchar	—	An area	Corresponding Region fields
2	Date	—	smalldatetime	—	2000/01/01	Above field as the primary key
3	Min temperature	℃	float	[-50, 40]	0.0	—
4	Max temperature	℃	float	[-45, 45]	25.0	Max temperature> Min temperature
5	Sunshine hours	h	float	[0, 24]	6.0	—
6	Rainfall	mm	float	[0, 500]	0.0	—
7	SmID	—	int	—	1	Non-null

4. EXPERIMENTAL RESULTS

4.1. The Evaluation Factors and Factor Selection

On the basis of evaluation of agricultural environmental quality model in national standard and the selection of high toxicity, easy accumulation of crop material as the evaluation factors in the evaluation principle, selection of factors are improved, put forward the concept of key evaluation factors and optional evaluation factor, can suit one's measures to local conditions in order to properly increase the number of optional evaluation factor at the same time to determine the critical evaluation factors the evaluation, has great rationality and superiority, the evaluation results more close to the o-bjective reality, constructs the evaluation factor set elements U:{U1, U2, U3} three single environmental factors set as shown, which underlined factor for optional evaluation factors.

$$U_{soil} = \{C_r, C_d, A_s, P_b, H_g, \\ HCH, DDT, Ni, Cu, Zn\} \tag{6}$$

$$U_{irrigating\ water} = \{BOD, Fluoride, Hg, Cr^{6+} \\ , Cd, Pb, As, Chloride, Total\ cyanide\} \tag{7}$$

$$U_{air} = \{SO_2, NO_x, Fluoride, TSP\} \tag{8}$$

The prediction based on the fuzzy theory, by reducing half echelon distributing method of partial small fuzzy distribution of each single evaluation factors, establish the membership degree of each class standard, the membership calculation formula is as follows:

When the quality level of j=1:

$$r_{ij} = \begin{cases} 1 & (C_i < S_{i1}) \\ (S_{i2} - C_i)/(S_{i2} - S_{i1}) & (S_{i1} \le C_i < S_{i2}) \\ 0 & (C_i > S_{i2}) \end{cases} \tag{9}$$

When the quality level of j=m:

$$r_{ij} = \begin{cases} 0 & (C_i < S_{im-1}) \\ (C_i - S_{im-1})/(S_{im} - S_{im-1}) & (S_{im-1} \le C_i < S_{im}) \\ 1 & (C_i > S_{im}) \end{cases} \tag{10}$$

When the quality level of 1<j<m:

$$r_{ij} = \begin{cases} 0 & (C_i < S_{ij-1}) \\ (C_i - S_{ij-1})/(S_{ij} - S_{ij-1}) & (S_{ij-1} \le C_i \le S_{ij}) \\ (S_{ij-1} - C_i)/(S_{ij+1} - S_{ij}) & (S_{ij} \le C_i \le S_{ij+1}) \\ 1 & (C_i > S_{ij+1}) \end{cases} \tag{11}$$

In the formula , r_{ij} for the i evaluation factor of the j grade membership, C_i for the i evaluation factor of measured value, S_{i1}, S_{i2} , S_{ij-1}, S_{ij}, S_{ij+1}, S_{im-1}, S_{im} respectively for the i evaluation factor of level $1,2$,$j-1,j+1,m-1,m$ standard value, other by analogy.

4.2. The Development and Realization of System

The main meteorological data meteorological data tables are stored in different years throughout the year, including area SmID, name, date, the daily maximum temperature, daily minimum temperature, sunshine duration and rainfall, the table structure as shown in Table **1**:

4.3. Application and Analysis of the Instance System

Agricultural cultivation management plan model effecti-vely save labor, technology, management cost, improve production efficiency, the management level and the quality of agricultural products; accurate environmental monitoring in production factor, improve the scientific and technological content of agricultural production process, implementation of science and technology in the agricultural production pro-cess of full penetration of informationization, industrializa-tion of agriculture, comprehensive rural management infor-mation, the realization of green, organic agricultural production, the planting scale and industrialization of

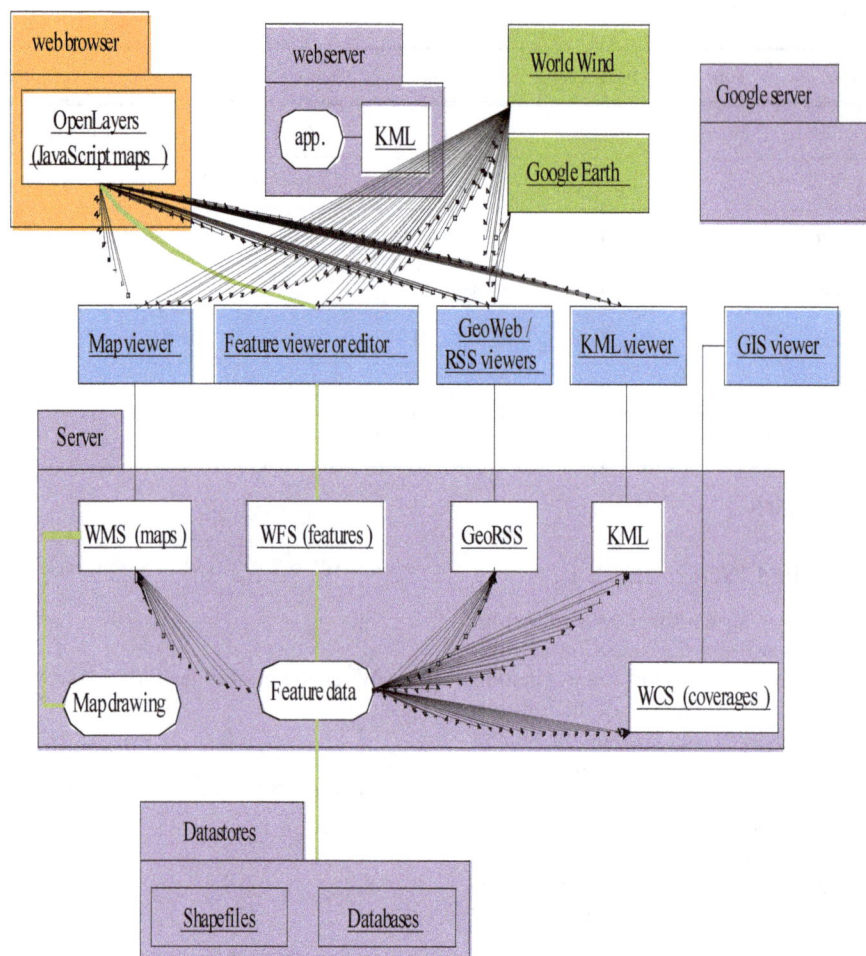

Fig. (3). The full width and Hawkeye map display function chart.

agricultural technology extension; broaden the channels and channels of communication, the realization of agricultural production guidance, such as remote diagnosis, inadequate supplement agricultural technology service; implementation of safety traceability of agricultural products; through the environmental monitoring in production, achieve the varie-ties of cross regional planting, enhanced weakening varieties of environmental adaptability; to crop growth conditions, growth trend, production is predicted or simulated analysis, provides the basis for the agricultural product market macro-control; promote green, organic agricultural production, planting scale and factory.

The system has realized the basic operating functions map GIS system has, including the map to enlarge, narrow, roaming, eagle eye area and distance calculation, display control, map switch. Fig. (**3**) is the full width and Hawkeye map display function chart.

production, post natal in integrated management, market forecast: agricultural products, planting guide, information market of agricultural means of production; production: a-gricultural regionalization, the expert system to the needs of agricultural production based on information, through infor-mation technology to realize collection, storage, monitoring, warning, analysis and display, control of environmental pa-rameters in the production process, so as to realize the gui-dance and control for the whole production process of plan-ting industry; postpartum: agricultural product information and market of agricultural products, agricultural products logistics. The agricultural management, production, management and decision plays a guidance and support role. To further promote the agricultural production, agricultural resources information, precision agriculture to achieve scien-tific decision-making.

CONCLUSION

Prenatal to the chain of ACMP based on Internet of things technology really realize agriculture antenatal,

ACKNOWLEDGEMENTS

We gratefully acknowledge financial support from Science and Technology Research Project of Liaoning,

China (No.2013301008); the National Science and Technology Support Project of China (No. 2014BAD12B04) and the Special Scientific Research Fund of Agricultural Public Welfare Profession of China (No.201303125); the Hunan Provincial Natural Science Foundation of China (Grant No. 2015JJ5025)

REFERENCES

[1] Li Hang, Chen Houjin, "Key technology and application prospect of the internet of things", Forum on Science and Technology in China, January 2011.

[2] Du Hongli, Wu Jun, Yu Hong, "The research of enterprise's SCM based on internet of things". Logistics Sci-Tech, March 2011.

[3] Xing Zhiqing, Fuxing, "Study on the internet of things in modern agricultural production", Agricultural Technology & Equipment, August 2010.

[4] S. C. Kim, I. Song, S. Yoon and S. R. Park, "DOA estimation of angle-perurbed sources for wireless mobile communications", IEICE Trans. Communication, vol. E83-B, No.11, 2000, pp.2537-2541.

[5] J.X.Wu, T.Wang, Z.Y.Suo, et al, "DOA estimation for ULA by spectral Capon rooting method", Electronics Letters, vol.45, No.1, 2009, pp.84-85.

[6] J.M.Xin and S.A, "Linear prediction approach to direction estimation of cyclostationary signals in multipath environment", IEEE Transactions on Signal Processing, vol.49, No.4, 2001, pp.710-720.

[7] E.Grosicki, K. Abed-Meraim and K.Y.Hua, "A weighted linear prediction method for near-field source localization ", IEEE Transactions on Signal Processing, vol.53, No.10, 2005, pp.3651-3660.

[8] T.B.Lavate, V.K.Kokate and A.M.Sapkal, "Performance analysis of MUSIC and ESPRIT DOA estimation algorithms for adaptive array smart antenna in mobile communication", International Journal of Computer Networks, vol.2, No.3, 2010, pp. 152-158.

[9] A.Hirata, T.Morimoto and Z.Kawasaki, "DOA estimation of ulta-wideband EM waves with MUSIC and interferometry", IEEE Antennas and Wireless Propagation Letters, vol.2, No.1, 2003, pp.190-193.

[10] F.Taga, "Smart Music algorithm for DOA estimation", Electronics Letters, vol.33, No.3, 1997, pp.190-191.

[11] W.Sun, J.L.Bai and K.Wang, "Novel method of ordinal bearing estimation for more sources based on obique projector", Journal of Systems Engineering and Electronics, vol.20, No.3, 2009, pp.445-449.

23

Alternating Magnetic Field Prior to Cutting Reduces Wound Responses and Maintains Fruit Quality of Cut *cucumis melo* L. cv Hetao

Jin Jia, Xiujuan Wang, Junli Lv, Shan Gao and Guoze Wang[*]

School of Mathematics, Physics and Biological Engineering, Inner Mongolia University of Science and Technology, no. 7, Aerding Street, Kundulun District, Baotou, CN-014010 Inner Mongolia, China

Abstract: The quality of fruit targeted for human consumption is affected greatly by the methods used in processing. The objective of this study was to assess a method of treating cantaloupe melons to reduce the wound responses incurred by cutting while maintaining fruit quality. After being treated with an alternating magnetic field (AMF) at the strength of 2mT for 0, 5, 10, 15, 20, or 25 min, *Cucumis melo* fruits were cut longitudinally into five pieces and stored at 5°C until analysis. The rates of decomposition, electrolyte leakage and respiration, were determined along with fruit firmness, soluble solids and titratable acids. Cutting the melons damaged the quality of fruits, as indicated by lower firmness soluble solids, higher electrolyte leakage, decomposition, respiration and titratable acids in cut slices compared to intact melons. Treatment with AMF before cutting influenced all parameters above, compared to untreated samples, and treatment with 2mT for 15 min resulted in reduced respiration rates, lower electrolyte leakage, delayed softening and decomposition, and reduced flows of soluble solids and titratable acid. Therefore, application of AMF treatment prior to cutting is beneficial for reducing wound responses and maintaining quality and flavor of cut *Cucumis melo* L. cv, which could provide a useful means to increase the market share of this popular fruit crop species.

Keywords: Alternating magnetic field (AMF), Freshly cut, *Cucumis melo* L. cv Hetao, fruit quality.

1. INTRODUCTION

The cantaloupe melon (*Cucumis melo* L. cv Hetao) is a popular and commercially important fruit crop of the Cucurbitaceae family that grows commonly in Inner Mongolia, where it is served freshly cut. Sales trends for longitudinal cuts of the fruit suggest that consumers are willing to pay for the convenience, if quality is perceived to be better than or equal to the quality of uncut melons [1], and if nutritional measures such as the contents of sugars, amino acids, and vitamins are maintained after processing. However, mechanical handling such as cutting, slicing and peeling may cause physical damages during processing, which causes the quality of fruit to deteriorate rapidly [2]. Any such deterioration when melons are cut into individual portions for sale clearly affects the purchasing choices of consumers in stores, and can lead to substantial market losses [3]. Thus, long shelf life has become an important factor in the marketing and modern use of this species. Development of better treatment methods to enable cutting melons during processing could alleviate the damage to fruits, thus enhancing their quality and ultimate food safety. Improving the shelf life of this fruit would also increase its popularity, enabling cantaloupe growers to achieve a larger share of the produce market across China [4].

Cantaloupes are generally valued for their intense aroma and taste, so they have a high commercial value but are highly perishable during their storage life [5]. To broaden the market for them while maintaining the good quality of fruits, several methods have been investigated. For example, the use of an edible coating made of pectin has been tested as a new method to maintain quality while prolonging the shelf life of fresh-cut melon, however that pretreatment masked melon taste and reduced the sensory acceptance scores at the end of the study [6]. Another method that has been tested is the combination of hot water pasteurization of whole cantaloupe with low-dose irradiation of packaged fresh-cut melon to maintain quality [7], but when the temperature was returned to normal, the benefits of staying fresh were no longer maintained [8] so shelf life was not effectively prolonged. Amaro *et al*. [9] showed that 1-methylcyclopropene (1-MCP) can preserve the soluble solids, total phenolics, total carotenoids and carotene contents of fresh-cut cantaloupe, yet some softening occurred, and the treatments were less effective when 1-MCP was applied at a low temperature (0°C) [10].

Thus, there is a need for development of new methods to protect fruits during the cutting process and thus extend their shelf life. Alternating magnetic field (AMF) has potential for this purpose, they are naturally occurring, easy to work with and create no residue or environmental pollution. The use of electromagnetic fields to stimulate the growth and metabolic cascades and control biochemical pathways had been researched, and report had been found that electromagnetic fields have an important role in eliciting the effects of many biological systems [11], and may affect organisms in both

School of Mathematics, Physics and Biological Engineering, Inner Mongolia University of Science and Technology, no. 7, Aerding Street, Kundulun District, Baotou, CN-014010 Inner Mongolia, China

negative and positive manner such as acceleration of growth and metabolism. Experiments using oscillating magnetic fields have uncovered new effects in the living systems of growth [12]. AMFs are also capable of changing a plant's growth rates [13], so their action on various biological systems can be used as effectively as a means of regulating the biological activity of plants and fruits [14]. For example, in the previous study, magnetic fields of different intensities had varying influences on the biology of PC12 cells [15].

Justo *et al.* [16] observed that the *Escherichia coli* had 100 times greater viability when treating with magnetic field for 6.5h than the others without treatment. Magnetic field can change the range of molecular organization in living organisms, as to cell membrane, it changed the membrane permeability, the distribution of protein and lipid domains, and the internal molecular distribution of electronic charge inside lipid molecule, all these changes can affect the rate of biochemical reactions, also the magnetic field have some influence on chloroplast, nucleus, proteins, protoplasm and whole cell [17]. Magnetic treatments may also protect plants against adverse environmental conditions such as stress caused by drought [18]; for instance, they are reportedly effective in determining the distribution of apparent microporosity on apple and tomato fruit [19].

As part of the continuous research efforts to identify useful applications of magnetic fields in food processing, the objective of this study was to conduct the first investigation of the effects of AMF on physiological responses and quality of cut cantaloupe. To do so, several parameters were measured, including electrolyte leakage, respiration rate, fruit firmness, content of soluble solids and titratable acid, and rate of decomposition.

2. MATERIAL AND METHODS

2.1. Plant Materials and Treatments

Cantaloupes were purchased from the city of BaMeng of Inner Mongolia in July. They were carefully selected to have a uniformity of size, color, absence of damage and defects, and with a ripeness of 90%. The instrument used to generate AMF was made in our laboratory. It was comprised of two parallel couples of Helmholtz coil, with the radius of each coil being equal to the distance between them. The two couples were connected in series, and connected to a potential transformer that could transform the direct current power into alternating current power. By adjusting the compensation current of the coils, an AMF was generated with a volume of $40 \times 40 \times 20$ cm (Fig. **1**). A table between the two coils provided a surface where fruits were placed for the treatments (Fig. **2**).

A total of 240 melons were used, with 40 intact melons undergoing AMF treatment at a strength of 2mT, in each of six time durations: 0, 5, 10, 15, 20, or 25 min. After being treated, they were cut longitudinally into six pieces, and the sliced fruits were used in the following experiment. The cut pieces were placed on a plastic tray and covered with 40 μm-thick polyethylene film. All trays, holding six pieces each, were stored at 5°C until the fruit had decomposed. During this time, the physical variables were measured daily.

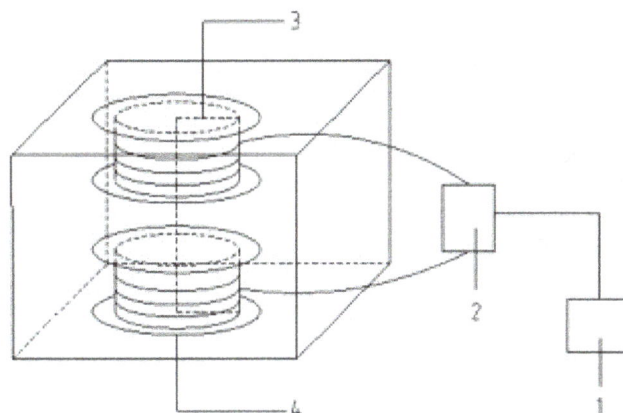

Fig. (1). Schematic drawing of the AMF-generating instrument; 1 is a power source, 2 is a potential transformer, and 3 and 4 are the Helmholtz coils.

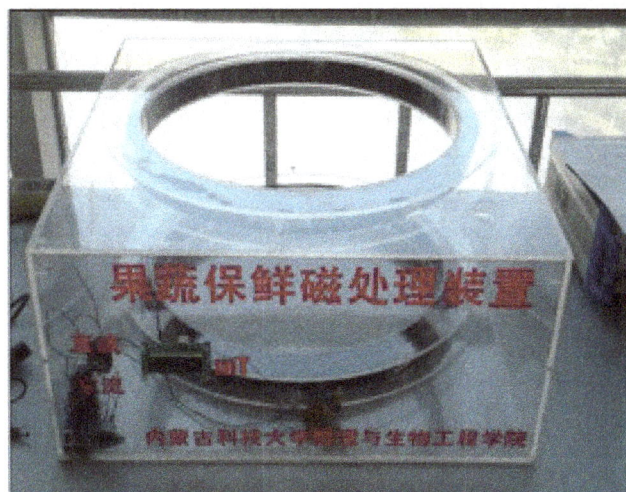

Fig. (2). The finished AMF instrument: 3 intact melons were placed on the table homogeneous for each round of AMF treatment at a strength of 2mT.

2.2. Fruit Firmness and Soluble Solids

The firmness of fruit was measured as the force required to puncture the outer cortex using a Texture Analyzer (QTS-25 Brookfield, USA) fitted with a of 5-mm diameter probe that had a maximum penetration depth of 10 mm. We set the probe to a target value of 5 mm penetration into the fruits at a speed of 30 mm/min. The content of soluble solids was determined using an ATAGO hand-held refractometer.

2.3. Respiration Rate

Three pieces of melon slices were placed in 20 mL of 0.4 mol/L NaOH in a hermetic dryer. The respiration rate of the fruits was calculated according to the method of Alves *et al.* [20] with a minor modification. The amount of CO_2 produced by the melons was calculated by measuring the consumption of NaOH. The respiration rate was expressed as micrograms of CO_2 produced per kilogram per hour (mL kg^{-1}h^{-1}).

2.4. Electrolyte Leakage

For determining electrolyte leakage, a modification of the method reported by Campos *et al.* [21] was used. A stainless steel cork borer of 0.5 cm diameter was used to obtain cylinders of melon tissue. Two 4-mm thick samples were cut from each cylinder. After being rinsed 3 times (2-3 min) with demineralised water, ten pieces were put into 50 ml of demineralised water and shaken at 100 cycles per min for at room temperature. The electrolyte leakage in the solution was measured after 180 min using a Conductance Bridge (DDS-11A, Yamei Electron Instrument Factory, Shanghai, China). Total conductivity was obtained after keeping the flasks in an oven (90°C) for 2 h. Results were expressed as percentage of total conductivity.

2.5. Titratable Acids

The amount of titratable acids was determined following established methods [22]. In brief, we used a titration of 0.1 mol/L NaOH, with phenolphthalein as the indicator of the end point. As 20 g of fluid was extracted from tissue was transferred into a 250 mL volumetric flask with a constant volume, then 25 mL of the solution from the volumetric flask was used as the unit for the statistical analysis of titration results.

2.6. Rate of Decomposition

The rate of decomposition of melon slices was measured after being removed from 5°C to room temperature at 20°C for 24h by visual estimation, we counted the number of melon slices that were decomposed, and then divided by the total number of melon slices in the treatment, and expressed this as a rate out of 100%.

2.7. Statistical Analysis

All experiments were performed in triplicate. Analysis of variance (ANOVA) between treatment means was carried out with the SPSS 17.0 software using Duncan's one-way test at $P<0.05$.

3. RESULTS AND DISCUSSION

3.1. Fruit Firmness and Soluble Solids

The firmness of melon slices that were treated with AMF was higher than the firmness of untreated slices, throughout the entire storage period (Fig. **3**). We found that firmness was highest for all fruits within 3 days, then firmness loss was observed by indicating the lower firmness in untreated slices than that in treated fruit. Further, firmness in fruit treated with AMF for 15 min was 54.9% higher than fruit without AMF treatment on the 4th day. These combined results showed that AMF retarded the softening of fruit tissue. Furthermore, applying AMF for 15 min was the best treatment for maintaining high firmness of melons.

Indeed, firmness is an important parameter that is closely related to fruit maturity, harvest time and quality grade [23]. A decrease in firmness during storage is mainly related to increases in degradative metabolism [24]. Therefore, higher firmness clearly equates to better quality melons. In our

experiment, the softening of melons was effectively delayed when samples were treated with AMF.

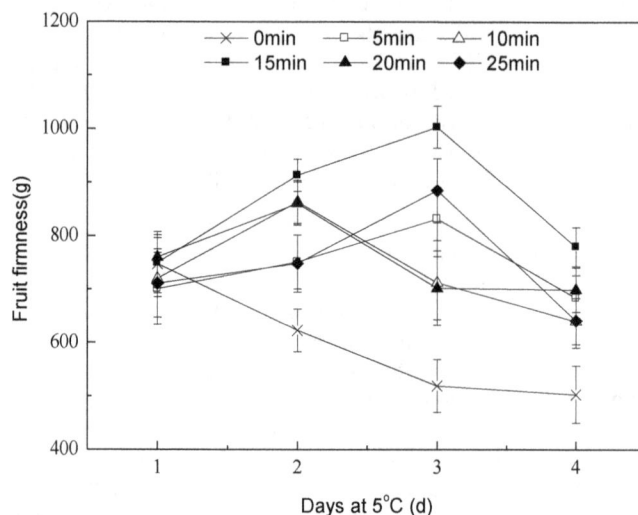

Fig. (3). Changes in the firmness of flesh of cantaloupe slices stored at 5°C. Before storage, melons were treated with an AMF of 2mT for 0, 5, 10, 15, 20, or 25 min, and then were cut longitudinally into six pieces.

Soluble solids content, which provides a reflection of the carbohydrate content of fruits, can be divided into soluble and insoluble components, with the soluble component consisting largely of soluble sugars [25]. Sugar content is an important determinant of fruit quality, for example it is a key factor that determines the eating quality of watermelons [26]. We detected a rapid decrease in soluble solids appeared in all samples during the storage period (Fig. **4**). Melon slices treated for 20 and 25 min showed little change in soluble solids compared to those without treatment. However, when the treatment time was 5 or 10 min, the soluble solids decreased rapidly. The highest levels of soluble solids were found in cut melon treated with AMF for 15min (Fig. **4**). Thus, the treatment of 2 mT for 15 min was able to maintain a high level of soluble solids in melon slices.

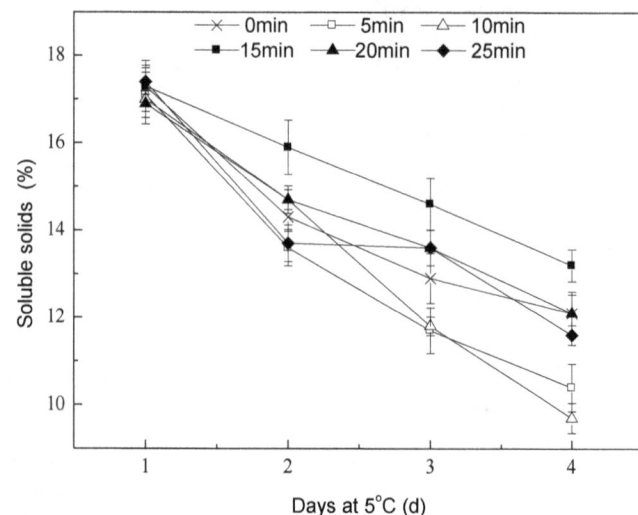

Fig. (4). Changes in the concentration of soluble solids in cantaloupe slices stored at 5°C. Before storage, melons were treated with an AMF of 2mT for 0, 5, 10, 15, 20, or 25 min, and then were cut longitudinally into six pieces.

3.2. Respiration Rate

The pretreatment of soybean seeds with magnetic fields is used in agriculture to achieve enhanced crop growth and increased yields [13]. According to Wills *et al.* [27], the respiration rate of fruits, vegetables and ornamental plants is an excellent indicator of the metabolic activity of tissue, and is therefore an important determinant of a potential product's shelf life.

In our study, treatment times of 5, 10 and 20 min led to higher respiration rates than those measured in untreated melon slices on the 2nd day. CO_2 production in melon slices increased even when they were treated with AMF, which suggested that the quality of melons deteriorated slowly. However, samples treated for 15 min had an obviously lower respiration rate than those that were untreated, until the 4th day (Fig. **5**). All of these results indicated that wounding caused by cutting increased the rate of respiration, but AMF of 2mT for 15 min effectively reduced this wound response.

Some studies have reported that the respiration rate of selected fresh-cut produce is inversely related to shelf life. That is, when the respiration rate of freshly cut products was higher, the shelf life would be shorter, and vice versa [28], such as has been demonstrated with fresh-cut carrots [20]. We have demonstrated that the use of AMF treatment can significantly prolong the shelf life of cantaloupes by slowing respiration.

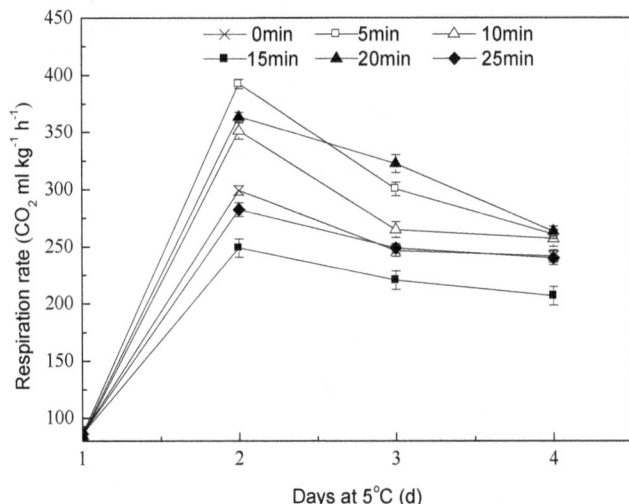

Fig. (5). Changes in the respiration rate of cantaloupe slices stored at 5°C. Before storage, melons were treated with an AMF of 2mT for 0, 5, 10, 15, 20, or 25 min, and then were cut longitudinally into six pieces.

3.3. Electrolyte Leakage

Sliced fruit, treated with AMF for 0 min and 20 min, had the highest electrolyte leakage during the storage period. However, the lowest electrolyte leakage was observed when the treatment time was 15 min during the storage period (Fig. 6). In our study, results showed that electrolyte leakage was stimulated by cutting, which supports previous results that electrolyte leakage is a useful parameter to indicate physical damage, because leakage increases to the plasmalemma when wounds to fruit tissue are incurred [29].

Also, Gerasopoulos *et al.* [30] showed that adverse conditions such as low temperature cause considerable losses of quality during the prolonged shelf life of fruits, apparently related to factors that affect membrane function, and thus, to the more electrolyte leakage. Less electrolyte leakage is consistent with better fruit quality. In this study, AMF of 2mT for 15 min was able to effectively reduce electrolyte leakage and maintain good melon quality.

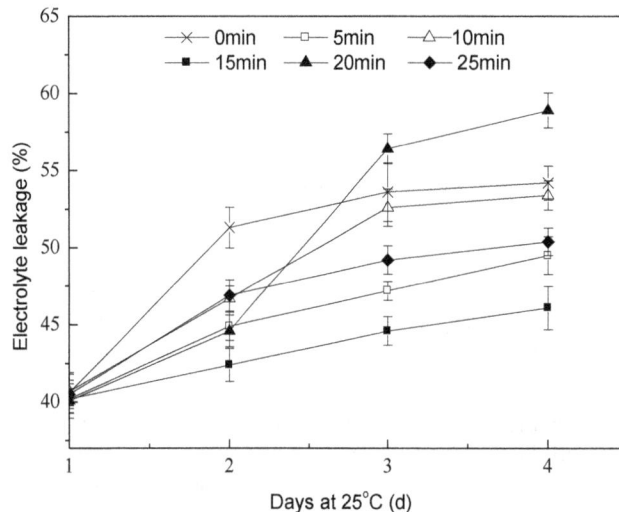

Fig. (6). Changes in the electrolyte leakage from cantaloupe slices stored at 5°C. Before storage, melons were treated with an AMF of 2mT for 0, 5, 10, 15, 20, or 25 min, and then were cut longitudinally into six pieces.

3.4. Titratable Acid

The amount of titratable acid is another one of the key characters that refelct the taste of fruits [31], lower acidity with little changes in that measure may help to maintain the consistency of fruit flavor. In our report, the cut melons treated with AMF all had lower titratable acid than their untreated counterparts during the first day, even though a rise occured at the cut treat group afterwards. In all of the fruits tested, the amount of titratable acid had a greater range when the treatment time was 10, 20 or 25 min. The lowest amount of titratable acid was found when treatment time was 5 min, but it also changed rapidly in that group. However, samples treated for 15 min not only showed smaller changes, but also had the lowest initial measures of titratable acid (Fig. **7**). Similar beneficial results were obtained by Martiñon *et al.* [4]., who found that the titratable acid changed little in fresh-cut cantaloupe, and was able to delay the postharvest ripening process effectively. These findings indicated that AMF of 2mT for 15 min can maintain the desired taste of melons.

3.5. Rate of Decomposition

Tripathi *et al.* [32] reported that fruits with higher moisture content and nutrient composition were highly susceptible to attack by pathogenic fungi, which might make them unsuitable for eating. Clearly, fruit quality is better when the rate of decay is lower. The use of AMF might be able to address this problem by changing enzyme activity,

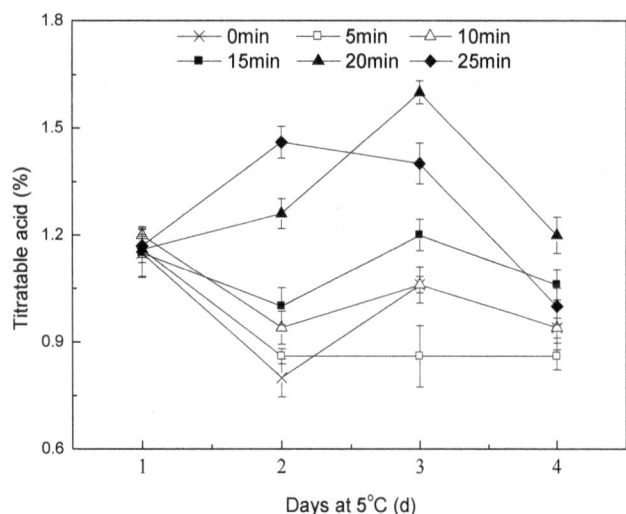

Fig. (7). Changes in the titratable acid in cantaloupe slices stored at 5°C. Before storage, melons were treated with an AMF of 2mT for 0, 5, 10, 15, 20, or 25 min, and then were cut longitudinally into six pieces.

gene expression, and the release of calcium from intracellular storage sites [33]. In our experiment, the rate of decomposition in untreated, cut melons increased faster than the rate in treated, cut melons, within 4 days (Fig. **8**). Samples treated with AMF for 5 and 10 min had a lower decomposition rate, but not significantly so (P > 0.05). A possible reason may be that the periods of AMF treatment were not long enough. When treatment lasted for 15 min, the rate of decomposition was obviously lower. Prolonging the treatment time further, to 20 or 25 min, did not lower the decomposition rate further (Fig. **8**). These results showed that an optimum length of AMF treatments for cantaloupe would be 15 min, to achieve a much lower decomposition rate.

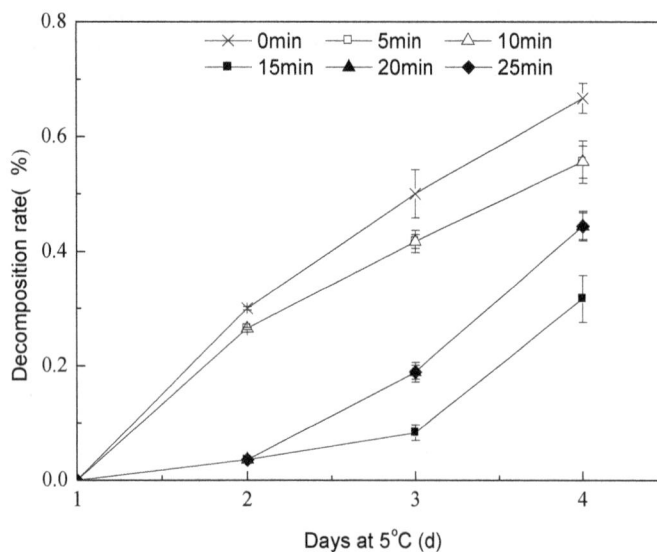

Fig. (8). Changes in the rate of decomposition of cantaloupe slices stored at 5°C. Before storage, melons were treated with an AMF of 2mT for 0, 5, 10, 15, 20, or 25 min, and then were cut longitudinally into six pieces.

CONCLUSION

In conclusion, we demonstrated that AMF treatment for 15 min was able to reduce the respiration rate and electrolyte leakage, delay softening and decomposition rates, and also reduce the flow of soluble solids and titratable acids in cut cantaloupes. Reductions in all of these parameters are related to good quality in cantaloupe fruits. Therefore, the application of AMF at 2mT for 15 min before cutting is the best treatment to reduce wound responses and maintain the quality and good flavor of fruits in this important crop species.

ACKNOWLEDGEMENTS

This research was supported by the National Natural Science Foundation of China (NNSFC-31260406) and Inner Mongolia university of science and technology Innovation Foundation for Young Core Instructor (2014QNGG06).

REFERENCES

[1]   Beaulieu J, Lea JM. Quality changes in cantaloupe during growth, maturation, and in stored fresh-cut cubes prepared from fruit harvested at various maturities. J Am Soc Hortic Sci 2007; 132(5): 720-8.

[2]   Watada AE, Ko NP, Minott DA. Factors affecting quality of fresh-cut horticultural products. Postharvest Biol Tec 1996; 9: 115-25.

[3]   Manohar SH, Murthy HN. Estimation of phenotypic divergence in a collection of Cucumis melo including shelf-life of fruit. Sci Hortic-Amsterdam 2012; 148: 74-82.

[4]   Martiñon ME, Moreira RG, Castell-Perez ME, Gomes C. Development of a multilayered antimicrobial edible coating for shelflife extension of fresh-cut cantaloupe (Cucumis melo L.) stored at 4°C. LWT - Food Sci Tec 2014; 56: 341-50.

[5]   Huang CH, Zong L, Buonanno M, Xue X, Wang T, Tedeschi A. Impact of saline water irrigation on yield and quality of melon (Cucumis melo cv. Huanghemi) in northwest China. Eur J Agr 2012; 43: 68-76.

[6]   Cristhiane C, Ferrari-Claire IGL, Sarantópoulos SM, Carmello-Guerreiro MDH. Effect of osmotic dehydration and pectin edible coatingson quality and shelf life of fresh-cut melon. Food Bio Tec 2013; 6: 80-91.

[7]   Fan XT, Annous BA, Sokorai KJB, Burke A, Mattheis JP. Combination of hot-water surface pasteurization of whole fruit and low-dose gamma irradiation of fresh-cut cantaloupe. J food protect 2006; 69(4): 912-9.

[8]   Chan HT, Tam S, Seo ST. Papaya polygalacturonase and its role in thermally injured ripening fruit. J Food Sci 1981; 46: 190-7.

[9]   Amaro AL, Fundo JF, Oliveira A, Beaulieu JC, Fernandez-Trujillo JP, Almeida DPF. 1-Methylcyclopropene effects on temporal changes of aroma volatiles and phytochemicals of fresh-cut cantaloupe. J Sci food Agr 2013; 93(4): 828-37.

[10]   Villalobos-Acuña MG, Biasi WV, Mitcham EJ, Holcroft D. Fruit temperature and ethylene modulate 1-MCP response in 'Bartlett' pears. Postharvest Biol Tec 2011; 60(1): 17-23.

[11]   Pilla AA, Markov MS. Bioeffects of weak electromagnetic fields. Rev environ health 1994; 10: 155-69.

[12]   Moore RL. Biological effects of magneticfields: studies with microorganisms. Can J Microbiol 1979; 25: 1145-51.

[13]   Radhakrishnan R, Kumari BDR. Pulsed magnetic field: A contemporary approach offers to enhance plant growth and yield of soybean. Plant Physiol Bioch 2012; 51: 139-44.

[14]   Cakmak T, Cakmak ZE, Dumlupinar R, Tekinay T. Analysis of apoplastic and symplastic antioxidant system in shallot leaves: Impacts of weak static electric and magnetic field. J Plant Physiol 2012; 169(11): 1066-73.

[15]   Zhang Y, Ding J, Duan W. A study of the effects of flux density and frequency of pulsed electromagnetic field on neurite outgrowth in PC12 cells. J Biol Phys 2006, 32(1): 1-9.

[16]   Justo OR, Pérez VH, Alvarez DC, Alegre RM. Growth of Escherichia coliunder extremely low-frequency electromagnetic fields. Appl Biochem Biotech 2006; 134: 155-63.

[17]   Hunt RW, Zavalin A, Bhatnagar A, Chinnasamy S, Das KC. Electromagnetic Biostimulation of Living Cultures for Biotechnology, Biofuel and Bioenergy Applications. Int J Mol Sci 2009; 10(10): 4515-58.

[18]   Selim AH, El-Nady MF. Physio-anatomical responses of drought stressed tomato plants to magnetic field. Acta Astronaut 2011; 69: 387-96.

[19]   Musse M, Guio FD, Quellec S, Cambert M, Challois S, Davenel A. Quantification of microporosity in fruit by MRI at various magnetic fields: comparison with X-ray microtomography. Magn Reson Imaging 2010; 28(10): 1525-34.

[20]   Alves JA, Júnior RAB, Boas EVdBV. Identification of respiration rate and water activity change in fresh-cut carrots using biospeckle laser and frequency approach. Postharvest Biol Tec 2013; 86: 381-6.

[21]   Campos PS, Quartin V, Ramalho JC, Nunes MA. Electrolyte leakage and lipid degradation account for cold sensitivity in leaves of Coffea sp.plant. J Plant Physiol 2003; 160: 283-92.

[22]   Xie LJ, Ye XQ, Liu DH, Ying Y. Prediction of titratable acidity, malic acid and citric acid in bayberry fruit by near-infrared spectroscopy. Food Res Int 2011; 44: 2198-204.

[23]   Peng YK, Lu RF. Prediction of apple fruit firmness and soluble solids content using characteristics of multispectral scattering images. J Food Eng 2007; 82(2): 142-52.

[24]   Mao LC, Wang GZ, Que F. Application of 1-methylcyclopropene

prior to cutting reduces wound responses and maintains quality in cut kiwifruit. J Food Eng 2007; 78: 361-5.

[25]   Burdon J, Lallu N, Pidakala P, Barnett A. Soluble solids accumulation and postharvest performance of 'Hayward' kiwifruit. Postharvest Biol Technol 2013; 80: 1-8.

[26]   Jie DF, Xie LJ, Rao XQ, Ying YB. Using visible and near infrared diffuse transmittance technique to predict soluble solids content of watermelon in an on-line detection system. Postharvest Biol 2014; 90: 1-6

[27]   Wills R, McGlasson B, Graham D, Joyce D. Postharvest. An introduction tothe physiology and handling of fruit, vegetables and ornamentals. 4th ed. Adelaide, Australia, 1998; p. 262.

[28]   Waghmare RB, Mahajan PV, Annapure US. Modelling the effect of time and temperature on respiration rate of selected fresh-cut produce. Postharvest Biol Tec 2013; 80: 25-30.

[29]   Antunes MDC, Sfakiotakis EM. Changes in fatty acid composition and electrolyte leakage of 'Hayward' kiwifruit during storage at different temperatures. Food Chem 2008; 110: 891-6.

[30]   Gerasopoulos D, Chlioumis G, Sfakiotakis E. Non-freezing points below zero induce low-temperature breakdown of kiwifruit at harvest. J Sci Food Agric 2006; 86(6): 886-90.

[31]   Vallone S, Sivertsen H, Anthon GE, Barrett DM, Mitcham EJ, Ebeler SE, Zakharov F. An integrated approach for flavour quality evaluation in muskmelon (Cucumis meloL. reticulatus group) during ripening. Food Chem 2013; 139: 171-83.

[32]   Tripathi P, Dubey NK. Exploitation of natural products as an alternative strategy to control postharvest fungal rotting of fruit and vegetables. Postharvest Biol Technol 2004; 32: 235-45.

[33]   Uckun FM. Exposure of B-lineage lymphoid cells to low energy electromagnetic fields stimulates Lyn kinase. J Biochem Chem 1995; 270: 27666-670.

24

A Short Sequence Splicing Method for Genome Assembly using a Three-Dimensional Mixing-Pool of BAC Clones and High-throughput Technology

Xiaojun Kang[1,2,3], Cheng Yang[2], Xuguang Zhao[2], Weiwei Chen[2], and Sifa Zhang[2,*], Yaping Wang[3,*]

[1]State Key Laboratory of Biogeology and Environmental Geology, Wuhan Hubei 430074, P.R. China
[2]School of Computer Science, China University of Geosciences (Wuhan), Wuhan Hubei 430074, P.R. China
[3]State Key Laboratory of Freshwater Ecology and Biotechnology, Institute of Hydrobiology, Chinese Academy of Sciences, Wuhan Hubei 430072, P.R. China

Abstract: Current genome sequencing techniques are expensive, and it is still a major challenge to obtain an individual whole-genome sequence. To reduce the cost of sequencing, this paper introduced a high-throughput sequencing strategy using a three-dimensional mixing-pools based on the cube. Following the strategy, BAC clones were injected into each vertex of the cube, and sequencing of each plane provided information about multiple clones, thereby significantly reducing the cost of sequencing. In addition, Velvet was used to assemble the sequencing data. The scaffold generated from Velvet contained a number of contigs, which were orderless. Therefore, to address this problem, a scaffold assembly algorithm based on multi-way trees was used. The algorithm used a multi-way tree to build the framework of chromosomes, and subsequently, the frame was filled to complete the scaffold assembly. This algorithm alone outperformed Velvet in the assembling of a scaffold.

Keywords: BAC clone data, scaffold assembly algorithm, three-dimensional mixing-pool.

1. INTRODUCTION

Next generation sequencing (NGS) technology has a had a tremendous impact on the biological sciences as it has allowed scientists to perform whole-genome scatter (WGS) sequencing on almost any organism. Compared to Sanger sequencing technology [1], NGS technology is faster, with millions of reads generated at once [2], and cheaper, yet the cost is still expensive for an individual study. Meanwhile, other factors such as short read lengths (usually only 35~100bp [3]), large amount of generated data, and a high error rate make WGS sequence splicing and assembly difficult [4, 5].On the other hand, the complex structure of a genome sequence itself also makes short sequence splicing difficult [6]. To obtain better assembly, the current software [7] requires data for sequencing, from previously established databases of multiple insertion length, or even from partial Sanger sequencing, to make the assembly cost effective.

In order to reduce the cost of sequencing, a high-throughput sequencing strategy using three-dimensional mixing-pool, named as Cube, was designed and implemented along with the zebrafish chromosome data that was used to verify the effectiveness of the strategy. The results

showed that the strategy can effectively obtain sequencing data and reduce the cost of sequencing. However, to precisely assemble contigs obtained by Velvet splicing [8, 9], a scaffold assembly algorithm based on multi-way tree was utilized, and multiple sets of data were used for the actual measurement and analysis of the algorithm. The results demonstrate that the algorithm showed an excellent assembly performance.

2. CONSTRUCTION AND ANALYSIS OF THREE-DIMENSIONAL MIXING-POOL

2.1. Construction and Sequencing of Three-Dimensional Mixing-Pool

To verify the effectiveness of a sequencing program, a three-dimensional mixing-pool was utilized, as shown in Fig. (**1**). One pool was constructed using roughly 850 Mb of sequence from the zebrafish genome. Based on preliminary findings, the study design required a three-dimensional mixing-pool of 18×18×18, with a total of 5832 BAC clones, each occupying a vertex on a three-dimensional cube. The three-dimensional cube was constructed with 54 (18×3) planes, and each plane (mixed 18×18=324 clones) was used as a DNA sample for sequencing. By sequencing each plane, the sequencing data of 5832 BAC clones was obtained, thus greatly reducing the cost of sequencing. The specific cube preparation steps are described as follows:

(1) Firstly, 54 planes were prepared which were divided into X, Y, Z.

*Address correspondence to these authors at the School of Computer Science, China University of Geosciences (Wuhan), Wuhan Hubei 430074, P.R. China; E-mail: zhangsifa@cug.edu.cn and State Key Laboratory of Freshwater Ecology and Biotechnology, Institute of Hydrobiology, Chinese Academy of Sciences, Wuhan Hubei 430072, P.R. China; E-mail: wangyp@ihb.ac.cn

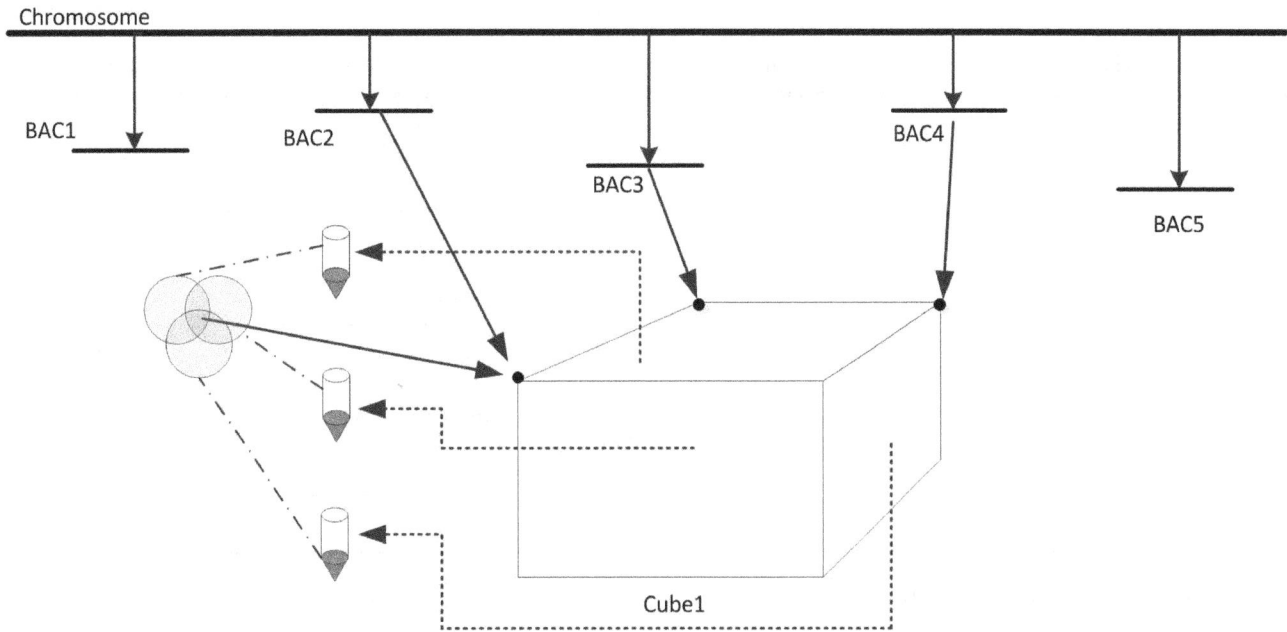

Fig. (1). BAC pooling and DNA extraction. DNA is extracted after pooling each plane(X,Y,Z),then sequence tags are assigned to individual BACs based on presence of three mutually perpendicular planes.

(2)   The 5832 BAC clones were placed respectively into the 5832 filling slots of the X-axis, and then respectively into the 5832 filling slots of the Y-axis, and finally, the 5832 filling slots of the Z-axis.

(3)   Paired-end sequencing was performedfor each plane of BAC clones.

As each BAC clone was simultaneously present in each plane, the sequence data of each clone could be obtained by calculating the intersection of three mutually perpendicular planes. This paper obtained two sequencing databases when sequencing the cube: a paired-end database with the insertion length of 1 kb and a paired-end database with the insertion length of 8 kb. The sequence length of each database was 100 bp. Finally, the two databases were used for contig splicing. The 8 kb database was used to build a scaffold frame. Then 1 kb database was used to fill the scaffold framework and finalize the splicing of clones.

2.2. Analysis of Single BAC Clone

To reduce the cost of sequencing, multiple BAC clones were placed on one plane at a time for sequencing. Therefore, it was necessary to analyze the cube after the completion of sequencing and each BAC clone was extracted to ensure accuracy . Extraction of each BAC clone from the cube was based on each vertex being an intersection of three mutually perpendicular planes. Therefore, each BAC clone could be obtained by calculating three mutually perpendicular planes in 54 planes. However, BAC clones obtained in this way contained many impurities. For example, a number of paired-end reads that did not belong to a particular clone

might also be included. Therefore, after extracting BAC clones from the cube, it was necessary to purify them. During the purification operation, all paired-end reads that existedin multiple clones were deleted, thus all preserved paired-end reads only appeared once in the vertices of one cube. The specific methods of extraction and purification are detailed in the following sections.

2.2.1. Preliminary Extraction of BAC Clones

There is a basic principle of analysis for single BAC clones in a three-dimensional mixing-pool: three mutually perpendicular planes intersect at a point, and three planes of data (a mixing-pool) intersect at single data set (single BAC clone). Therefore, a clone that simultaneously exists on three associated planes belongs to a data set of these corresponding clones. Based on this principle, all clone data can be obtained from X0, Y0, Z0 to X17, Y17, Z17 using the intersection calculation for each of the mutually perpendicular planes.

2.2.2. Purification of BAC Clones

The previous section described the preliminary extraction of BAC clones, namely the tags included in each BAC clone which were parsed out of the cube. For each BAC clone, the operation is guaranteed to extract all the tags contained in it, but the extraction operation also brings a serious problem: for a particular BAC clone, some tags that do not belong may also be extracted from the data set, and this contamination will negatively affect the subsequent cloning assembly operations. To avoid this contamination phenomenon, the initially extracted clones were purified.

Table 1. Analysis on extraction results of BAC clone.

Cube Number	Number of the Initial Tags	Number of the Parsed Tags	Proportion
1	17,496,000	1,289,659	31.77%
2		1,951,817	54.33%
3		1,483,333	38.75%
4		1,966,490	48.67%
5		1,945,673	44.26%
6		3,463,959	36.69%
7		2,843,583	47.48%
8		3,520,113	43.77%

Table 2. Proportion of the effective clones in cube.

Cube Number	Theoretical Number of the Clones	Number of the Parsed Clones	Number of Effective Clones	Proportion
1	5832	5018	2861	49.06%
2		5326	3005	51.53%
3		5213	3046	52.23%
4		5089	2879	49.37%
5		4981	2914	49.97%
6		5032	2977	51.05%
7		5107	3004	51.51%
8		4924	3102	53.19%

This quality checkpoint deletes all tags appearing in multiple clones. The steps implemented for all tags in the 5832 BAC clones are as follows:

(1) For each tag, the sum N of ASCII code of each character in the tag was calculated, followed by the calculation of number of file mapping by the tag (k).

(2) All tags were divided into 200 small files followed by step (1). The record form of tag in each file contained following data: tag sequence, and the clone number where the tag existed.

(3) The data in each small file was sorted out according to the tag sequence placing common tag sequences adjacent to each other.

(4) A pairwise alignment operation between the sorted tags and the adjacent tags was then performed, and repeated tag numbers were removed, thereby purifying BAC clone data from being contaminated.

(5) The purified data were then restored to 5832 BAC clones based on tag location.

2.3. Test Results and Analysis

According to the extraction and purification methods described above, eight cubes were parsed successively. To assess the effect of data parsing, the proportions of parsed data acquired from the original data are as shown in Table 1. For each clone, the number of the parsed tags must be more than 500, for the assembly .

As shown in Table 1, the proportion of parsed tags for each cube was only about 40% of the tags after processing normalization, which were caused by the characteristics of the cube and repetitive sequences. Although parsing decreased the data set, the analytical results of eight cubes provided sufficient coverage of the test data: each cube initially covered 80% of the test data and the parsed data of each cube covered more than 30% of the test data. As a result, eight cubes provided 2.5X coverage of the test data.

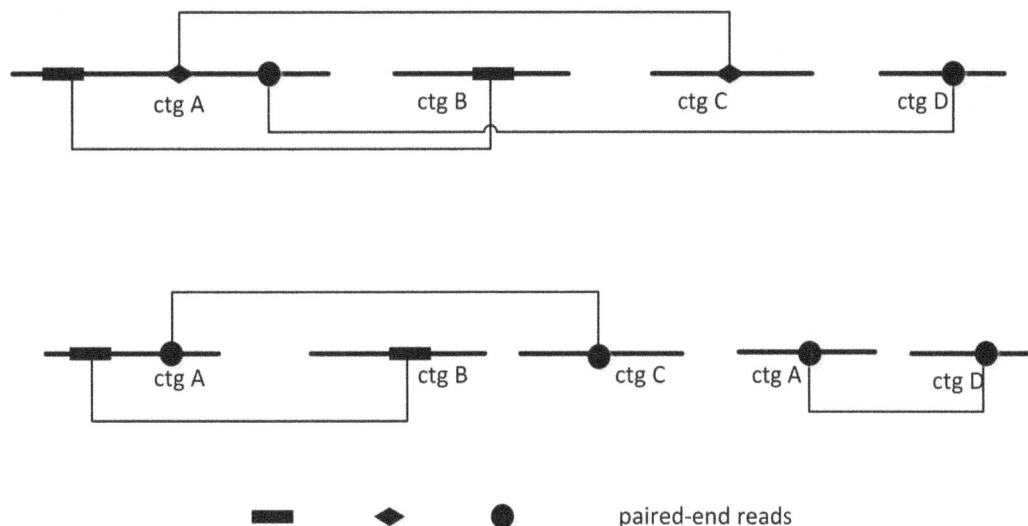

Fig. (2). Scaffold assembly when affected by repetitive sequence. The correct order of 4 contigs is A-B-C-D. They are connected by some paired-end reads. When there are repetitive sequences, the order may chaotic, and the result of assembly is A-B-C-A-D.

On the other hand, approximately 50% of the effective clones were parsed out of each cube, implying that half of the cube data covered 80% of the test data with each cube covering roughly 40% of the test data (Table **2**). Thus, eight cubes provided 3.2X coverage of the original test data, which was sufficient to assemble the test data. Therefore, high-throughput sequencing using a three-dimensional mixing-pool could effectively restore the test data at a significantly reduced cost.

3. SCAFFOLD ASSEMBLY ALGORITHM BASED ON MULTI-WAY TREE

3.1. Problem Analysis

Paired-ends are used to assemble contigs into a scaffold. This process identifies shared sequences to splice overlapping contigs together. One contig (A) may have a number of paired-end corresponding to multiple contigs (B, C, D) as shown in Fig. (**2**). The order of these contigs corresponding with A must be determined. Under normal conditions, the order is determined by other sequence in B, C, and D compared to A. Sometimes, however, repetitive sequences can lead to errors in alignment. The order of four contigs A, B, C and D in the chromosome is A-B-C-D, butbecause ctg A had repetitive sequences, incorrect results could be obtained (A-B-C-A-D) if these were spliced according to normal conditions. To avoid this error, we utilized a scaffold assembly algorithm based on a multi-way tree.

3.2. Establishment of Multi-Way Tree

To avoid splicing errors caused by repetitive sequences, scaffold assembly was completed by creating a multi-way tree to establish the overall framework of chromosomes and then fill the frame. Two paired-end databases were used during the process of contig stitching and scaffold assembly. The insertion length of the first database was 1KB, whereas the insertion length of the second database was 8KB.

However, the sequence length in each database was 100bp. The 8KB insertion length database was used to create the multi-way tree, and the 1KB insertion length database was used to fill the frame. A flow chart describing the establishment of the multi-way tree is shown in Algorithm **1**.

Algorithm 1:

Input:

paired-end reads which insertion length is 1KB and 8KB

Output:

Scaffold established with multi-way tree

1 Select the longest contig as a initial node

2 Search the child node that has a distance of 8kb from the initial node

3 **while** there is a child node **do**

4 Search the child node of the next level for each child node

5 **end**

Each contig functioned as a node. To begin, the longest contig was selected as an initial node. Following this, a child node roughly 8KB from the initial one was chosen through paired-end search. An ordinal relation between these child nodes and the initial nodes was observed and, some may be located towards the left or right of an initial node. For this study, only the nodes located towards the right of the initial one were used for extension. This process continued for each of the child nodes, continuing selection for extension towards the right. Eventually, a subsequent child node could not be identified, and the path was terminated. Circulation, search, and extension were conducted in this way until all the paths were terminated. Following this, the same method was used to search and extend the paths to the left of the initial nodes. The path with the maximum number of nodes (*i.e.* the longest path) was chosen as an overall framework of the chromosome.

Table 3. Contrast experiment between scaffold assembly algorithm based on multi-way tree and velvet algorithm.

Data Sources	Ctg Quantity	Assembly Length of Algorithm in this Article	Assembly Length of Velvet
Chromosome 22	108	117,573	30,795
Chromosome 22	86	108,798	29,865
Chromosome 22	87	122,379	59,884
Chromosome 24	95	120,893	40,702
Chromosome 24	113	121,759	32,716
Chromosome 24	107	122,678	59,937
Chromosome 24	99	119,374	21,329
Chromosome 25	64	988,595	60,016
Chromosome 25	97	100,987	49,057
Chromosome 25	106	122,336	38,467

3.3. Filling of Multi-Way Tree

After completing the overall framework, it was filled in with further extensions where, each node in the framework was referred to as a stub. For each stub, paired-end was used to find contigs roughly 1KB from the stub. If multiple contigs were found, the ordinal relationship between these contigs and stub was determined. By filling out each stub in this process, the assembly of whole chromosome was completed.

3.4. Test Results and Analysis

To validate the correctness and validity of the scaffold assembly algorithm using a multi-way tree, four sets of data in chromosome 22, 24, and 25 of zebrafish were selected for analysis. The test data and results generated by Velvet are shown in Table **3**. In addition, a total of ten sets of data from multiple fish chromosomes were used to compare assembly by velvet with multi-way tree scaffold assembly. The "Ctg quantity" column indicates the number of contigs used to assemble the scaffold, which were obtained by velvet splicing paired-end. Most contig lengthswere very short because a high number were obtained by velvet splicing; therefore, only those contigs longer than 100 bp were selected for scaffold assembly.

Of the 10 sets of test data, the assembled length almost reached 100KB which is very close to the original. Therefore, the algorithm used in this paper sufficiently assembled a correctly sequenced scaffold. However, the algorithm does have some shortcomings, for example, chromosome ends may not be assembled. This problem will require subsequent work and may be solved using multiple paths from multi-way tree analysis rather than simply selecting the longest path.

The assembled length using velvet was substantially less than 60KB while the algorithm developed in this study generally assembled a construct greater than 100KB. Therefore, this algorithm not only corrected splicing error created by Velvet software but also greatly improved the assembled length and demonstrated good splicing performance.

CONCLUSION

This study described the design and implementation of a high-throughput sequencing strategy using a three-dimensional mixing-pool and scaffold assembly with a multi-way tree-based algorithm. The experimental results show that the strategy was feasible, and the method reduced the cost of sequencing while improving assembly. This provides a new direction for the improvement of genome sequencing and might greatly reduce costs, thus contributing to the development of genome sequencing projects.

ACKNOWLEDGEMENTS

The project was supported by the National Natural Science Foundation of China under Grant 31372573, State Key Laboratory of Freshwater Ecology and Biotechnology under Grant 2013FB04, State Key Laboratory of Biogeology and Environmental Geology China University of Geosciences (No.GBL31506), and the National Science Foundation for Post-doctoral Scientists of China under Grant 2014M552119, Supported by Wuhan Branch, Super computing Center, CAS, China.

REFERENCES

[1]   Sanger F, Nicklen S. Coulson AR. DNA sequencing with chainterminating inhibitors. Proc Natl Acad Sci USA 1977; 74(12): 5463-7.

[2]   Metzker ML. Sequencing technologies-the next generation. Nat Rev Genet 2010; 11(1): 31-46.

[3]   Steven L, Phillippy AM, Zimin A, *et al.* GAGE: A critical evaluation of genome assemblies and assembly algorithms. Genome Res 2012; 22: 557-67.

[4]   Li R, Fan W, Tian G. The sequence and de novo assembly of the giant panda genome. Nature 2010; 463: 311-7.

[5]   Star B, Nederbragt AJ, Jentoft S. The genome sequence of Atlantic cod reveals a unique immune system. Nature 2011; 477: 207-210.

[6]   Schatz MC, Delcher AL, Salzberg SL. Assembly of large genomes using secondgeneration sequencing. Genome Res 2010; 20(9): 1165-73.

[7]   Fan W, Li R. Test driving genome assemblers. Nature Biotechnol 2012; 30(4): 330-331.

[8]   Zerbino DR, Birney E. Velvet: algorithms for de novo short read assembly using de Bruijn graphs. Genome Res 2008; 18: 821-9.

[9]   Pevzner PA, Tang H, Waterman MS. An Eulerian path approach to DNA fragment assembly. Proc Natl Acad Sci 2001; 98: 9748-53.

25

Risk Assessment of Cold Damage to Maize based on GIS and a Statistical Model

Zhewen-Zhao[a,b,c], Jingfeng-Huang[a,b,c,*], Zhuokun-Pan[a,b,c] and Yuanyuan-Chen[a,b,c]

[a]*Institute of Remote Sensing and Information Application, Zhejiang University, Hangzhou, 310058, China*

[b]*Key Laboratory of Polluted Environment Remediation and Ecological Health, Ministry of Education, College of Natural Resources and Environmental Science, Zhejiang University, Hangzhou, 310058, China*

[c]*Key Laboratory of Agricultural Remote Sensing and Information System of Zhejiang Province, Hangzhou, 310058, China*

Abstract: Cold damage to maize is the primary meteorological disaster in northwest China. In order to establish a comprehensive risk assessment model for cold damage to maize, in this study, risk models and indices were developed from average daily temperature and maize yield and acreage data in 1991-2012. Three northwest provinces were used to calculate the temperature sum during the growth period, temperature departure over the years and relative meteorological yield in order to obtain the climate risk index, risk sensitivity index and damage assessment index. Using the geographic information system (GIS) and cold damage risk indices obtained from the statistical assessment model, the studied area was divided into four risk regions: low, medium, medium-high and high. Northeast and southwest Gansu were grouped to the high-risk region; west Shaanxi and north NHAR were grouped into to the low-risk region; all other areas fell into medium and medium-high risk regions. Our results can help growers avoid cold damage to maize using local climate data and optimize the structure and layout of maize planting. It is of significance in guiding the agricultural production in the three northwest provinces in China and also can serve as a reference in modeling risk assessment in other regions.

Keywords: Maize, climate, cold damage, risk assessment, geographic information system (GIS).

1. INTRODUCTION

Shaanxi, Gansu and the Ningxia Hui Autonomous Region (NHAR) are three major provinces in northwest China. Topographically, they include the Qinba Mountains in south Shannxi, the Guanzhong Plain, the Loess Plateau in north Shannxi, the Hexi Corridor in Gansu and Qilian Mountain in NHAR. Digital elevation model (DEM) data show that altitudes of this undulating terrain vary from over 3,000 meters to below 1,000 meters above sea level. Hence, temperature changes are relatively dramatic [1], and extreme weather may easily occur. Damage due to cold weather is of notable regionality [2]. North Shannxi, part of Gansu, and the majority of NHAR are important maize growing areas of China. According to data published on the website of the Farming Management Division of the Ministry of Agriculture of the People's Republic of China, the present total maize acreage of these three provinces is approximately 2 million hectare (ha), with a yield of 4500-7500 kg per ha; i.e., in a normal year, the total maize yield is approximately 10 million tons. If moderate cold damage occurs, the yield can be reduced by 5-15%, which is approximately 1 million tons of maize, equivalent in value to nearly 17 million dollar. If there is severe cold damage during the growth period, the yield will be reduced by 15-25% which is approximately 2 million tons of maize, equivalent in value to nearly 35 million dollar.

Due to the impact of global warming, in recent years, temperature conditions have improved in general, and the number of years with severe cold damage has been reduced. However, as extreme weather events tend to frequently occur, the amplitudes of temperature fluctuations in different regions increase. In the future, there will still be years of low or relatively low temperature. In addition, crop rotation, intercropping and other cultivation methods are being practiced in many places, and efforts are being made to cultivate late maturing varieties..Thus, once cold damage occurs, it will cause even more severe agricultural economic losses than ever before [3]. Therefore, it is necessary to study the risk of cold damage to maize.

Prior studies on cold damage to maize and rice in China have mainly focused in the northeast region. The studies on cold damage in the northwest of China are limited to the following reports. Li and Wang [4] classified Gansu and NHAR as provinces with serious freeze damage. Li *et al.* [5]

*Address correspondence to this author at the Institute of Remote Sensing and Information Application, Zhejiang University, Hangzhou 310058, China; E-mail: hjf@zju.edu.cn

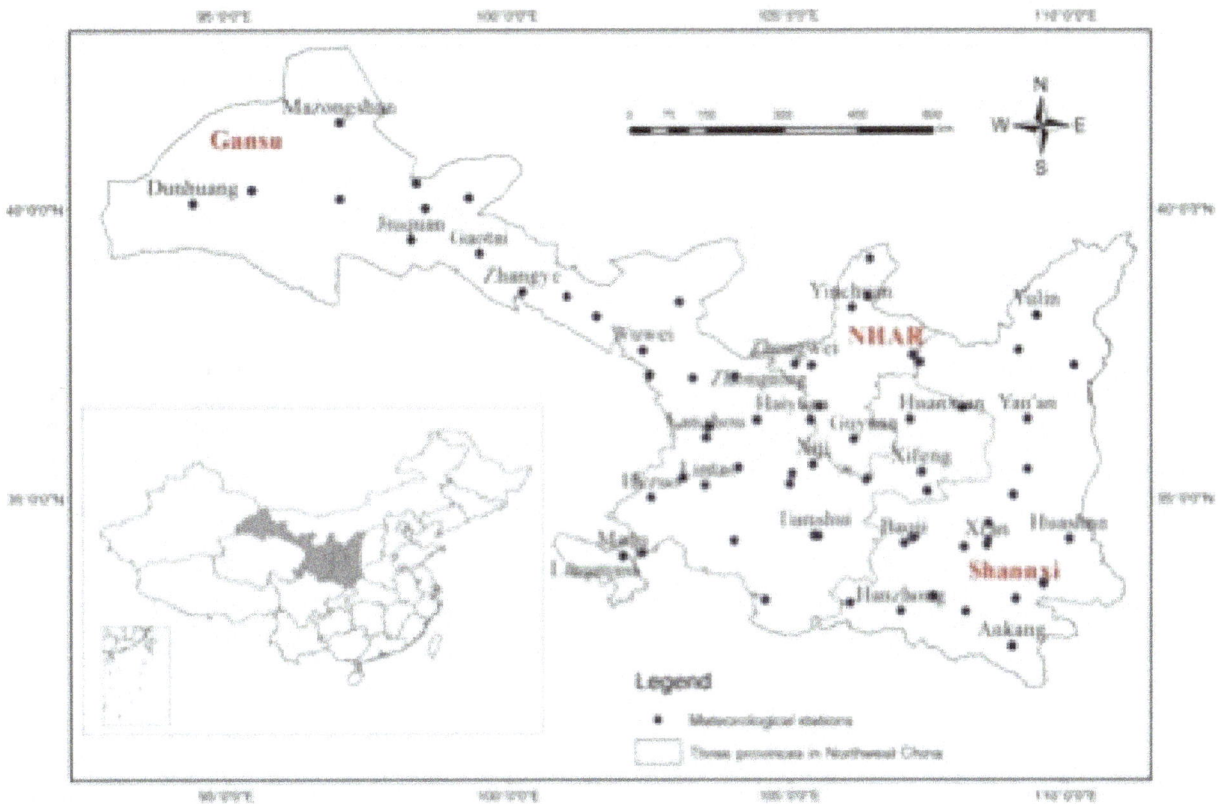

Fig. (1). Location of the studied area and spatial distribution of meteorological stations.

reported that the type of weather that causes cold damage was relatively common in Gansu; Li *et al.* [6] investigated cold damage to rice in NHAR. Liu *et al.* [7] designed a risk index of cold damage to apples in the Shannxi region according to the flowering time of apples. However, there has been no report on cold damage to maize in northwest China, its risk assessment or assessment indices.

In our investigation, meteorological data and maize yields and acreage in three northwest provinces (Gansu, NHAR, and Shannxi) were used to explore cold damage to maize using a mathematical model based on the temperature sum needed for crop growth. In order to effectively improve the intuitiveness of the risk assessment results, in addition, GIS technology was used in classifying the degree of risk. The goal of our study is to provide a theoretical basis for scientific measures of cold damage and prediction models to prevent cold damage. The objectives of this paper are to (1) build a model of a comprehensive risk assessment for cold damage to maize and (2) classify the degree of risk.

2. MATERIALS AND METHODS

The three northwest provinces are located at 31°42'-42°57'N, 92°13'-111°15'E, expanding about 2350 km from west to east and about 1000 km from north to south (Fig. 1). They include Gansu Province, the NHAR and Shannxi Province, with a total area of about 725,500 km², approximately 7% of the total area of China. Average daily temperature data collected from 73 meteorological stations (28 in Shannxi, 35 in Gansu and 10 in NHAR) from 1951 to 2012 were used. Data on maize growth and development from 1991 to 2012 were provided by the China Meteorological Data Sharing Network (http://cdc.cma.-gov.cn/). Data on the total yield of maize, yield per unit area and total acreage were obtained from provincial rural statistical yearbooks.

Identification of cold damage to maize in the three northwest provinces was based on the meteorological industry standard "level of cold damage to rice and maize"(QX/T101-2009which is a National Industry Standards) Table **1** lists indicators of level of cold damage to maize during growth periods; T (°C) is the average of sum temperature from May to September over all the years examined and ΔT (°C) is the temperature departure of the average monthly temperature of May-September of a particular year from that of the entire years. For data collected in each meteorological station, average daily temperatures were used to calculate the average monthly temperature of each year. Then, based on the indicator table (Table **1**), the years, frequency and sites at which general and severe cold damage occurred during growth period were determined.

Table 1. Indicators of maize yield reduction due to cold damage.

Degree of Damage	T≤80	80<T≤85	85<T≤90	90<T≤95	95<T≤100	100<T≤105	Yield Reduction Rate
General cold damage	-1.7<ΔT≤-1.1	-2.4<ΔT≤-1.4	-3.1<ΔT≤-1.7	-3.7<ΔT≤-2.0	-4.1<ΔT≤-2.2	-4.1<ΔT≤-2.2	5-15%
Severe cold damage	ΔT ≤-1.7	ΔT ≤-2.4	ΔT ≤-3.1	ΔT ≤-3.7	ΔT ≤-4.1	ΔT ≤-4.4	>15%

Multiple methods are available for simulating trend yield [8]. In the present study, the linearly weighted moving average method was used because with this method, no sample data are lost, and the stimulation result is relatively good [9, 10]; the step size was set to 11, and meteorological yields were obtained.

2.1. Comprehensive Indicators of Cold Damage Assessment

A. Climate Risk Index

Climate risk is a natural attribute of agricultural disasters. In this study, changes in the temperature during crop growth and the frequency of low temperature were analyzed to obtain the climate risk index. The amount of total heat can be used to indicate whether cold damages occurs in a particular year in a certain region. The inter-annual stability of temperatures also directly relates to the risk of cold damage to crops. Thus, temperature data from 1951-2012 were used to calculate the coefficient of variation (CV) of T in meteorological stations in Gansu, NHAR and Shannxi provinces.

$$CV = \frac{1}{T} \times \sqrt{\frac{\sum_{i=1}^{n}(T_i - \overline{T})^2}{n-1}} \tag{1}$$

where CV is the coefficient of variation, $\overline{T}$ is the mean of T_{5-9} over the period of 1951-2012, T_i is the value of a certain year in a certain region, and n is the total number of years.

B. Probability of Cold Damage

Compared to the frequency of cold damage, when the sample size is large enough, the probability of cold damage is not subject to change as the number of years increases, and thus is of higher objectivity and stability. Before calculating the probability of cold damage, a skewness-kurtosis test on climate samples determines if the data fits a normal distribution. If the distribution is skewed it is transformed into a normal distribution [11].

$$\text{Skewness: } C_s = \frac{\frac{1}{n}\sum_{i=1}^{n}(x_i - \overline{x})^3}{[\frac{1}{n}\sum_{i=1}^{n}(x_i - \overline{x})^2]^{\frac{3}{2}}} \tag{2}$$

$$\text{Kurtosis: } C_e = \frac{\frac{1}{n}\sum_{i=1}^{n}(x_i - \overline{x})^4}{[\frac{1}{n}\sum_{i=1}^{n}(x_i - \overline{x})^2]^2} \tag{3}$$

$\overline{x}$ is the mean of T_{5-9} over the period of 1951-2012, x_i is the T_{5-9} value of a certain year in a certain region, and n is the total number of years.

Based on meteorological data collected in Shannxi, Gansu and NHAR from 1951-2012, the skewness-kurtosis values over 50 years were calculated. By comparing the results with the values on the critical value table [12], it was concluded that the data all fit a normal distribution. Thus, the risk probability at corresponding levels of cold damage can be obtained based on the assumption of a normal distribution. The probability density function is as follows:

$$f(x) = \frac{1}{\sigma\sqrt{2\pi}}e^{\frac{-(x-\mu)^2}{2\sigma^2}} \tag{4}$$

where x is the value of ΔT sequence, μ is the mathematical expectation and can be replaced by the mean in a large sample sequence (sample size ≥ 30), and σ is the standard deviation. Integration of the probability density function was performed to obtain climate risk probability at different cold damage indicators:

$$F_1 = \int_{\Delta T_2}^{\Delta T_1} f(x)dx \tag{5}$$

$$F_2 = \int_{-\infty}^{\Delta T_2} f(x)dx \tag{6}$$

where F_1 and F_2 are the probabilities of moderate and severe cold damage, respectively, and ΔT_1 and ΔT_2 denote ΔT in moderate and severe cold damage, respectively (see Table 1 for reference).

C. Climate Risk Index of Cold Damage

The climate risk index of cold damage is a comprehensive index combining the intensity and frequency of cold damage [13] and can objectively reflect the degree of the risk of cold damage. The years with cold damage were divided into two groups: the moderate and severe cold damage groups. In each group, the frequency of the corresponding level of cold damage occurring over the years D_i and the median value H_i were calculated. The climate risk index k was calculated according to the following formula:

$$k = \sum_{i=1}^{2}\frac{D_i}{n} \times H_i \tag{7}$$

where D_i is the number of years in which the corresponding level of cold damage occurred (Table 1), H_i is the mean of the corresponding maximum and minimum values and n is the total number of years.

D. Risk Sensitivity Index

Risk sensitivity index is determined by the production layout of the crop and includes physical exposure index (S_1) and disaster resilience index (S_2). Physical exposure refers to the ratio of crop acreage to the geographical area of the studied area, i.e., larger planting area and higher proportion of the total land area will increase the sensitivity to risk correspondingly. In addition, planting density also affects physical exposure. S_1 is calculated using the following equation:

$$S_1 = \frac{S_i}{C_i} \qquad (8)$$

where S_i is regional maize acreage in units of square hectometers (hm^2), C_i is the land area of this region in hm^2, and i stands for different areas.

The disaster resilience index (DRI) mainly depends on three factors: genetics, productivity and yield. The DRI is positively correlated with the resistance of the crop itself, which is determined by the genetic characteristics. For example, resistant varieties have high resistance. The DRI is also positively correlated with the level of agricultural productivity; for example, taking agricultural cultivation measures such as optimization of sowing time and transplanting time, ground cover and artificial irrigation and using special equipments can improve disaster resilience. The DRI is positively correlated with agricultural yield [14]. Currently, the DRI is often defined by yield [15]. In this study, relative maize yield per unit area in each region was used as the DRI.

$$S_2 = \frac{P_i}{P} \qquad (9)$$

where P_i is the maize yield per unit area in one region in kg and P is the average maize yield per unit area of an entire province in kg.

Prior methods for assessment on agricultural loss due to meteorological risk mainly involve the establishment of risk indicators of disaster intensity and the loss risk assessment model [16]. In this study, the loss assessment index of cold damage was obtained from 1) the mean yield reduction rate, 2) the CV of yield reduction rates in years when cold damage occurred, 3) the probability of yield reduction rates in different ranges, and 4) the yield reduction risk index and other indicators for comprehensive assessment of the maize yield loss.

1) Average Yield Reduction Rate

The difference between trend yield and the actual yield per unit area in percentage is called "yield reduction rate" (i.e., relative meteorological yield). The relative meteorological yield is a relative value, indicating the amplitude of the deviation of actual yield from trend yield [17]. The normalized relative yield data can be treated as normal data, and the mean and variance were calculated as characteristic parameters. These data were combined with relative yield data that passed the normality test, and risk assessment was performed based on the assumption that all relative yield data fit a normal distribution.

Average yield reduction rate can reflect the level of average yield reduction caused by cold damage in a certain region. As shown in Table 1, years in which the yield reduction rate was greater than 5% were considered as years of disaster. The average yield reduction rates in years of disaster in different counties (cities) of the studied area were calculated using Eqns. 10 and 11.

$$Y_1 = \sum_{0.15 > y_i > 0.05} \frac{y_i}{k_1} \qquad (10)$$

$$Y_2 = \sum_{y_i \geq 0.15} \frac{y_i}{k_2} \qquad (11)$$

where Y_1 and Y_2 are the average yield reduction rate in moderate and severe yield reduction years, respectively, y_i is the yield reduction rate sequence, k_1 and k_2 are the total number of samples in years of moderate and severe cold damage, respectively.

2) CV of Yield Reduction Rate

According to Eqn. 1, the CV of the maize yield reduction rate in years of disaster, CV_z, was calculated to describe the amplitude of fluctuation of maize yield loss in years affected by disaster. High CV_z indicates that the region experiences a relatively large number of years of moderate and severe yield reduction, suggesting that the growth environment is relatively fragile and subject to relatively high risk of yield reduction due to external conditions.

3) Risk Probability of Yield Reduction

Fluctuation in crop yield is very clear due to the influence of climate and temperature. Because the constructed yield reduction rate sequence fits a normal distribution, risk probability of yield reduction can be calculated. The risk probabilities of maize yield reduction in different counties (cities) of the three northwest provinces were calculated using Eqns. 12 and 13:

$$P_1 = \int_{0.05}^{0.15} f(x)dx \qquad (12)$$

$$P_2 = \int_{0.15}^{\infty} f(x)dx \qquad (13)$$

where P_1 and P_2 are the risk probability of moderate and severe yield reduction, respectively, x is time sequence of yield reduction rate that passed the normality test or had been subject to normalization.

4) Loss risk Index of Cold Damage

Like climate risk index of cold damage, loss risk index is also a comprehensive index reflecting the amplitude and frequency of yield reduction and can objectively reflect the

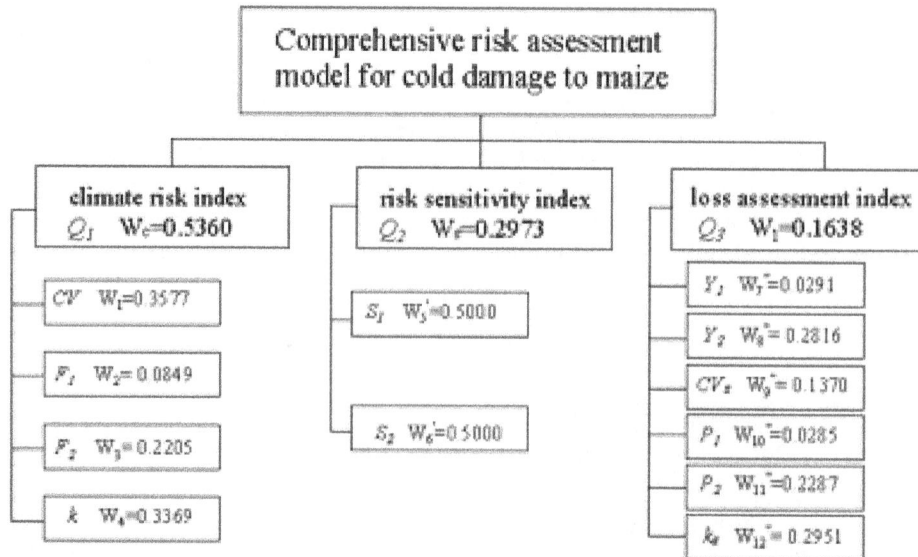

Fig. (2). Comprehensive risk assessment model for cold damage to maize and weight coefficients. CV is the coefficient of variation. F1 and F2 are the probability of moderate and severe cold damage, respectively. k is risk index. S1 and S2 are physical exposure and disaster resilience indices, respectively. Y1 and Y2 are average yield reduction rates in moderate and severe yield reduction years, respectively. CVz is the CV of the maize yield reduction rate in years of disaster. P1 and P2 are the risk probability of moderate and severe yield reduction. kz is the loss risk index of cold damage.

degree of the loss risk of cold damage. According to Eqn. 7, the product of frequency of moderate yield reduction and group median value was added to the product of frequency of severe yield reduction and group median value to obtain the loss risk index of cold damage k_z.

2.2. Comprehensive Risk Assessment Model for Cold Damage

To comprehensively reflect the risk of cold damage, all indices were combined with corresponding weight coefficients. After range standardization, the values of all indices were between 0-1, and the influence of different range of magnitude of indices was eliminated. Range standardization was performed using Eqn. 14:

$$x_i = \frac{x - x_{min}}{x_{max} - x_{min}} \quad (14)$$

where x and x_i are index values before and after standardization, respectively; x_{max} and x_{min} are the maximum and minimum value of the same index in all regions in the studied area, respectively.

Weighted composite indices were calculated using Eqns. 15- 17, including climate risk index (Q_1), risk sensitivity index (Q_2) and loss assessment index (Q_3), where W, W' and W'' are corresponding weight coefficients (Fig. 2). These indices were then combined to establish the comprehensive assessment model using Eqn. 18 where W_c is the weight coefficient of climate risk index Q_1, W_s is the weight coefficient of risk sensitivity index Q_2 and W_l is the weight coefficient of loss assessment index Q_3.

$$Q_1 = W_1 \times CV + W_2 \times F_1 + W_3 \times F_2 + W_4 \times k \quad (15)$$

$$Q_2 = W_5' \times S_1 + W_6' \times S_2 \quad (16)$$

$$Q_3 = W_7'' \times Y_1 + W_8'' \times Y_2 + W_9'' \times CV_z + W_{10}'' \times P_1 + W_{11}'' \times P_2 + W_{12}'' \times k_z \quad (17)$$

The comprehensive risk model is:

$$R = W_c \times Q_1 + W_s \times Q_2 + W_l \times Q_3 \quad (18)$$

The AHP was used to determine index weight. The importance of three indices is empirically rated as climate risk index > risk sensitive index > loss assessment index, and their weights W_c, W_s and W_l were determined as shown in Fig. (2). Taking into account that the physical exposure (S_1) can be reflected with regional disaster resilience index (S_2), the weights (W_5 and W_6) for these two indices were given the same. Determination of all weight coefficients was completed with the AHP analysis software yaahp V6.0 (http://www.yaahp.com/). The consistency test statistic was 0.0331, less than 0.1, i.e., the result passed the consistency test.

3. RESULTS

Regionalization

Using the spatial and temporal variation of temperatures, the amplitudes of risk of cold damage to maize during the growth period in different regions was analyzed. ArcGIS10 was used to plot the calculated CVs into elements on a map. As the results shown, the largest portion of the CVs were

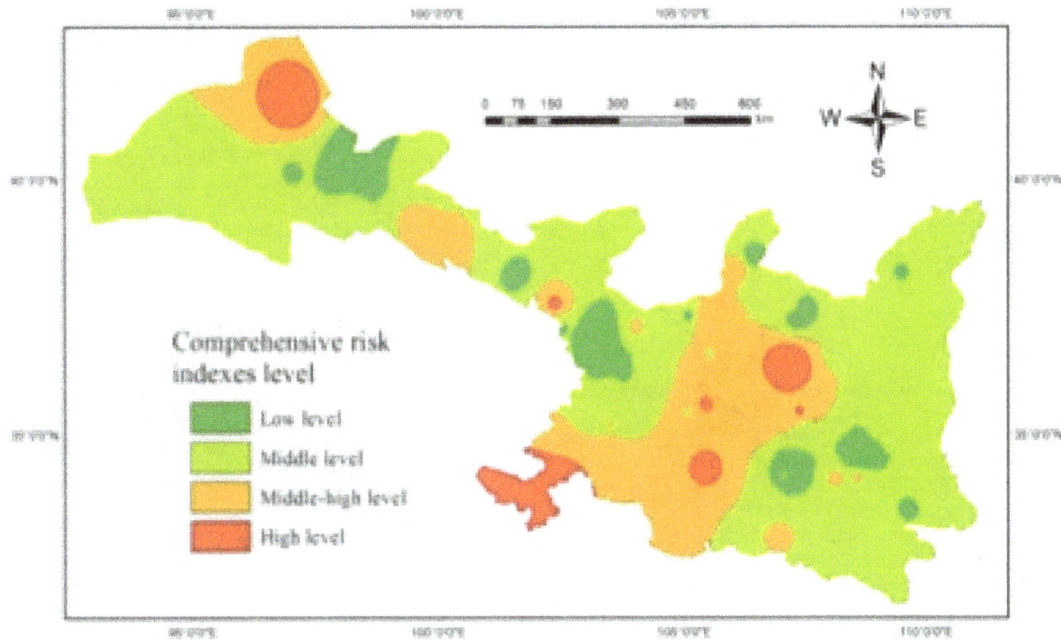

Fig. (3). Distribution of comprehensive risk index of cold damage to maize in the three northwest provinces.

above 0.04, followed by those between 0.03 and 0.04; those 0.03 below accounted for the lowest proportion of all CVs. Regarding spatial distribution, in northwest Gansu there was ample heat, and temperature was relatively stable, whereas in southwest Gansu, temperature variation was relatively large. Results show that CVs in Maqu and Langmusi reached a maximum, and variation in the Hexi Corridor in Gansu was also relatively large; heat variation in central and west Shannxi was relatively small and heat variation in NHAR was medium. In addition, the CV of moderate cold damage index and that of severe cold damage index were also calculated; a higher value indicated more severe disaster and higher risk (data not shown).

According to the industry standard of cold damage to maize (Table 1), the calculated risk probabilities of cold damage of different regions in the three northwest provinces were used for zoning analysis. As illustrated in Fig. (3), sites with high risk probability of cold damage such as Zhangye (18%) and Wuwei (17%) were also regions of relatively high CV. It was found that the risk index of Xifeng in Gansu was the highest, reaching 1.49. The risk indices of Yan'an, Wuwei and Zhangye were all at about 1.4. At all other sites except Linxia, the risk indices were between 0.86-1.17; at Linxia, the risk index was 0.05, almost negligible.

The comprehensive risk index R was calculated using Eqn. 18 and the weight coefficients shown in Fig. (2). With the natural break classification method, the studied area was divided into four risk areas: low, medium, medium-high and high (Fig. 3). The regions of highest risk to cold damage included Mazongshan, Maqu, Langmusi and Huanxian. The regions of lowest risk included Xi'an, Lanzhou, and Yinchuan *et al.*

4. DISCUSSIONS AND CONCLUSION

According to the studied area data in 1951-2012 of meteorological disasters information, choose 1962, 1967, 1968, 1977, 1984, 1989, 1993, 1996, 2001, 2004 and 2006 as typical cold damage years, perform the statistics over the average yield reduction rate, and analysis of the correlation between average yield reduction rate and the Chilling average risk index during typical years in the counties (cities), which is to verify the feasibility and applicability of the comprehensive risk assessment model for cold damage.

By analyzing the regression relationship between the comprehensive risk index and maize average yield reduction rate in typical chilling years, we can know that they meet the 0.05 level of significance related. So it is reliable to the comprehensive risk assessment model for cold damage which is established by climate risk index (Q_1), risk sensitivity index (Q_2) and loss assessment index (Q_3). We can infer from the maize yield reduction degree affected by chilling damage.

Three provinces in northwest China were selected for a study on risk of cold damage to maize. The risk assessment index of cold damage and a crop damage assessment system were established using a combination of meteorological data. The weight of each index from the Comprehensive Weight Method was determined based on AHP method and overcame the limitations of deflection caused by single-weighted weighting method in study of conventional risk assessment, and thus increasing the credibility and reliability of risk assessment in index analysis. These results can be used to help avoid the risk of cold damage to maize, and to optimize the planting structure and layout using local climate resources, thereby ensuring the safety of maize in these growing areas. In the future, selection of assessment indices and establishment of a comprehensive assessment system still needs further in-depth exploration.

ACKNOWLEDGEMENTS

This study was supported by the National Science and Technology Support Program, project number 2012 BAH29B00.

REFERENCES

[1]   Deng B. Statistical analysis of test data. Tsinghua University Press, Beijing, 1994.

[2]   Huo ZG, Li SK, Wang SY. Study on the risk evaluation technologies of main agrometeorological disasters and their application. J Nat Resour 2003; 18: 692-703.

[3]   Li SK, Huo ZG, Wang SY. Risk evaluation system and models of agrometeorological disasters, J Nat Disaster 2004; 13: 77-87.

[4]   Li YJ, Wang CY. An integrated maize chilling damage forecast model based on the multi-prediction model. J Catastrophol 2006; 21: 1-7.

[5]   Li YJ, Wang CY. Research on comprehensive index of chilling damage to corn in Northeast China. J Nat Disaster 2007; 16: 15-20.

[6]   Li YJ, Wang YH, Zhang XF, Wang CY. Research on chilling damage of maize in northeast China, J Nat Disaster 2011; 20: 74-80.

[7]   Liu XZ, He WL, Li YL, Bai QF, Liang T, Zhang T. A Study on the risk index design of agricultural insurance on apple florescence freezing injury in shaanxi fruit zone. Chin J Agrometeorol 2010; 31: 125-129.

[8]   Ma SH, Liu YY, Wang Q. Dynamic prediction and evaluation method of maize chilling damage. Chin J Appl Ecol 2006; 17: 1905-1910.

[9]   Wang CY. Studies on cold damage to crops in Northeast China. China Meteorological Press, Beijing, 2008.

[10]   Wang RC, Huang JF. Rice yield estimation using remote sensing, China Agriculture Press, Beijing, 2009.

[11]   Wang SL. Review of the progress in methods of agrometeorological disaster prediction in China. J Appl Meteorol Sci 2003; 14: 574-582.

[12]   Wang SL, Ma YP, Zhuang LW. Improvement study on prediction and assessment model for chilling damage of maize in Northeast China. J Nat Disast 2008; 17:12-18.

[13]   Wu DL, Wang CY, Xue HX. Drought risk map for winter wheat in North China. Acta Ecol Sin 2011; 31: 760-769.

[14]   Wu WL. Changes in cold damage to rice in NHAR under the background of global warming. NHAR Agric For Sci Technol 2008; 14(1), 54-59.

[15]   Tang YJ, Pan JY. Characteristics of agro-meteorological and agrobiological disasters in China in recent years. J Nat Disaster 2012; 21: 26-30.

[16]   Yang AP, Feng M, Liu AG. Research on sensitivity of rice to chilling injury in summer in hubei province, Chin J Agrometeorol, 2009; 30: 324-327.

[17]   J.M. Zhang, Q.F. Shang, Y.A. Qi, and X.Z. Wang, Analysis of agricultural risk factors in Gansu Province. J Nat Disaster 2006; 15: 144-148.

26

Extraction and Antioxidant Activity of Soybean Saponins from Low-temperature Soybean Meal by MTEH

Liu Zhong-Hua*, Ge Hong-Lian, Luo Rui-Ling and Zhao Jin-Hui

College of Life Science and Agronomy, Zhoukou Normal University, Zhoukou, Henan, 466000, China

Abstract: The research aimed at developing an optimal procedure for the extraction of soybean saponins from low-temperature soybean meal with microwave treatment combined with enzymatic hydrolysis (MTEH), and studied the antioxidant activity of soybean saponins. The result shows that the optimal parameters of microwave treatment, determined with the orthogonal array design method, were the medium fire of microwave power, 1.5 min of microwave time, 80% of ethanol, and 1:25 ratio of material to water, moreover, the optimal conditions of enzymatic hydrolysis, determined with the response surface experiments, were 50 minutes of hydrolysis time, 51°C of hydrolysis temperature and 1.5% dosage of cellulase, with which the optimal extraction ratio of the soybean saponins reached 0.916%. The saponins extracted from soybean meal exhibited antioxidant activity and the effect of scavenging superoxide anion radicals (SAR) and hydrogen peroxide (H_2O_2).

Keywords: Soybean saponina, microwave, enzymatic hydrolysis, antioxidant activity.

1. INTRODUCTION

Recently, physiologically active substances from soybean, such as soybean saponins, and soybean isoflavone, have become a hot topic in the researches based on industry of health foodstuff. Soybean saponins have various physiological functions, such as anti-lipid oxidation, anti-free radical, immune regulation, anticoagulation, antithrombotic, anti-diabetic, antitumor and antivirus [1, 2]. Low-temperature soybean meal, produced as a by-product in the process of extracting oil from soybean, with low price and large yield in China, contains rich soybean saponins. The research studied the method of extraction of soybean saponins from the raw materials, and soybean meal, to make full use of the potential value of soybean meal.

Currently, the researches mainly focus on the physiological activity of soybean saponins and less focus on the extraction technology to improve the extraction ratio of soybean saponins, which is the key to extract the high-purified products [3]. Microwave extraction, is efficient, highly selective and has no pollution hazards as compared to the traditional solvent extractions. It is one of the new technologies used to extract the biological active substances [4, 5]. The method of biological enzyme hydrolysis, with strong selectivity and high efficiency, is widely used for extracting effective components from plants [6].

The researches based on the methods of microwave treatment combined with enzymatic hydrolysis to extract the soybean saponins from low-temperature soybean meal are less reported. The research for improving efficiency and making full use of soybean meal, studied the process condition of microwave treatment combined with enzymatic hydrolysis to extract the soybean saponins from low-temperature soybean meal to provide a more efficient method for extracting soybean saponins and obtain basic data for large-scale industrial production. In addition, the research studied the antioxidant activity of soybean saponins extracted from low-temperature soybean meal to provide scientific basis for its further exploration.

2. MATERIALS AND METHODS

2.1. Materials and Reagents

Low-temperature Soybean meal was offered by Henan Zhoukou Yihai Grain and Oil Co Ltd; Cellulase enzyme (20000U/g). Oleanolic acid of standard substance, ethanol, petroleum ether were analytically pure.

2.2. Methods

Pretreatment of raw material. The grinded soybean meal was screened through 80 mesh sieve and treated with solvent of petroleum ether to carry out soxhlet extraction for about 3 h, with the ratio of solid-liquid at the l:15; the defatted soybean meal was dried for reserve.

*Address correspondence to this author at the College of Life Science and Agronomy, Zhoukou Normal University, Zhoukou, Henan, 466000, China; E-mail: lzh.hzl@163.com

Drawing of standard curve of soybean saponins. With the oleanolic acid used as a standard acid, the standard curve was drawn [7]. The oleanolic acid methanol solution was drawn with the following volume of 0 mL, 0.2 mL, 0.4 mL, 0.6 mL, 0.8 mL, and 1 mL respectively, poured into different volumetric flasks of 10 mL, and diluted with the 95% ethanol to 10 mL; the control group had 95% ethanol. The ODs were measured at a wavelength of 210 nm, and the quality of concentration by absorbance was regressed with the regressed equation: A=88.350 C+0.0611 (R^2=0.9930).

Extraction and measurement of soybean saponins. The defatted soybean meal was treated by cellulase, and then saponins in the cellulase-lyzed solution were extracted in the solvent of ethanol with microwave treatment . The extracted solution was diluted to 20 folds with 95% ethanol for measuring its OD at a wavelength of 210 nm. The content of soybean saponins was calculated with the regressed standard curve.

Calculation of the ratio of soybean saponins. The ratio of soybean saponins (%) = the content of soybean saponins (g) / sample mass (g) ×100

Optimization on extraction conditions of soybean saponins. Different microwave treatments, including concentration of ethanol, ratio solid-liquid, microwave power and microwave time, and the parameters of enzymatic hydrolysis, including enzyme usage, enzymolysis time, hydrolysis temperature and pH value, were studied in single-factor experiments to investigate their effect on the extraction ratio of soybean saponins. Based on the single-factor experiments, the orthogonal experiments and response surface experiments were applied to optimize the extraction condition of soybean saponins.

Measurement of antioxidant activity of soy saponins from soybean meal. The scavenging effect of soybean saponins on SARs was detected with the methods of pyrogallol autoxidation [8]. The scavenging effect of soybean saponins on H_2O_2 was detected with the method of spectrophotometry [9].

3. RESULTS AND ANALYSIS

3.1. Single-factor Experiments of Microwave Treatment

Microwave time. Soybean meal treated with cellulase was extracted with 60% ethanol and the ratio of solid-liquid was 1:20, and the solution was treated with microwave at medium fire for 0 min, 0.5 min, 1.0 min, 1.5 min, 2.0 min, and 2.5 min, and the experiment was repeated for three times. The results areshown in Fig. (1).

Fig. (1) shows that the longer the microwave time, the higher the extraction ratio of soybean saponins. The extraction ratio increased to a peak value with the treatment time of 1.0 min, but then decreased with the extension of time. The possible reason for the decrease in ratio might be protein denaturation of the soybean meals caused by increased microwave time, which blocked the infiltration and diffusion of soybean saponins. Therefore, the optimal time of microwave treatment was 1.0 min for extracting soybean saponins among the experiments.

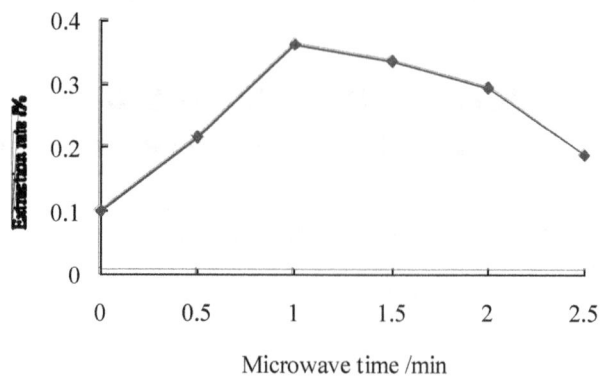

Fig. (1). Effect of different microwave time on the extraction rate of soybean saponina.

Ratio of solid-liquid. Soybean meal treated with cellulase was extracted with 60% ethanol and the ratio of solid-liquid was 1:10, 1:15, 1:20, 1:25, and 1:30, respectively, and the solution was treated with microwave at medium fire for 1.0 min. The experiment was repeated for three times. The result is shown in Fig. (2).

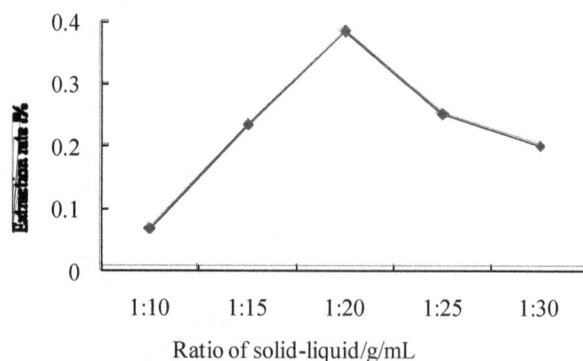

Fig. (2). Effect of different ratio of solid-liquid on the extraction rate of soybean saponina.

Fig. (2) shows that the bigger the ratio of solid-liquid, the higher the extraction ratio of soybean saponins. The extraction ratio increased to a peak value at 1:20, but then decreased with the increase in the ratio of solid-liquid. Thus, the optimal ratio of solid-liquid was 1:20 for extracting soybean saponins among the experiments.

Concentration of ethanol. Soybean meal treated with cellulase was extracted with 50%, 60%, 70%, 80%, 90%, and 100% ethanol and the ratio of solid-liquid was 1:20, and the solution was treated with microwave at medium fire for 1.0 min. The experiment was repeated three times. The result is shown at Fig. (3).

Fig. (3) shows that the bigger the concentration of ethanol, the higher the extraction ratio of soybean saponins. The extraction ratio increased to the peak value with the concentration of ethanol at 80%, but decreased with the increase inthe concentration. The possible reason for the decrease in the ratio might be the change in the polarity of extraction solution, which blocked the infiltration and diffusion of soybean saponins, thus, the optimal concentration of ethanol was observed to be 80 % for extracting soybean saponins among the experiments.

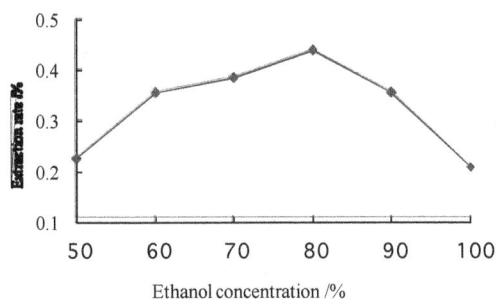

Fig. (3). Effect of different ethanol concentration on the extraction rate of soybean saponins.

Microwave power. Soybean meal treated with cellulase was extracted with 80% ethanol and the ratio of solid-liquid was 1:20, and the solution treated with microwave treatment at low-power, thawing, medium fire, medium-high fire, high fire for 1.0 min. The experiment was repeated three times. The result isshown in Fig. (**4**).

Fig. (**4**) shows that the stronger the microwave power, the higher the extraction ratio of soybean saponins. The extraction ratio increased to a peak value with the medium fire, but decreased with the increase in the microwave power. The possible reason for the decrease in the ratio might be the decomposition of the soybean saponins caused by microwave power. Thus, the optimal microwave power was the medium fire for extracting soybean saponins among the experiments.

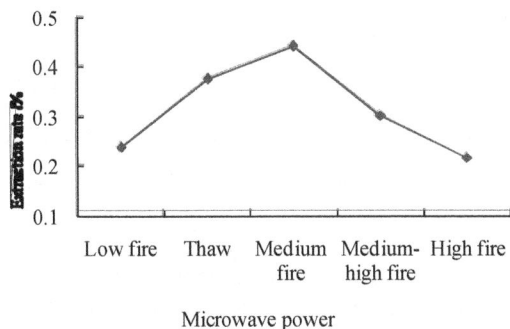

Fig. (4). Effect of different microwave power on the extraction rate of soybean saponins.

3.2. Orthogonal Experiment of Microwave Conditions

Based on experimental results of the single-factor microwave treatment, the experiment of $L_9(3^4)$ orthogonal (Table **1**) was conducted to optimize the parameters in microwave treatment. The results of orthogonal experiment are shown in Table **2**.

Table 2 shows that the factors influencing the extraction ratio of soybean saponins from soybean meal were; microwave time (C), microwave power (B), ethanol concentration (A), ratio of solid-liquid (D) in the order from the strongest to the weakest; the optimal combination of treatment for extraction was $A_2B_2C_3D_3$: ethanol 80%, microwave power at medium fire, microwave treatment time of 1.5 min, and 1:25 g/mL ratio of solid-liquid.

3.3. Single Factor Experiment of Enzymolysis Condition

Enzyme dosage. With the cellulase enzymolysis buffer of pH 6.0 and the enzymolysis temperature at 45°C, the solution was extracted with 0%, 0.5%, 1.0%, 1.5%, 2.5%, and 3.0% of cellulase enzymolysis dosage for 30 min, respectively, and subsequently treated with the above mentioned optimal condition for microwave treatment.

With an increasein the enzyme dosage, the extraction ratio of soybean saponins also increased gradually (Fig. **5**). At 1.5% of the enzyme dosage, the extraction ratio reached the maximum, but gradually decreased with the continued increase in the enzyme dosage. This suggested that the optimal enzymolysis dosage was 1.5% for the extraction of soybean saponins among the experiments.

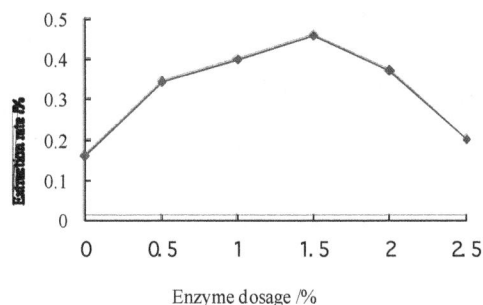

Fig. (5). Effect of different enzyme dosage on the extraction rate of soybean saponins.

PH value. With cellulase dosage of 1.5%, extraction solution was treated with the cellulase enzymolysis buffer liquid of pH 2.0, 3.0, 4.0, 5.0, 6.0, and 7.0, respectively, and was hydrolysed for 30 min at 45°C to extract the saponins with the above mentioned optimal condition for microwave treatment. The experiment was repeated three times and the results are shown in Fig. (**6**).

Fig. (**6**) shows that at pH 4.0 of the cellulase enzymolysis buffer liquid, the extraction ratio of soybean saponins reached the peak, and at other pH values, the extraction ratios were lower. The most suitable pH of cellulase enzymolysis buffer for extracting soybean saponins was 4.0. Thus, the most optimal pH value was 4.0 and dosage of the enzymatic hydrolysis was 1.5% for extracting soybean saponins among the experiments.

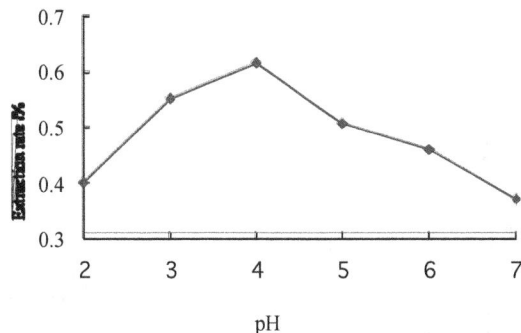

Fig. (6). Effects of different ph on the extraction rate of soybean saponins.

Table 1. Factor and levels in orthogonal array design.

Levels	Factors			
	A **Ethanol Concentration/%**	**B** **Microwave Power**	**C** **Microwave Time/min**	**D** **Ratio of Solid-Liquid /(g/mL)**
1	70	thaw	0.5	1:15
2	80	medium fire	1.0	1:20
3	90	medium-high fire	1.5	1:25

Table 2. The results of orthogonal experiment.

Experiment Number	Factor				Soybean Saponins Extraction/ (%)
	A	**B**	**C**	**D**	
1	1	1	1	1	0.435
2	1	2	2	2	0.317
3	1	3	3	3	0.551
4	2	1	2	3	0.353
5	2	2	3	1	0.765
6	2	3	1	2	0.482
7	3	1	3	2	0.517
8	3	2	1	3	0.611
9	3	3	2	1	0.298
R	0.099	0.129	0.288	0.066	

Cellulase hydrolysis temperature. With the cellulase enzymolysis buffer liquid at pH 4.0 and 1.5 % of enzyme dosage, the solution of enzymatic hydrolysis was treated at different temperatures 40°C, 45°C, 50°C, 55°C, and 60°C for 30 min to extract the saponins, respectively, with the above mentioned optimal condition for microwave treatment. The experiment was repeated three times and the results are shown in Fig. (7).

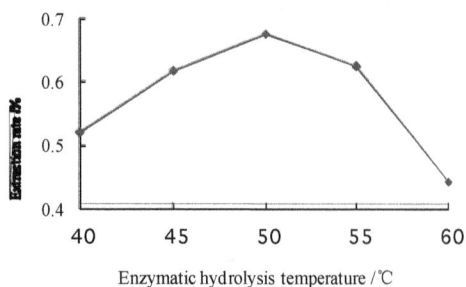

Fig. (7). Effects of different enzymatic hydrolysis temperature on the extraction rate of soybean saponins.

Fig. (7) shows that with the rise in the temperature of enzymatic hydrolysis, the extraction ratio of soybean saponins increased gradually. At 50°C of the enzymatic hydrolysis temperature, the extraction ratio reached the maximum, but gradually decreased with the continued rise in the temperature of enzymatic hydrolysis. The possible reason for the decrease in the extraction ratio might be the inhibition or deactivation of the cellulase activity, which was the result of the high temperature. Thus, the optimal enzymatic hydrolysis temperature was 50°C for extracting soybean saponins among the experiments.

Enzymatic hydrolysis time. With the cellulase hydrolysis buffer liquid at pH 4.0 and the dosage of cellulase at 1.5%, the extraction solution was hydrolysed for 20 min, 30 min, 40 min, 50 min, and 60 min at 50°C to extract the saponins with the above mentioned optimal condition for microwave treatment, respectively. The experiment was repeated three times and the results are shown in Fig. (8).

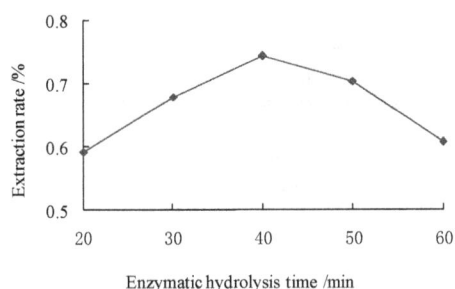

Fig. (8). Effect of different enzymatic hydrolysis time on the txtraction rate of soybean saponins.

Fig. (8) shows that with increased enzymolysis time, the extraction ratio of soybean saponins also increased. At 40 min of the hydrolysis time, the extraction ratio reached the maximum, but decreased with the hydrolysis time over 40

minutes. Therefore, the most suitable hydrolysis time was 40 minutes for the extraction of soybean saponins among the experiments.

3.4. The Response Surface Experiments of Enzymatic Hydrolysis Conditions

Based on the principle of Box-Benhnken central composite design, and with the most suitable condition of microwave treatment for extraction and the results of single-factor enzymatic hydrolysis experiments, the enzymatic hydrolysis condition of soybean saponins was optimized with the method of response surface (RSM) and the factor level of RSM isshown in Table 3. The experimental results of RSM areshown in Table 4 and Table 5.

Based on the test data represented in Table 4, the effect of cellulase dosage, and the temperature and time of enzyme

Table 3. Factor and levels in Box-Behnken central composite design.

Levels	Factors		
	A Enzyme Dosage/%	B Enzymatic Hydrolysis Temperature /°C	C Enzymatic Hydrolysis time/min
1	1.0	45	30
2	1.5	50	40
3	2.0	55	50

Table 4. Results of Box-Behnken central composite design.

Experiment Number	A Enzyme Dosage/%	B Enzymatic Hydrolysis Temperature /°C	C Enzymatic Hydrolysis time/min	Soybean Saponins Extraction/ (%)
1	1.0	50	50	0.861
2	2.0	50	50	0.896
3	1.5	50	40	0.890
4	1.0	45	40	0.573
5	1.5	50	40	0.887
6	1.5	45	50	0.695
7	2.0	45	40	0.660
8	1.5	50	40	0.874
9	1.5	55	30	0.729
10	1.5	50	40	0.879
11	2.0	50	30	0.882
12	1.0	55	40	0.722
13	1.0	50	30	0.717
14	2.0	55	40	0.731
15	1.5	55	50	0.769
16	1.5	50	40	0.882
17	1.5	45	30	0.737

Table 5. ANOVA for the regression response surfuce model.

Source	DF	Sum of Squares	Mean Square	F value	Prob > F	
Mode 1	9	0.15	0.017	15.52	0.0008	**
A	1	0.011	0.011	9.99	0.0159	*
B	1	0.01	0.01	9.32	0.0185	*
C	1	0.00304	0.00304	2.77	0.1398	
A^2	1	0.011	0.011	10.46	0.0144	*
B^2	1	0.11	0.11	96.68	< 0.0001	**
C^2	1	0.000326	0.000326	0.3	0.6025	
AB	1	0.00152	0.00152	1.39	0.2774	
AC	1	0.00423	0.00423	3.85	0.0905	
BC	1	0.00168	0.00168	1.53	0.2556	
Residual	7	0.00768	0.0011			
Lack of fit	3	0.00752	0.00251	2.17	0.058	
Pure error	4	0.000161	0.0000403			
Cor total	16	2.16				

hydrolysis during on the extraction of soybean saponins wereanalyzed with the software of Design Expert to do multiple regressions, and the acquired ternary quadratic response surface regression equation was:

$$Y = -16.01810 + 1.35040A + 0.63725B - 0.015840C - 0.20880A^2 - 0.006348B^2 + 0.000088C^2 - 0.0078AB - 0.0065AC + 0.00041BC \quad (R^2 = 0.9523)$$

Table 5 shows that the regression model was very significant (P < 0.01). In terms of R2 = 0.9523, the linear relationship between the dependent variable and the examined variable was significant. The model adjusted coefficient R^2Adj was 0.9109, indicating that the model could explain 91.09% variation of the response values with a higher fitting degree. The results of variance analysis show that the one degree of enzyme of dosage (A), and the temperature of enzyme (B), were significant while the time of enzyme hydrolysis (C) was not significant. Moreover, the quadratic terms A^2 and B^2 were very significant. The factors influencing the extraction ratio of soybean saponins from soybean meal were A, B and C from being the strongest to the weakest, while A and B had a stronger effect on the response values and the relationship of the factors for extraction of soybean saponins with the response value in the experiments not being purely linear while some were nonlinear.

The acquired regression model predicted the conditions for enzyme extraction to obtain the maximum response values which included , 50 min of enzymolysis time, enzymolysis temperature at 50.88 °C and 1.50% dosage of cellulase. With these conditions, the extraction ratio of soybean saponins reached 0.916%. Based on the actual operating conditions, the most suitable condition for extracting soybean saponins was 50 min of enzymatic hydrolysis time, 51°C of enzymolysis temperature and 1.5% dosage of cellulase enzyme.

To test the reliability of the response surface method, three parallel experiments were implemented under the most suitable extraction conditions. The average extraction ratio of saponins for the three parallel experiments was 0.905% which was slightly different from the theoretical predicted value, indicating that the response surface method could be applied to optimize condition forenzymatic hydrolysis.

3.5. Antioxidant Activity of Saponins Extracted from Soybean Meal

Scavenging effect of saponins extracted from soybean meal on SARs. Fig. (9) shows that more radicals were scavenged with increased extracted solution of soy saponins, indicating that the scavenging effect of the extracted solution of soy saponins on SARs was significantly associated with the content of soy saponins. It was concluded that the extracted soy saponins from soybean meal had the ability to scavenge SARs, suggesting that the scavenging activity of the extracted solution on SARs might be related to the extracted soy saponins.

Scavenging effect of saponins from soybean meal to H_2O_2. Hydroxyl free radicals produced by the oxidation of H_2O_2 through metal ions attack the DNA in cells, cause the tissue injury, and accelerate the process of lipid oxidation reaction. Thus, the strong antioxidant ability of saponins could be proved by the scavenging effect on H_2O_2.

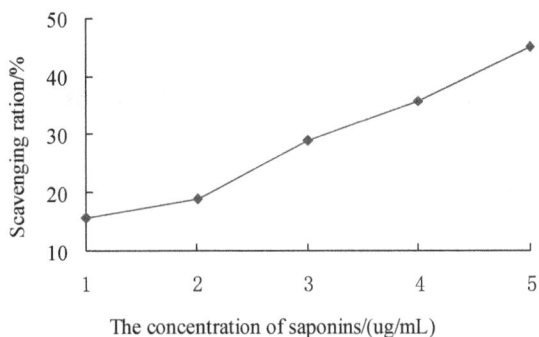

Fig. (9). Scavenging effect of saponins from soybean meal on SARs.

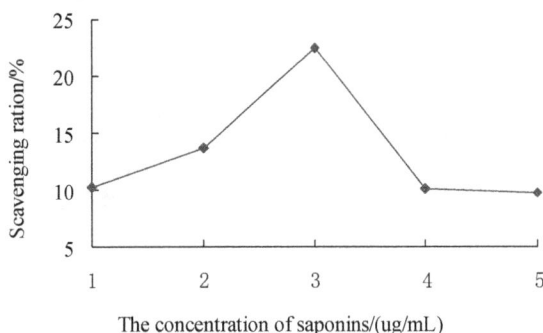

Fig. (10). Scavenging effect of saponins from soybean meal on H_2O_2.

Fig. **(10)** shows that more radicals were scavenged with increase in the concentration of soy saponins. It could be concluded that the extracted soy saponins from soybean meal had the ability to scavenge H_2O_2. The highest scavenging ratio of the extracted solution on H_2O_2 was 22.5% with the concentration of soy saponins at 3.0 ug/mL, and the scavenging ratio of H_2O_2 decreased with the increase in soy saponins concentration, and the reason for the decrease inscavenging ratio with more soybean saponins remained unclear and needs to be studied further.

CONCLUSION

This research studied the extraction and antioxidant activity of soy saponins from low-temperature soybean meal by MTEH. The most suitable process condition to extract the saponins with MTEH was 80% ethanol, medium fire of microwave, 1.5 min of microwave time, 1:25 g/mL of the ratio material to liquid, 50 min of cellulase hydrolysis time, 51°C of hydrolysis temperature, and 1.5% of cellulase dosage. Under this condition, the ratio of extraction of soybean saponins reached 0.905%, and the extraction time was markedly shortened, which facilitated their industrial production, while the difference in the function and structure of the extracted soybean saponins from traditional MTEH need to be studied further. It was proved that the extracted solution of soybean saponins had the ability to scavenge SARs and H_2O_2, indicating that they had clear antioxidant activity, and that the more radicals were scavenged with the increase in the concentration of the extracted solution of soy saponins, suggesting that its scavenging activity on SARs might be associated with saponins. The more hydrogen peroxide was scavenged with increased concentration of soy saponins, but the ratio of scavenging decreased with increased concentration of the extracted solution of soy saponins, the reason of which needs to be further researched.

ACKNOWLEDGEMENTS

This work wassupported by the Key Project of Scientific and Technological Research of the Education Department of Henan, China (No. 13A416110).

REFERENCES

[1] Wang ZY, Ai QJ, Gan JQ, Cheng JW. Research and development of soyasaponins. Cereal Food Indust 1955; 12: 31-34.

[2] Liu SJ, Guan NP, Yan HB. Nutritional and sensory evaluation of soybean meal. Mod J Anim Husbandry Vet Med 2009; 10: 30-32.

[3] Guan WJ, Gu KR. Recent progress of soyasaponin research. Sci Technol Cereals Oils Foods 2007; 15: 42-45.

[4] Upadhyay R, Ramalakshmi K, Rao LJM. Microwave-assisted extraction of chlorogenicacids from green coffee beans. Food Chem 2012; 130: 184-88.

[5] Yang M, Huang FH, Liu CS. Influence of microwave treatment of rapeseed on minor components content and oxidative stability of oil. Food Bioprocses Technol 2013; 6: 3206-16.

[6] Wang JW, Xu YF, Zhou JQ, Chen JL. Advances in enzyme-assisted extraction of Chinese traditional medicinal herbs." Chin J Bioprocess Eng 2008; 6: 6-11.

[7] Teng YP, Zhang YM. Determination of soybean saponins by the method of spectrophotometry. Chin J Food Hygiene 2000; 12: 10-13.

[8] Zhang LX, Zhao LJ. Study on antioxidant activity of Flammulina velutipes polysaccharide. Southwest China J Agri Sci 2014; 27: 240-3.

[9] Wang HB. The study of antioxidant activity of soyasaponin and isoflavone. Food Res Dev 2008; 29: 9-12.

27

The Growth Study of *Perinereis aibuhitensis* in Airlift Recirculating Aquaculture System

Dazuo Yang[a,b,c], Chenchen Cao[b,c], Gang Wang[b,c], Yibing Zhou[b,c] and Zhilong Xiu[a,*]

[a]*College of Life Science and Technology, Dalian University of Technology, Dalian, 116021, China*
[b]*Key Laboratory of Marine Bio-resources Restoration and Habitat Reparation in Liaoning Province, Dalian Ocean University, Dalian,116023, China*
[c]*Key Laboratory of North Mariculture, Ministry of Agriculture, Dalian Ocean University, Dalian 116023, China*

Abstract: For sustainable development of aquaculture industry, it is important to apply biological method and biotechnology to treat the waste of aquaculture. In this paper, an airlift recirculating aquaculture system was designed and polychaete worms were cultured in it. According to the different food level experiments, the growth of *Perinereis aibuhitensis* was tested in each system with different feeding ratios according to food/total worms' weight percentage. It was marked as M1, M2, M3, M4 and M5 respectively. The water parameters were also tested. The results showed that in the M3 groups, the production of *P.aibuhitensis* was the highest being 2.36 g/m^2, while other groups exhibited negative production. Fed on the residual feeds and feces of flounder fish, the mean weight of worms increased to a maximum in M4, which was 1.570g and in M3, it was 0.986g in 40 days. The results provided a novel method to biologically utilize the waste of aquaculture.

Keywords: *Perinereis aibuhitensis*, Growth, Airlift recirculating, Aquaculture system.

1. INTRODUCTION

China promotes aquaculture at large. In the recent years, its aquaculture production has been observed to be the highest in the world [1]. In 2012, China's total aquaculture production was 4,288.36 million tons, of which, mariculture production was 3,033.34 million tons, accounting for more than 70% of the world's mariculture production. However, with the continuous development of aquaculture, environmental problems have been increasingly focused, which have become a major problem restricting the sustainable development of China's aquaculture. Therefore, it is important for healthy development of aquaculture and sustainable use of resources to research the methods of effectively eliminating the rising environmental pollution, and to restore and optimize the breeding environment as soon as possible [2].

Polychaete belongs to Annelida. Polychaete is one kind of benthic polychaetes, which are widely distributed in the coastal beaches and estuarine areas. It has high economic value not only as the best marine bait [3], but also has wide prospects for drug development in the ocean and other areas [4]. As a typical marine benthic fauna of deposited diet, polychaete eats detritus and transforms it into body's energy [5]. It is not only a secondary producer of marine ecosystem, but also is an important part in marine detritus food chain [6]. *Perinereis aibuhitensis*, a kind of dominated benthic fauna, which is widely distributed in China's coast, is used as the fishing bait and bioremediation species to polluted marine

sediment according to their absorption and transformation of organic pollutants detritus [7].

In this study, we constructed an Airlift Recirculating Aquaculture System (ARAS). Polycheate worm *Perinereis aibuhitensis* was cultured with no water exchange. The purpose of this research was to evaluate the role of *P. aibuhitensis* in bioremediation in the marine organic detritus. At the same time, the experiment was also a good trial for the sample recirculating aquaculture system. It was effective for the sustainable development of aquacultural industry and for maintaining the health of marine ecosystem.

2. MATERIALS AND METHOD

2.1. Establishment of the ARAS

The system was constructed with polyvinyl chloride (PVC) materials; its size was 30cm(W)×20cm (H)×25cm(D). In each tank, there were two different layers from the bottom to upwards, which were, the sand layer and the water layer. Two L style PVC pipes were installed in the tanks with one side of the L type pipe inserted in the sand layer and the other side of the L type pipe was kept 3cm higher than the surface of water. The pipe surface in the sand had many holes which were covered with nylon net(diameter:210μm). An air stone was put in the PVC tube with the pipe having a wide mouth, and the water in the pipes was lifted by the air lifting effects through sand layer to the water layer. When the water in the pipes was lifted and outflowed, the water in the tanks flowed in the pipes through the holes. In this way, a self-circulation system was constructed (see Fig. **1**). During the experiment, the circulation velocity of water was 0.2-

*Address correspondence to this author at the College of Life Science and Technology, Dalian University of Technology, Dalian, 116021, China; E-mail: zhilxiu@dlut.edu.cn

0.3L/min and the actual content in the tanks was 5L. Before the experiment, all the pipes and sand were disinfected with sodium hypochlorite. During the experiment period, no water was exchanged.

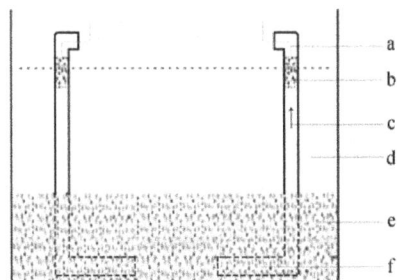

a. Gas-filled tube; b. Air stone; c. Water (flow) direction; d. Water layer; e. Sand layer; f. L style PVC pipe.

Fig. (1). Diagram of the Airlift Recirculating Aquaculture System (ARAS).

2.2. Animals and Aquaculture for Experiment

The animals for the experiment of *P.aibuhitensis* were selected from their natural intertidal habitats near Dalian city. Prior to the experiment, the worms were temporarily cultured in tanks about 15 days, and the worms in good condition were selected for the experiment. The density of *P.aibuhitensis* was 100 ind/m^2 in each tank. The seawater was collected from worms habitat and the water quality parameters included were; temperature 19±1℃, salinity 30-32, and pH 8.01-8.03. During the experiment, air pump was used continuously and the temperature was controlled by air conditioner. The illumination was 50-100Lx.

The food of *P.aibuhitensis* in the experiment included feces and residual feeds of the Japanese flounder fish farm. The feces and residual feeds were mixed evenly with a small amount of water and sodium alginate, which were then turned to 1-2mm diameter particles by feed mechanism.The particles were preserved in the refrigerator at 4°C, whose nutrient composition is shown in Table **1**.

2.3. Grouping

The worms were fed with five feeding ratio according to the food/worms' weight percentage (0%, 6%, 12%, 18%, 24%). The M1, M2, M3, M4 and M5 were marked in different feeding groups. Each feeding group performed four repeated experiments. The experimental period was 40 days and 20 worms were dug up from the sediment and weighed in each ten days.

2.4. Test Methods of Water Chemistry Parameters

The methods of water chemistry parameters used the Chinese specification for marine monitoring—Part 4: Seawater analysis [8] in which the total nitrogen was tested by potassium persulfate oxidation-ultraviolet spectrophotometry method, ammonia was determined using the hypobromite method. Nitrate and nitrite were tested by cadmium column reduction and spectrophotometry method and hydrochloric acid naphthyl ethylenediamine photometric method, respectively.

2.5. Calculation and Statistical Analysis

Specific Growth Rate (SGR) was calculated by the following formula:

$$SGR = 100 \cdot (\ln DW2 - \ln DW1) \cdot T^{-1}$$

where SGR is the specific growth rate (%·d^{-1}), DW1 and DW2 are the worms' body weight (g) in the beginning and in the final stage of experiment, respectively. T is the experimental time(d).

The production and biomass of worms in tanks were calculated by the methods in Gray J S,(1987) [9].

Statistical analyses were performed using SPSS (version 13.0). One-way ANOVA was applied to determine the difference of worms' growth in each tank.

3. RESULTS

3.1. Worms' Growth

The growth in the worms with different diet ratio under the temperature of 20℃ is shown in Fig. (**2-5**). From Figs. (**2-5**), it is obvious that the mean-weight of worms of different groups increased following the experimental time. The M4 group was the most significant with a mean weight of 1.117±0.221g/ind in the beginning of the experiment. After 40 days of aquaculture, the weight of worms was 1.570±0.352g/ind, while the mean-weight of worms in the other groups was 1.374±0.208g/ind, 1.215±0.332g/ind, 0.986±0.211g/ind, 1.448±0.361g/ind. Through statistical analysis, the effect of different diets ratio on the worms' growth was significantly analyzed (F=23.05, P<0.01).

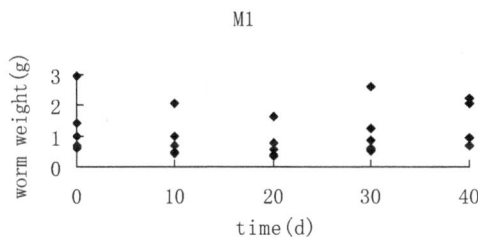

Fig. (2). the weight change of worms in M1.

Table 1.   The nutritional composition of food.

Index	Dry Matter (%)	Organic Matter (%)	Crude Protein (%)	Crude Fat (%)	P (%)	C (%)
Food	90	71.22	35.60	17.79	0.09	35.11

M2

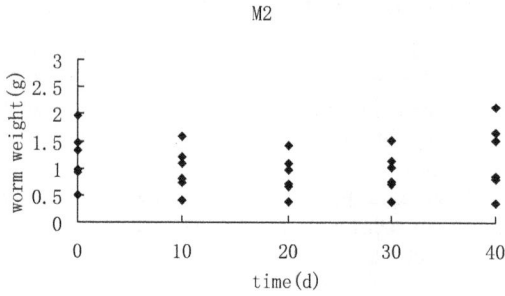

Fig. (3). The weight change of worms in M2.

M3

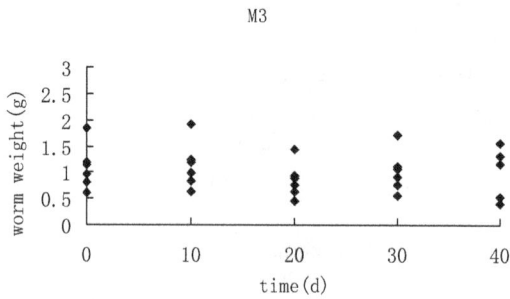

Fig. (4). The weight change of worms in M3.

M4

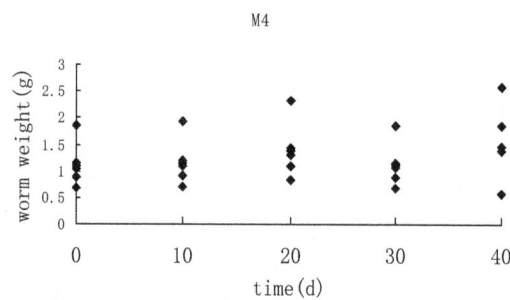

Fig. (5). The weight change of worms in M4.

The SGR of different groups is shown in Fig. (7). It is obvious that the growth and SGR of worms with the different food levels changed differently. In the ARAS, the aquaculture period was 10-20 days, the SGR of M2, M4 and M5 increased, but it was decreased in groups M1 and M3. When aquaculture period was 20-30 days, the SGR of M5 decreased, while the SGR of the rest of the groups increased. Throughout the experiment period, the SGR of M2 increased with special growth rate of M4 being 0.97. The values of SGR of the rest of the groups were 0.56, 0.32 and 0.20. The SGR of the group M3 was negative, being -0.37.

M5

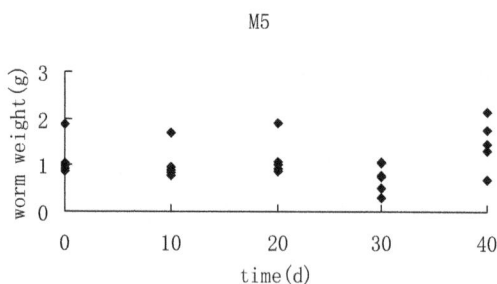

Fig. (6). The weight change of worms in M5.

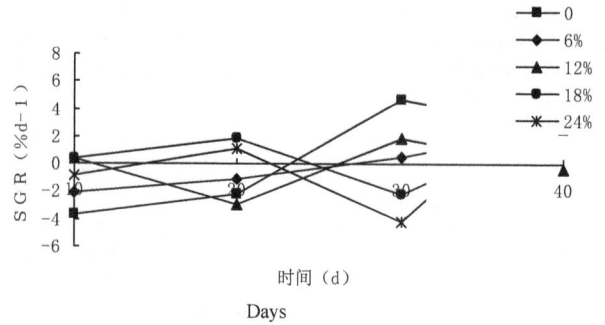

Fig. (7). The SGR of worms in each group.

In Table **2**, it is shown that the production of M1, M2, M3, M4 and M5 decreased at the beginning of the experiment, which was -82.56 g/m^2, -62.73 g/m^2, 2.63 g/m^2, -40.92 g/m^2 and -33.77 g/m^2, respectively. The production of each group increased markedly after 30 days, with the highest production of M3 as 2.36 g/m^2, while the production of M1 was the lowest as -82.56, and the production of M4 was close to M5.

3.2. The Change in Water Chemistry Parameters in Each Group

Figs. (**8-11**) show change in the water chemistry parameters in each group. It could be seen from Figures 7-10 that the change in Total Nitrogen (TN) was similar in each group, which increased and decreased smoothly throughout the experiment. However, the overall trend inclined towards its increment. In the early 20 days of experiment, the ammonia nitrogen (NH$_4^+$-N) concentration significantly changed. It first reached the peak in the experiment of the larger group M3 and subsequently the groups M5, M2, M4 and M1 reached the peak. In 20-40 days, ammonia nitrogen concentration in all the groups declined to the lowest point. The concentration of nitrite nitrogen (NO$_2^-$-N) increased rapidly at first, then decreased to maintain a stable level, and again increased at the end of the experiment. The nitrate nitrogen (NO$_3^-$-N) concentration overall showed an increasing trend.

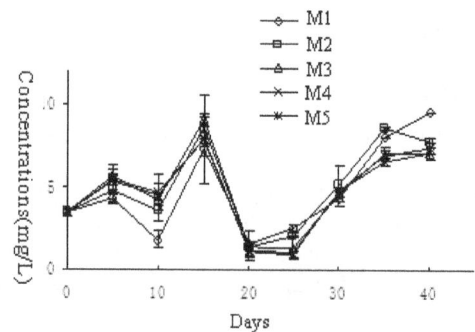

Fig. (8). The total nitrogen change in each group.

4. DISCUSSION

Polychaetes can reduce the pollution from aquaculture and wastewater, therefore, they can promote sustainable aquaculture development [10]. Different polychaete was used to reuse the aquaculture solid waste [11, 12]. In this experiment, *P.aibuhitensis* not only acted well (the worms average

Table 2. The biomass and production of worms in the experiment.

Index	Time (d)	Mean Weight (g)	Density (ind/m^2)	Biomass (g/m^2)	Production (g/m^2)
M1	1	1.594	100	159.43	
	10	1.152	100	115.16	-44.27
	20	0.832	100	83.18	-31.98
	30	1.153	66.67	76.87	-6.31
Total				434.64	-82.56
M2	1	1.468	100	146.82	
	10	1.017	100	101.73	-45.09
	20	1.129	100	112.95	11.22
	30	1.261	66.67	84.08	-28.86
Total				445.58	-62.73
M3	1	1.616	100	161.55	
	10	1.409	100	140.92	-20.63
	20	0.928	100	92.81	-48.80
	30	1.907	83.33	158.92	71.46
Total				591.10	2.63
M4	1	1.221	100	122.12	
	10	1.185	100	118.50	-3.62
	20	0.779	100	77.91	-40.59
	30	1.218	66.67	81.20	3.29
Total				399.73	-40.92
M5	1	1.441	100	144.08	
	10	0.979	100	97.88	-46.21
	20	0.641	100	64.09	-33.79
	30	1.324	83.33	110.32	46.23
Total				416.37	-33.77

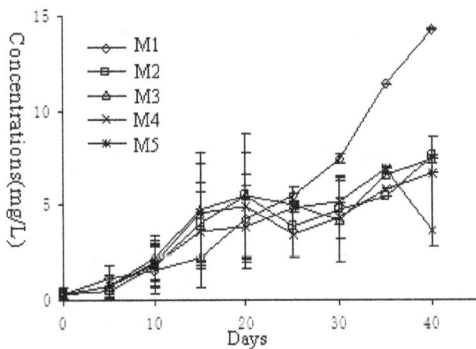

Fig. (9). The change of nitrate in each group.

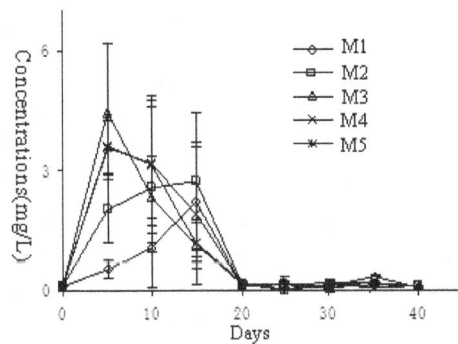

Fig. (10). The change of ammonia in each group.

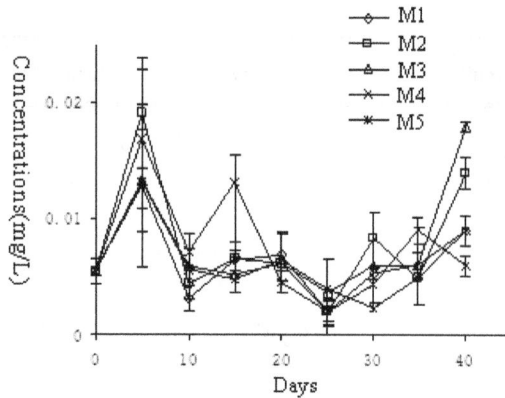

Fig. (11). The change of nitrite in each group.

survival rate was more than 70%), but it also showed growth, depending on the feces and residual food of the fish in the ARAS. In spite of the SGR, under different feeding levels, the growth and production were low. This phenomenon can be explained by several reasons. Firstly, the nutrition of food was very low, and the crude protein was 35.60%, which resultantly made the growth limited. At the same time, the density of worms in the tank became high, therefore, the food was not enough in most of the tanks. The food involved feces and residual food with inadequate viscosity, which were easily dispersed in water after feeding, therefore the worms did not get sufficient food. The particularity of the distribution food also affected the absorbing efficiency of worms. Secondly, there was a consanguineous relation between the creature and the space, where it lived. Most of the studies on aquatic-organism indicated that space is a limiting factor to the growth of the creature. Ali (2003) studied fish in the containers of different sizes in the same feeding condition. It was shown that the fishes which were fed in bigger containers grew rapidly than other fishes [13]. The feeding to shrimp, polliwog and carp gave the same results [14]. The tank used in this experiment was small with a capacity of 5L, therefore, the growth of worms was restricted due to limited space. Thirdly, the substrate was the important habitation environment of polychaete. In the natural ecosystem, the surface of sediment can accumulate a large amount of organic matter by sedimentation. It is the major food resource of polychaete. In the natural sediments, the content of organic matter is 4-5% [15]. Compared to this, the density of worms was not low in the experiment [16], but in the ASRS, the organic matter was very low at the beginning of the experiment, being lower than 2%. As a result, the serious shortage of surface organic matter deposition also affected the growth and survival of worms. Our results show that in the ARAS, under the condition, in which, the feces and residual feeds of fish farm were the only food for *P.aibuhitensis* and without water exchange, most of the worms could survive. The ratio of food/total worms' weight was 12% (M3 group), and the biomass and production of worms were the highest. From the water chemistry results, we also observed that the ARAS could maintain the water quality, especially the concentration of ammonia-N could be kept at lower level in spite of substantial wasted food given to the worms. The nitrite was not stable in water and would be transferred to the nitrate by

microorganisms living in the sediments. The concentration of nitrite was high during the experiment, but the worms tolerated and showed growth in high concentration nitrate [17]. The experimental results provided us a confident trial to reuse the waste of aquaculture by the polychaete airlift recirculating aquaculture system.

ACKNOWLEDGEMENTS

This work was funded by the National Marine Public Welfare Research Project (No. 201305002, 201305043), and the project of marine ecological restoration technology research to the Penglai 19-3 Oil Spill Accident and Dalian Natural Foundation (No. 2012J21DW014).

REFERENCES

[1] Chen J, Xu H, Ni Q, Liu H. The study report on the development of China industrial recirculating aquaculture. Fishery Modernization 2009; 36: 1-7.

[2] Bunting SW. Principles of Sustainable Aquaculture, Promoting Social, Economic and Environmental Resilience. Routledge, Abingdon, UK, 2013; pp. 301.

[3] Carvalho AN, Vaz ASL, Sérgio TIB, Santos PJ. Sustainability of bait fishing harvesting in estuarine ecosystems–Case study in the Local Natural Reserve of Douro Estuary, Portugal. Revista da Gestão Costeira Integrada, 2013; 13: 157-168.

[4] Roscia G, Falciani C, Bracci L, Pini A. The development of anti-microbial peptides as new antibacterial drugs. Curr Protein Pept Sci 2013; 14: 641-649.

[5] McArthur M, Brooke B, Przeslawski R, Ryan D, Lucieer V, Nichol S, McCallum A, Mellin C, Cresswell I, Radke LC. On the use of abiotic surrogates to describe marine benthic biodiversity. Estuar Coast Shelf S 2010; 88: 21-32.

[6] Gerlach SA. Food-chain relationships in subtidal silty sand marine sediments and the role of meiofauna in stimulating bacterial productivity. Oecologia 1978; 33: 55-69.

[7] Zhao H, Zhou Y, Li Y, Li S, Yang D. Molecular cloning and expression of the gene for G protein alpha subunit induced by bisphenol A in marine polychaete *Perinereis aibuhitensis*. Environ Toxicol Phar 2014; 37: 521-528.

[8] The Specification for Marine Monitoring-Part 4: Seawater Analysis, General Administration of Quality Supervision, Inspection and Quarantine of the People's Republic of China, Standardization Administration of the People's Republic of China, 2008.

[9] Gray, JS. In: Gee, JHR and Giller, PS, Ed.Organization of Communities Past and Present.Species-abundance patterns,Blackwell Science, Oxford 1987; 53-68.

[10] Aguado-Giménez F, Piedecausa M, Carrasco C, Gutiérrez J, Aliaga V, García-García B. Do benthic biofilters contribute to sustainability and restoration of the benthic environment impacted by offshore cage finfish aquaculture? Mar Pollut Bull 2011; 62: 1714-1724.

[11] Honda H, Kikuchi K.Nitrogen budget of polychaete *Perinereis nuntia* vallata fed on the feces of Japanese flounder. Fisheries SCI 2002; 68: 1304-1308.

[12] Palmer PJ. Polychaete-assisted sand filters. Aquaculture 2010; 306: 369-377.

[13] Ali M, Nicieza A, Wootton RJ. Compensatory growth in fishes: a response to growth depression. Fish Fish 2003; 4: 147-190.

[14] Mascaró M, Rodríguez-Pestaña L, Chiappa-Carrara X, Simões N. Host selection by the cleaner shrimp *Ancylomenes pedersoni*: Do anemone host species, prior experience or the presence of conspecific shrimp matter? J Exp Mar Biol Ecol 2012; 413: 87-93.

[15] Heiri O, Lotter AF, Lemcke G. Loss on ignition as a method for estimating organic and carbonate content in sediments: reproducibility and comparability of results. J Paleolimnol 2001; 25: 101-110.

28

Molecular Character, Phylogeny and Expression of Tomato *LeNHX3* Gene Involved in Multiple Adverse Stress Responses

Jing Fan[1,*], Jianping Hu[2], Lanyang Gao[3], Jinhua Liao[1] and Mingyuan Huang[1]

[1]*College of Life Science, Leshan Normal University, Leshan, 614004, P.R. China*

[2]*College of Chemistry, Leshan Normal University, Leshan, 614004, P.R. China*

[3]*Sichuan Academy of Botanical Engineering, Chengdu, 641200, P.R. China*

Abstract: Crop production is severely affected by high salt stress. To obtain more salt-tolerant crops by genetic modification, it is crucial to explore some key genes associated with salt tolerance. *LeNHX3* gene is considered one putative Na^+/H^+ antiporter with the ability of improving plant salt tolerance by maintaining intracellular ionic balance in tomato, however, limited information about it has been reported. Here, we report the structure, phylogenetic evolution and expression of *LeNHX3* gene from wild type tomato (*Lycopersicon esculentum Mill* cv. Ailsa Craig). Sequence analysis showed that *LeNHX3* encodes a protein containing 10 transmembrane domains, with a typical conserved amiloride binding domain presented in the third transmembrane domain. An interesting discovery also showed that sequence of LeNHX3 was more conserved than its allele protein collected by GenBank (designated as LeNHX3-GB in this study) when compared with others Na^+/H^+ antiporters. Homology modeling results showed that the structure of LeNHX3 protein consists mainly of α-helix and random coil, it has similar tertiary structure to that of LeNHX3-GB, however, inter-residue interactions were found to be further strengthened in LeNHX3. Phylogenetic analysis showed LeNHX3 was clustered with vacuolar Na^+/H^+ antiporters and has distant relationship to plasma membrane Na^+/H^+ antiporters. Expression profiles analysis indicated *LeNHX3* gene was constitutively expressed in roots, stems and leaves, its expression was also induced by salt, low temperature and abscisic acid. The results presented in this work provide new insights into *LeNHX3* gene, it is particularly important that one new *LeNHX3* allele from wild tomato was mined, which can serve as a candidate gene for improving plant stress tolerance by genetic engineering.

Keywords: Homology modeling, *LeNHX3* gene, Na^+/H^+ antiporters, Phylogenetic evolution, Salt tolerance.

1. INTRODUCTION

Soil salinization has been one of the severest negative environmental constraints, nearly 7 percent of the total land, 20 percent of the cultivated area and 50 percent of the irrigated lands in the world are adversely affected by salinity stress [1-3], it disrupts the normal photosynthesis and carbohydrate metabolism of corps, with a consequence of plant growth retardation and yield reduction. Global agricultural sustainability is largely dependent on the improvement of crop salt tolerance [4]. Tomato is one of the most widely grown and consumed vegetables in the world [5], however, most of the cultivated tomatoes are highly or moderately sensitive to soil salinity, which results in substantially reducing the yields under salt stress [6, 7]. Wild tomatoes are more salt-tolerant than cultivated tomatoes [8], they are suitable as the germplasms for mining genes for genetic improvement of salt tolerance in cultivated tomatoes.

In order to avoid occurrence of high salt toxicity in plants, Na^+ should be transported outside the cytosol or inside the vacuoles, all the processes can be mediated by Na^+/H^+ antiporter, a protein conferring salt tolerance for plant by maintaining ion homeostasis in cells [9]. To date, many Na^+/H^+ antiporters have been cloned and characterized. *AtNHX1* was the first vacuolar Na^+/H^+ antiporter isolated from *Arabidopsis thaliana*, its over-expression led to increased salt tolerance of *Arabidopsis thaliana*, peanut and maize [10, 11]. Na^+/H^+ antiporters from other species also confer salt tolerance in plants, for example, the vacuolar Na^+/H^+ antiporter *SbNHX1* gene from extreme halophyte *Salicornia brachiata* conferred salt tolerance for Jatropha curcas [12]. In tomato, several Na^+/H^+ antiporters have also been reported. *LeNHX2*, one Na^+/H^+ antiporter located in vacuole, is an important determinants for salt tolerance of tomato [13, 14]. *LeNHX3* is another Na^+/H^+ antiporter in tomato, a positive correlation was found between its expression level and salt tolerance in tomato [15], however, more molecular information about it is still lacking. In this study, structure, phylogeny and expression profiling of *LeNHX3* from wild type tomato (*Lycopersicon esculentum Mill* cv. Ailsa Craig) were analyzed. This work is helpful for us to explore more information of *LeNHX3* and improve plant abiotic stress tolerance by genetic engineering in the future.

*Address correspondence to this author at College of Life Science, Leshan Normal University, Leshan, Sichuan, 614004, P.R. China; E-mail: fanjing972001@126.com

Table 1. Comparison of Na$^+$/H$^+$ antiporters from different species.

Protein Name	Species	GenBank Accession Number	Numbers of Amino Acid	Molecular Weight (Da)	Theoretical Isoelectric Point
LeNHX3	*Solanum lycopersicum* (wild type, Ailsa Craig)	—	537	59421.5	8.55
LeNHX3-GB	*Solanum lycopersicum*	CAK12754.1	537	59443.5	8.54
InNHX2	*Ipomoea nil*	BAD91200	536	59317.6	7.17
CmNHX1	*Chrysanthemum x morifolium*	ABN71591	550	61085.3	6.46
AgNHX1	*Atriplex gmelini*	BAB11940	555	61504.8	6.70
BnNHX2	*Brassica napus*	ACZ92142	542	59931.1	7.67
ZmNHX2	*Zea mays*	NP001105531	540	59808.2	8.25
MzNHX1	*Malus zumi*	ADB80440	544	60474.0	8.85
VvNHX1	*Vitis vinifera*	AAV36562	541	60137.2	7.24
SbNHX1	*Salicornia brachiata*	ACA33931	560	62322.7	6.43
PeNHX3	*Populus euphratica*	ACU01854	545	60293.7	8.13
TaNHX2	*Triticum aestivum*	AAK76738	538	59082.4	8.41
GhNHX1	*Gossypium hirsutum*	AAM54141	543	60089.5	7.20
KcNHX2	*Karelinia caspia*	ABC18331.1	550	61079.5	6.36
HtNHX1	*Helianthus tuberosus*	ABM17091.1	549	60744.2	6.90
PtNHA1	*Puccinellia tenuiflora*	EF440291	1137	125500.2	6.48
CsSOS1	*Cucumber*	AFD64618.1	1144	127272.0	6.30
OsSOS1	*Oryza sativa*	AAW33875.1	1148	127917.8	6.77

Note: LeNHX3-GB indicates the LeNHX3 protein collected in Genbank.

2. MATERIALS AND METHODS

2.1. Plant Materials

Mature seeds of *Lycopersicon esculentum Mill.* cv. Ailsa Craig were sanitized with 5% sodium hypochlorite and then germinated on 1/2 Murashige and Skoog (MS) medium, after grown at 25°C in complete darkness for one week, seeds were incubated under 16h light and 8h dark photoperiod cycles until seedlings reached the height of about 8 centimeters, after treated with 200mM NaCl, 150mM mannitol, low temperature (4°C) and 10μM Abscisic acid for 6h, all samples were then collected and frozen immediately in liquid nitrogen and stored at -80°C refrigerator for RNA extraction.

2.2. Protein Sequences

The *LeNHX3* gene has been cloned from wild type tomato (*Lycopersicon esculentum* Mill. cv. Ailsa Craig) and sequenced in our previous study [16]. In this study, amino acid sequence of *LeNHX3* was deduced from its cDNA sequence, sequences of other Na$^+$/H$^+$ antiporters were got from the protein database maintained by NCBI (http://www.ncbi.nlm. nih.gov/protein), for more details see Table 1.

2.3. Molecular Characteristics, Structure and Phylogenetic Relationship Analysis

Physical characteristics of LeNHX3 protein were deduced by protparam program at the ExPASy server (http://au.expasy.org/tools/protparam.html) with default parameters. Transmembrane analysis was performed by TMHMM server (http://www.cbs.dtu.dk/services/ TMHMM -2.0/) with default parameters. Multiple sequence alignments and amino acid sequence homology analysis between LeNHX3 and other Na$^+$/H$^+$ antiporters were performed by DNAMAN software (Lynnon corporation, Quebec, Canada) using the full alignment method. To construct three-dimensional structure of LeNHX3, fasta format sequence of which was submitted to the Swiss-model workspace (http://swissmodel.expasy.org/workspace/index. php), the template hits for LeNHX3 protein was then searched using template identification tool [17], the resulting structure with the largest sequence homology to LeNHX3 was used as template, homology modeling of LeNHX3 was then performed using alignment mode in Swiss-Model, the resulting structure was viewed using pymol software (version 0.99, DeLano Scientific LLC, South San Francisco, California, USA). On the basis of amino acid sequence alignments by Clustalx1.83 (EMBL-EBI, Cambridge, UK) using multiple alignment mode, phylogenetic evolutionary analysis was completed using MEGA software version 5.0 (www.megasoftware.net), the neighbor-joining (NJ) tree was generated using the p-distance method with complete deletion option and 1000 bootstrap replicates.

2.4. Tissue Specific and Stress Induced Expression of *LeNHX3* Gene

Total RNA for tissue specific expression was extracted from the mashed roots, stems and leaves. Total RNA for stress induced expression was extracted from the plantlets induced by salt, mannitol, low temperature and ABA using RNAprep Kit (Tiangen, Beijing, China), their cDNA were synthesised using cDNA synthesis kit (TaKaRa, Dalian, China). RT-PCR primers were designed with Primer 5.0 software, the primer sequences for *LeNHX3* gene amplification were as follows: 5'-GACTTATGCGAGG TGCTGTT-3' (forward primer) and 5'-CACTTGGTTCCG TTGGTGAT-3' (reverse primer). The housekeeping gene ubiquitin III (Ubi3) was assayed as an internal control (GenBank accession no. X58253.1), the primer sequences were 5'-AGAAGAAGACCTACACCAAGCC-3' (forward primer) and 5'-TCCCAAGGGTTGTCACATACATC-3' (reverse primer). The PCR amplification program was as follows: 94°C for 5min (initial denaturing), followed by 30 cycles of 94°C for 30s (denaturation), 55°C for 30s (annealing) and 72°C for 30s (extension), with a final extension at 72°C for 10 min, PCR products were then analyzed on 1.0% ethidium bromide-stained agarose gel.

3. RESULTS

3.1. Physiochemical Properties

To analyze the physicochemical parameters of different Na^+/H^+ antiporters, the protein sequences of 3 plasma membrane (PtNHA1, CsSOS1, OsSOS1) and 15 vacuolar type Na^+/H^+ antiporters were analyzed using ProtParam tool. The results showed that LeNHX3 protein encoding a polypeptide containing 537 amino acid residues with a predicted molecular weight of 59421.5 Da and theoretical isoelectric point of 8.55, it was smaller in molecular weight, but with larger theoretical isoelectric point than other Na^+/H^+ antiporters (Table **1**). To analyze the distribution of transmembrane domain, amino acid sequence of LeNHX3 was analyzed using TMHMM 2.0, the result indicated that LeNHX3 was consisted of ten transmembrane domains between the residues 21 to 43, 53 to 72, 77 to 99, 114 to 136, 218 to 240, 270 to 292, 304 to 326, 341 to 363, 384 to 402, 417 to 436, respectively (Fig. **1**).

3.2. Sequence Alignments and Homology Analysis

To identify the conserved domain presented in LeNHX3 protein and compare the sequence differences between LeNHX3 and other Na^+/H^+ antiporters, multiple sequence alignments of LeNHX3 against other 14 known vacuolar type Na^+/H^+ antiporters were performed using full alignment method of DNAMAN software. The result revealed that the conserved amiloride-binding domain LFFIYLLPPI was present in the third transmembrane domain at the N terminal of LeNHX3. Interesting, by comparing the amino acid

sequence of LeNHX3 with its allele-associated protein LeNHX3-GB, three amino acid substitutions between them were found, it was a Tyrosine to Histidine substitution at position 143 (Y143H), a Proline to Serine substitution at position 346 (P346S) and a Glycine to Alanine substitution at position 399 (G399A) in LeNHX3 protein, respectively. The Serine at position 346 and Alanine at position 399 were located in the eighth and ninth transmembrane domain of LeNHX3, both of them were more conserved than that of LeNHX3-GB (Fig. **1**), this indicated that the Serine-346 and Alanine-399 are important for transport function of LeNHX3.

3.3. Modeling the Three-Dimensional Structure of LeNHX3

To establish the tertiary structure of LeNHX3 and compare the structure differences between LeNHX3 and its allele associated protein LeNHX3-GB, their protein sequences were homology-modeled using the SWISS-MODEL server in alignment model, structures were constructed based on the sequence ranging from Phenylalanine at position 3 (Phe 3) to Isoleucine at position 386 (Ile 386) of LeNHX3 and LeNHX3-GB proteins, structure of NapA (PDB code, 4bwzA) was chosen as the template for modeling. The results showed that LeNHX3 and LeNHX3-GB proteins were primarily composed of α-helix and random coil (Fig. **2A, B**), the conserved amiloride-binding domain LFFIYLLPPI and the different residues at position 143 and 346 were located at the protein surface (Fig. **2C, D**). Although LeNHX3 showed high similarity with the structure of LeNHX3-GB, changes of inter-residue interactions were still found. In LeNHX3, one oxygen of Histidine residue at position 143 (H143) was found to interact with the nitrogen of Glycine residue at position 147 (G147), with a distance of 3.36 Å, two nitrogen atoms of H143 were involved in the interface with the oxygen and nitrogen of Asparagine residue at position 140 (N140), including formation of two sets of hydrogen bonds with separation of 2.86 and 3.10Å (Fig. **2E**). We also found the Serine residue at position 346 (S346) interacted with the Glutamine residue at position 343 (Q343) in LeNHX3, with a single polar contact of 3.36 Å (Fig. **2F**). However, only the residue Tyrosine at position 143 (Y143) formed two polar contacts with the Glycine at position 147 (G147) and the Asparagine at position 140 (N140) in LeNHX3-GB, resulting in two hydrogen bonds with distance of 3.34 and 2.86 Å, respectively (Fig. **2G**).

3.4. Phylogenetic Analysis of LeNHX3 and other Na^+/H^+ Antiporters

In order to ascertain the evolutionary relationships between the LeNHX3 and Na^+/H^+ antiporters from other plant species, a phylogenetic tree was constructed. The result showed that LeNHX3 has close phylogenetic relationship to the vacuolar type antiporters, it falls into the same clade with tomato LeNHX3-GB and InNHX2 from *Ipomoea nil*. However, LeNHX3 showed distant genetic relationship to plasma-membrane type Na^+/H^+ antiporters (Fig. **3**), this allows us to confirm that LeNHX3 is a typical vacuolar Na^+/H^+ antiporter.

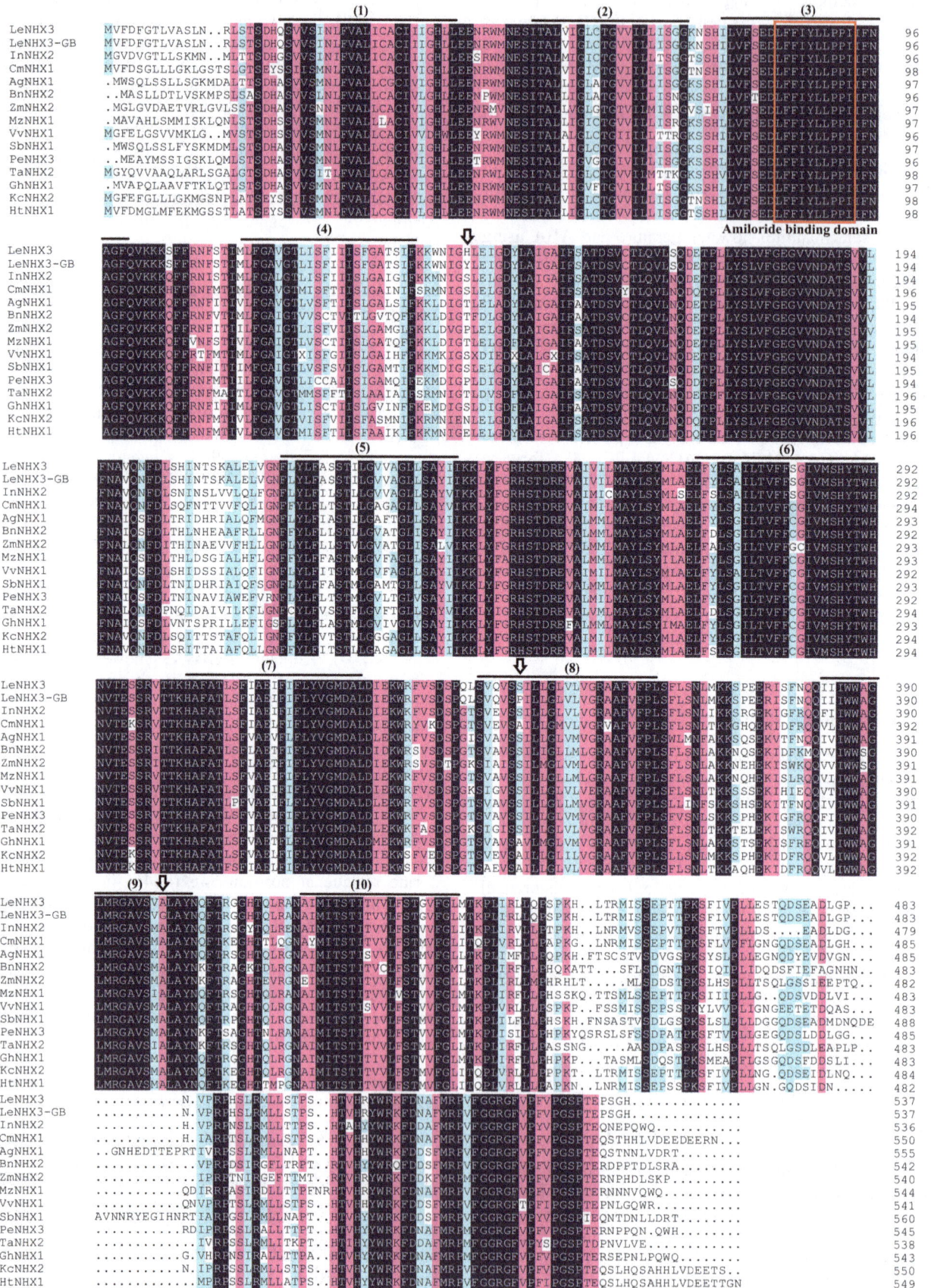

Fig. (1). Multiple sequence alignments of LeNHX3 with Na$^+$/H$^+$ antiporters from other species. The box indicates the conserved amiloride-binding site, the different amino acids in LeNHX3 and LeNHX3-GB are indicated by arrows, the 10 transmembrane domains are indicated by an overline respectively.

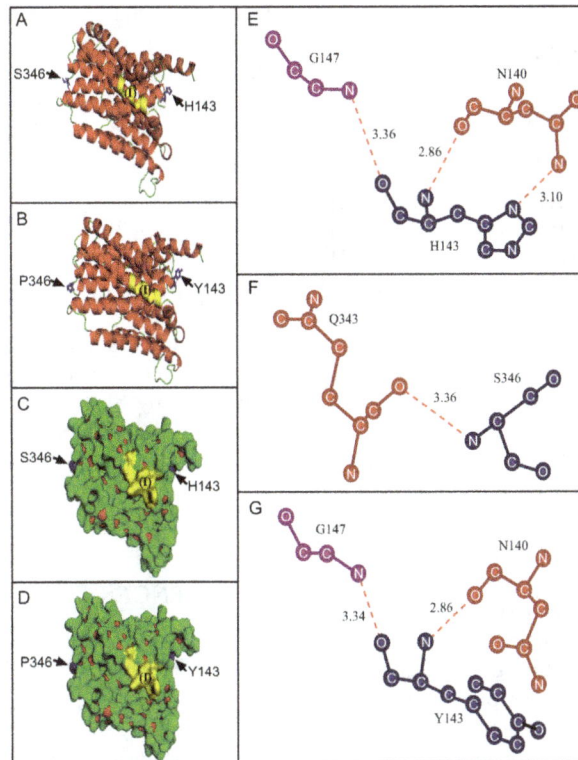

Fig. (2). Homology modeling and comparison of the interactions for different residues in LeNHX3 and LeNHX3-GB. Three-dimensional structures of LeNHX3 and LeNHX3-GB are shown in cartoon and surface representations, with α-helices colored red, coiled regions colored green, the conserved amiloride-binding domain colored in yellow and labeled as (I). Different residues at position of 143 and 346 of LeNHX3 and LeNHX3-GB were colored blue and indicated by arrows. (**A**) Cartoon representation of LeNHX3. (**B**) Cartoon representation of LeNHX3-GB. (**C**) Surface representation of LeNHX3. (**D**) Surface representation of LeNHX3-GB. The different residue interactions in LeNHX3 and LeNHX3-GB were shown in stick-sphere representations, hydrogen bonds were shown as dashed red lines. (**E**) Interaction of the Histidine residue at position 143 (H143) with the Glycine residue at position 147 (G147) and the Asparagine residue at position 140 (N140) in LeNHX3. (**F**) Interaction of the Serine residue at position 346 (S346) with the Glutamine residue at position 343 (Q343) in LeNHX3. (**G**) Interaction of the Tyrosine residue at position 143 (Y143) with the residue Glycine at position 147 (G147) and the residue Asparagine at position 140 (N140) in LeNHX3-GB.

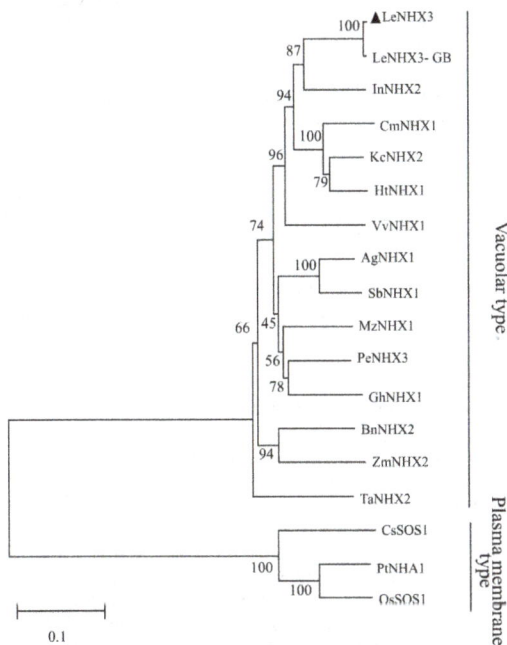

Fig. (3). Phylogenetic analysis of LeNHX3 and other Na^+/H^+ antiporters. Numbers near the nodes represent the percentage of 1000 bootstrap replicates that support each node, the scale bar corresponds to 0.1 amino acid substitutions per site.

3.5 Expression Patterns of *LeNHX3* Gene

To investigate the expression patterns of *LeNHX3* gene, specific tissue expressions of *LeNHX3* in various tissues were examined by RT-PCR. The results showed that *LeNHX3* was constitutively expressed in the leaves, stems and roots (Fig. **4A**), this suggested that *LeNHX3* is essential for the normal function of wild type tomato. Abiotic stress induced expression showed that *LeNHX3* was induced by salt, which demonstrated the potent role of LeNHX3 in salt tolerance in wild type tomato. Interesting, the transcript levels of *LeNHX3* were also up-regulated by low temperature and ABA, and the highest expression occurs at low temperature treatment, however, the transcript was not obviously induced by mannitol, this indicated that *LeNHX3* is involved in cross talk between salt, low temperature and ABA in tomato (Fig. **4B**).

4. DISCUSSION

Na^+/H^+ antiporter maintains a steady salt homeostasis by transporting Na^+ and H^+ ions across the cell membrane [18]. Due to wild genotype tomatoes are usually more salt tolerant than the cultivar, they are regarded as ideal gene donor for improving salt tolerance capacity of cultivated tomatoes [19]. The putative Na^+/H^+ antiporter LeNHX3 gene has been

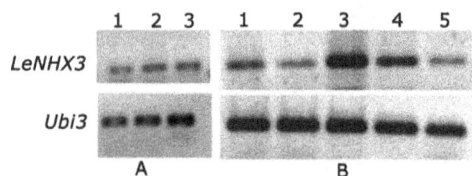

Fig. (4). Expression of *LeNHX3* gene. **(A)** RT-PCR analysis of *LeNHX3* gene (upper) and the internal control *Ubi3* gene (lower) in different tissues. Lane 1, roots; lane 2, stems; lane 3, leaves. **(B)** Abiotic stress induced expression of *LeNHX3 gene (upper) and the internal control Ubi3 gene (lower)*. Lane 1, salt; lane 2, mannitol; lane 3, low temperature; lane 4, ABA; lane 5, untreated.

previously cloned by us from wild type tomato [16], three amino acids were found different in comparison with its homologous LeNHX3-GB, and two of they were more likely to appear in most of the Na^+/H^+ antiporters (Fig. **1**). It has been previously reported that base substitutions can significantly alter gene function, for example, the Na^+/H^+ antiporter *SOS1* gene in *Arabidopsis thaliana* is essential for plant salt tolerance, however, a single base substitution in *SOS1* made plants show salt-hypersensitive and low K^+ affinità [20, 21], thus we speculate that the substituted amino acids in *LeNHX3* may confer plants more pronounced salt tolerance. It has been demonstrated that homology model is accurate enough to predict protein structures in wide ranging applications [22, 23], their folds are stabilized by inner residues contacts [24, 25]. Kozachkov and Padan have reported that two residues at position 136 and 399 of Na^+/H^+ antiporter NhaA in *Escherichia coli* were closely related to the conformational changes of protein [26]. In this study, no obvious conformational changes were observed between LeNHX3 and its homologues LeNHX3-GB, however, the substituted residues 143 (H143) and 346 (S346) in LeNHX3 strengthened inter-residue interactions in comparison with LeNHX3-GB, those make LeNHX3 conformation more stable than LeNHX3-GB (Fig. **2E-G**). Evolutionary tree is commonly used to infer phylogenetic relationships between species [27], our results showed LeNHX3 and LeNHX3-GB formed the same clade with vacuolar Na^+/H^+ antiporter InNHX2 [28], but with more distantly related to the plasma membrane Na^+/H^+ antiporters CsSOS1, PtNHA1 and OsSOS1 [29-31] (Fig. **3**), so we conclude that LeNHX3 and LeNHX3-GB play important roles in transporting excess cytoplasmic Na^+ into vacuoles. Na^+/H^+ antiporters can be induced by external stimuli, transcript level of *AtNHX1* has been reported to be up-regulated by NaCl and ABA, but not by cold [32]. The transcript of *OsNHX1* increased under condition of salt stress, but it was not induced by mannitol treatment [33]. In this study, the *LeNHX3* expression was improved by salt, ABA treatments, and the highest expression occurs at low temperature treatment, however, the expression was not obviously affected by mannitol (Fig. **4B**), this indicates that *LeNHX3* is involved in salt, cold and ABA stresses response.

CONCLUSION

Results of molecular character and phylogeny indicates residue substitutions in LeNHX3 make it more conserved as a typical vacuole Na^+/H^+ antiporter in compare to LeNHX3-GB. Expression analysis showed that *LeNHX3* gene is

involved in the cross-talk of salt, low temperature and ABA response in wild type tomato. All results in this study showed that the transcript form of *LeNHX3* in wild type tomato is suitable as an important target gene for improving plant adverse stress tolerance by genetic manipulation in the future.

ACKNOWLEDGEMENTS

The authors thank the anonymous referees for careful reading and constructive suggestions. This work was supported by Scientific research fund of Sichuan provincial education department (08ZB051) and Educational reform project of Leshan Normal university (JG14-YB14).

REFERENCES

[1]   Mishra S, Alavilli H, Lee Bh, Panda S, Sahoo L. Cloning and Functional Characterization of a Vacuolar Na^+/H^+ Antiporter Gene from Mungbean (*VrNHX1*) and Its Ectopic Expression Enhanced Salt Tolerance in *Arabidopsis thaliana*. PLoS ONE 2014; 9(10): e106678.

[2]   Yousofinia M, Ghassemian A, Sofalian O, Khomari S. Effects of salinity stress on barley (Hordeum vulgare L.) germination and seedling growth. Int J Agric Crop Sci 2012; 4(18): 1353-1357.

[3]   Hussain S, Khaliq A, Matloob A, Wahid M, Afzal I. Germination and growth response of three wheat cultivars to NaCl salinity. Soil Environ 2013; 32(1): 36-43.

[4]   Lv D, Subburaj S, Cao M, *et al.* Proteome and Phosphoproteome Characterization Reveals New Response and Defense Mechanisms of Brachypodium distachyon Leaves under Salt Stress. Mol Cell Proteomics 2014; 13(2): 632-652.

[5]   Kubo M, Augusto P, Cristianini M. Effect of high pressure homogenization (HPH) on the physical stability of tomato juice. Food Res Int 2013; 51(1): 170-179.

[6]   Bolarín M, Perez Alfocea F, Cano E, EstañM T, Caro M. Growth, fruit yield, and ion concentration in tomato genotypes after pre-emergence and post-emergence salt treatments. J Am Soc Horticult Sci 1993; 118(5): 655-660.

[7]   Chookhampaeng S, Pattanagul W, Theerakulpisut P. Screening some tomato commercial cultivars from Thailand for salinity tolerance. Asian J Plant Sci 2007; 6(5): 788-794.

[8]   Cano E, Pérez-Alfocea F, Moreno V, Caro M, Bolarín M. Evaluation of salt tolerance in cultivated and wild tomato species through in vitro shoot apex culture. Plant Cell, Tissue Organ Cult 1998; 53(1): 19-26.

[9]   Baltierra F, Castillo M, Gamboa MC, Rothhammer M, Krauskopf E. Molecular characterization of a novel Na^+/H^+ antiporter cDNA from Eucalyptus globulus. Biochem Biophys Res Commun 2013; 430(2): 535-540.

[10]   Banjara M, Zhu L, Shen G, Payton P, Zhang H. Expression of an Arabidopsis sodium/proton antiporter gene (*AtNHX1*) in peanut to improve salt tolerance. Plant Biotechnol Rep 2012; 6(1): 59-67.

[11]   Yin XY, Yang AF, Zhang KW, Zhang JR. Production and analysis of transgenic maize with improved salt tolerance by the introduction of *AtNHX1* gene. Acta Botanica Sinica 2004; 46(7): 854-861.

[12]   Jha B, Mishra A, Jha A, Joshi M. Developing transgenic Jatropha using the *SbNHX1* gene from an extreme halophyte for cultivation in saline wasteland. PLoS ONE 2013; 8(8): e71136.

[13]   Rodriguez-Rosales M, Jiang X, Gálvez F, Aranda M, Cubero B, Venema K. Overexpression of the tomato K^+/H^+ antiporter *LeNHX2* confers salt tolerance by improving potassium compartmentalization. New Phytologist 2008; 179(2): 366-377.

[14]   Huertas R, Rubio L, Cagnac O, *et al.* The K^+/H^+ antiporter *LeNHX2* increases salt tolerance by improving K^+ homeostasis in transgenic tomato. Plant Cell Environ 2013; 36(12): 2135-2149.

[15]   Gálvez F, Baghour M, Hao G, Cagnac O, Rodríguez-Rosales M, Venema K. Expression of *LeNHX* isoforms in response to salt stress in salt sensitive and salt tolerant tomato species. Plant Physiol Biochem 2012; 51(2): 109-115.

[16]   Fan J. Molecular cloning of *LeNHX3* gene from tomato and construction of the Over-expression Vector. Jiangsu Agric Sci 2010; 3(6): 47-49. (In chinese)

[17]   Arnold K, Bordoli L, Kopp J, Schwede T. The SWISS-MODEL workspace: a web-based environment for protein structure homology modelling. Bioinformatics 2006; 22(2): 195-201.

[18]   Zhu J. Regulation of ion homeostasis under salt stress. Curr Opin Plant Biol 2003; 6(5): 441-445.

[19]   Sun W, Xu X, Zhu H. Comparative transcriptomic profiling of a salt-tolerant wild tomato species and a salt-sensitive tomato cultivar. Plant Cell Physiol 2010; 51(6): 997-1006.

[20]   Shi H, Ishitani M, Kim C, Zhu J. The Arabidopsis thaliana salt tolerance gene *SOS1* encodes a putative Na^+/H^+ antiporter. Proc Nat Acad Sci USA 2000; 97(2): 6896-6901.

[21]   Wu S, Ding L, Zhu J. SOS1, a Genetic Locus Essential for Salt Tolerance and Potassium Acquisition. Plant Cell 1996; 8: 617-627.

[22]   Bordoli L, Schwede T. Automated protein structure modeling with SWISS- MODEL Workspace and the Protein Model Portal. Methods Mol Biol 2012; 857: 107-136.

[23]   Fraccalvieri D, Soshilov A, Karchner S, *et al*. Comparative analysis of homology models of the AH receptor ligand binding domain: verification of structure-function predictions by site-directed mutagenesis of a nonfunctional receptor. Biochemistry 2013; 52(4): 714-725.

[24]   Min J, Zhang Y, Xu R. Structural basis for specific binding of Polycomb chromodomain to histone H3 methylated at Lys 27. Genes Dev 2003; 17(15): 1823-1828.

[25]   Faure G, Bornot A, de Brevern A. Protein contacts, inter-residue interactions and side-chain modelling. Biochimie 2008; 90(4): 626-639.

[26]   Kozachkov L, Padan E. Site-directed tryptophan fluorescence reveals two essential conformational changes in the Na^+/H^+ antiporter NhaA. Proc Nat Acad Sci 2011; 108(38): 15769-15774.

[27]   Zhang G, Zhang X, Chen Z. Phylogeny of cryptogrammoid ferns and related taxa based on rbcL sequences. Nordic J Botany 2003; 23(4): 485-493.

[28]   Ohnishi M, Fukada-Tanaka S, Hoshino A, Takada J, Inagaki Y, Iida S. Characterization of a novel Na^+/H^+ antiporter gene *InNHX2* and comparison of *InNHX2* with *InNHX1*, which is responsible for blue flower coloration by increasing the vacuolar pH in the Japanese morning glory. Plant Cell Physiol 2005; 46(2): 259-267.

[29]   Wang S, Li Z, Rui R, Fan G, Lin K. Cloning and characterization of a plasma membrane Na^+/H^+ antiporter gene from Cucumis sativus. Russ J Plant Physiol 2013; 60(3): 330-336.

[30]   Wang X, Yang R, Wang B, Liu G, Yang C, Cheng Y. Functional characterization of a plasma membrane Na^+/H^+ antiporter from alkali grass (Puccinellia tenuiflora). Mol Biol Rep 2011; 38(7): 4813-4822.

[31]   Mahajan S, Pandey G, Tuteja N. Calcium-and salt-stress signaling in plants: shedding light on SOS pathway. Arch Biochem Biophys 2008; 471(2): 146-158.

[32]   Shi H, Zhu JK. Regulation of expression of the vacuolar Na^+/H^+ antiporter gene *AtNHX1* by salt stress and abscisic acid. Plant Mol Biol Rep 2002; 50(3): 543-550.

[33]   Fukuda A, Nakamura A, Tagiri A, *et al*. Function, intracellular localization and the importance in salt tolerance of a vacuolar Na^+/H^+ antiporter from rice. Plant Cell Physiol 2004; 45(2): 146-159.

29

Analysis of Correlations Between Climate and Molecular Adaptive Evolution of Wild Barley with Geographical Information Systems (GIS)

Lina Zhao, Tao Pei, Gege Yang and Chenghu Zhou[*]

State Key Laboratory of Resources and Environmental Information System, Institute of Geographical Sciences and Natural Resources Research, Chinese Academy of Sciences, China

Abstract: Climate is one of the most important factors determining the adaptive evolution of plants. In this study, 44 different populations of wild barley were used as materials to analyze the diversity of 17 genes (495 sequences), in order to study the influence of different climatic conditions on adaptive evolution of wild barley. A Geographical Information System (GIS) provided tools to visually present and analyse the geographical distribution of number of variable positions in the alignments and the differentiation index. 19 different bio factors were classified into 4 main component groups, and OLS Regression was used to analyze the differentiation coefficient of main climatic factors and the expression of gene differentiation degree. DHN family, a drought and temperature related gene family, showed significant spatial correlation with main climatic factors. Meanwhile, the results of other genes, which reported insignificant effects on drought or heat tolerance, were quite contrary. The phenotypes of plants were observed after the interaction of plant genotype with the environment. Simultaneously, the environment also has influence on some special genes. New analysis methods using GIS could be used to research the complex relationship between plant phenotype, plant genotype and the environment.

Keywords: Gene sequence analysis, biodiversity, molecular evolution, geographical information system (GIS), data meta-analysis, wild barley.

1. INTRODUCTION

Climate is one of the most important reasons for adaptive evolution in plants, and also leads to important selective factors determining intraspecific differentiation [1]. Yongfeng Zhou *et al.* used two closely related pine species growing in southeastern Chinese, Pinus massoniana Lamb and Pinus hwangshanensis Hisa as materials, to analyse 25 climate related genes.They found that variations in climate played an important role in the ecological divergence of the two species [2]. Not only on the woody perennial plants, climate also has a similar impact on the annual herbaceous plants such as Arabidopsis, barley and so on [3]. Wild barley (H. vulgare ssp. Spontaneum) is an annual, diploid grass species with 7 chromosomes and an estimated rate of self-fertilization of 98%. As the progenitor of cultivated barley, wild barley represents an important resource for the study of the adaptive evolution of plants' population [4].

The phenotypes of plants were observed after examining the results of the interaction of plant genotype and the environment. Meanwhile, the environment also had effects on some special genes. Alcohol dehydrogenases (ADH) is a group of Zn-binding enzymes which present in many organisms. The ADH family plays an important role in facilitating the interconversion between alcohols and aldehydes or ketones with nicotinamide adenine dinucleotide (NAD) as a coenzyme [5]. According to the previous reports, ADH family of plants has been observed to be involved in the biotic and abiotic stresses, such as disease resistance, drought tolerance, flooding resistance and so on [6]. Dehydrin (DHN) is a multi-family of proteins belong to the Group II Late Embryogenesis Abundant (LEA) family [7], and plays an important role in cold and drought stresses. Tommashi *et al.* found the expression levels of Dhn1, Dhn2, Dhn3, Dhn4, Dhn7, Dhn9 and Dhn10 in barley, to be highly increased in the germ, mesocotyl and the roots during drought stress. Meanwhile, Dhn5, Dhn8 and Dhn13 were significantly induced by cold and drought stresses [8]. In order to resist environmental change, and propagate the race, many factors such as natural selection, mating system, migration and genetic drift result in genetic diversity in the population. Even the same gene in the same population will have diversity because of the environmental differences. However, the relationship between genetic diversity and environmental differences could be analyzed through geographic information system (GIS) [9].

The base of Geographic Information System (GIS) is mainly the geographical space database. By collecting relevant information about the geographical space, appropriately processing the analysis, operating simulation, and using the geographic modeling, mutiple and dynamic geographic information resources were provided. Recently, GIS technology has been highly developed, and widely applied in many fields of life and production, such as for the

---

*Address correspondence to this author at the State Key Laboratory of Resources and Environmental Information System, Institute of Geographical Sciences and Natural Resources Research, Chinese Academy of Sciences, China; E-mail: 596295719@qq.com

Table 1. Loci, length of the aligned DNA sequences and numbers of accessions used in the study.

Locus	Abbrev.	Authors	Length of Alignment	No. of Accessions Investigated
Alcohol dehydrogenase 1	Adh1	Michael P. Cummings and Michael T. Clegg (1998)	1362	19
Alcohol dehydrogenase 2	Adh2	Jing-Zhong Lin *et al.* (2002)	1980	25
Alcohol dehydrogenase 3	Adh3	Jing-Zhong Lin *et al.* (2001)	1873	25
Alpha-amylase type B gene	α-AMYb	Peter L. Morrel *et al.* (2003)	856	25
C-repeat binding factor 3-like protein	Cfb3	Peter L. Morrell and Michael T. Clegg (2006)	1514	44
Dehydrin 1	Dhn1	Peter L. Morrel *et al.* (2004)	1538	23
Dehydrin 4	Dhn4	Peter L. Morrell and Michael T. Clegg (2006)	1047	24
Dehydrin 5	Dhn5	Peter L. Morrel *et al.* (2003)	1088	23
Dehydrin 7	Dhn7	Peter L. Morrel *et al.* (2004)	1400	27
Dehydrin 9	Dhn9	Peter L. Morrel *et al.* (2003)	1011	45
Glutathione-dependent formaldehyde dehydrogenase	Faldh	Peter L. Morrell and Michael T. Clegg (2006)	1092	25
Glyceraldehyde-3-phosphate dehydrogenase	G3pdh	Peter L. Morrel *et al.* (2003)	2010	26
Putative cleaveage stimulation factor	ORF1	Peter L. Morrell and Michael T. Clegg (2006)	1533	45
Phosphoenolpyruvate carboxylase	Pepc	Peter L. Morrel *et al.* (2003)	3173	25
Putative serine/threonine kinase	Stk	Peter L. Morrell and Michael T. Clegg (2006)	1057	25
MADS box transcription factor	Vrn1	Peter L. Morrel *et al.* (2004)	1262	19
Granule bound starch synthase	Waxy	Peter L. Morrel *et al.* (2003)	1232	26
Heat shock protein 17	Hsp17	Liu, Y. and Yang, Z.(2010)	1361	16

prediction of spatial distribution of variety of diseases and insect pests [10], topographic and geomorphic conditions, meteorological factors, and for the regional distribution of diseases [11]. In this article, GIS technology was used to analysz the relationship between climate and genetic diversity in wild barley, in order to reveal the important effect of gene evolution in the adaptive evolution of plants.

2. RELATED THEORY AND RESEARCH STATUS

2.1. The Data of Species Distribution

This study focused on examining the influences of climatic factors on the diversity of genes, especially factors like temperature and humidity . The wild barley was taken as the research object, and "Genetic diversity" and "Wild Barley" were chosen as key words. Web of Science (http://isiknowledge.com) and NCBI PubMed (http://www.ncbi.nlm.nih.gov/pubmed/) databases were used to search the related information. 422 papers were found; 17 genes (Table **1**) and 44 wild barley lines were selected. 495 sequences, related to the 17 genes were obtained from NCBI Genebank.

These sequences were aligned using CLUSTALX [12] and primarily analyzed by DNASP [13]. Geographical Information System Arc/Info was used for all geographical analyses.

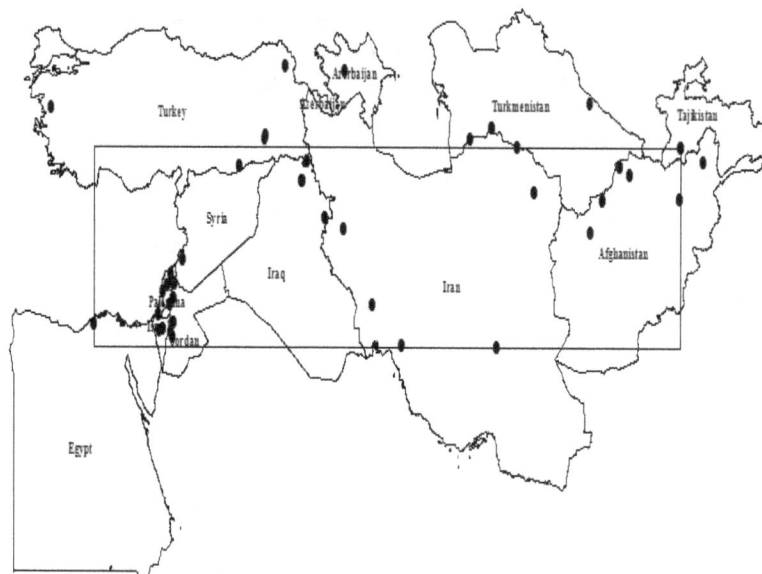

Fig. (1). The distribution of 44 wild barley populations.

The latitude and longitude of 44 wild barley populations were obtained through the GENESYS website. The distribution of these samples involved in this paper was graphically displayed by the GIS system (Fig. **1**). The rectangular area in the middle was chosen as the research scope according to finite sample frequency of 17 genes.

2.2. Climate Data Source

All climatic data were obtained from the world climate database (WORLDCLIM, http://www.worldclim.org/) [14]. From the year 1950 to 2000, climate data of 50 years from different weather stations around the world were pooled together. Using interpolation method, the global climate data was created in which, 19 biological climatic factors were observed to play important roles in the distribution of species . Detailed information is given as follows:

BIO1 = Annual Mean Temperature

BIO2 = Mean Diurnal Range (Mean of monthly (max temp - min temp))

BIO3 = Isothermality (BIO2/BIO7) (* 100)

BIO4 = Temperature Seasonality (standard deviation *100)

BIO5 = Max Temperature of Warmest Month

BIO6 = Min Temperature of Coldest Month

BIO7 = Temperature Annual Range (BIO5-BIO6)

BIO8 = Mean Temperature of Wettest Quarter

BIO9 = Mean Temperature of Driest Quarter

BIO10 = Mean Temperature of Warmest Quarter

BIO11 = Mean Temperature of Coldest Quarter

BIO12 = Annual Precipitation

BIO13 = Precipitation of Wettest Month

BIO14 = Precipitation of Driest Month

BIO15 = Precipitation Seasonality (Coefficient of Variation)

BIO16 = Precipitation of Wettest Quarter

BIO17 = Precipitation of Driest Quarter

BIO18 = Precipitation of Warmest Quarter

BIO19 = Precipitation of Coldest Quarter

A quarter is a period of three months (1/4 of the year)

3. MATERIALS AND METHODS

3.1. The Calculation of Differentiation Index

In a particular gene, the variable positions could be classified as singletons and nonsingletons, in which, singleton is a unique substitution ofa specific DNA sequence.

Considering a set of n aligned sequences, the differentiation index for the r-th sequence was calculated by

$$\pi_r = [\sum_{j=1}^{n}\pi_{rj}] / m_T(n-1) \tag{1}$$

Where , n is the number of homologous sequences in the specific gene. If one sequence r is considered for example, π_{rj} is the number of mismatches of sequence r compared with sequence j, and m_T is the length of the sequence r. Therefore, the differentiation index π_r represents the ratio between the sum of mismatched numbers and product of the length of the sequence and the homologous sequence number.

The interval of differentiation index was between 0 and 1, when the differentiation index was 0 which indicated that the sequence was totally identical with other homologous sequences, otherwise, 1 meant that it was completely different. The index was used in this paper to calculate the differentiation degree of the same gene in different wild barley populations.

3.2. Detection of Geographic Genetic Clines

All the investigated accessions of wild barley were sampled in 13 different countries in Middle-East, including Isreal, Jordan, Turkey *etc.* The Ordinary Kriging Interpolation, a spatial interpolation between sample points which uses information from the values of the sample points (z-axis in a Cartesian coordinate system) and the distances between them (x- and y-axes), was applied to quantify the spatial variation of the 17 genes [15]. In this study, the geographic genetic clines were classified through the Kriging maps, which were created by ArcGIS 10.

3.3. Analysis of the Correlations Between Gene Differentiation Index and Climate

Firstly, the regular meshes (Fishnet) were created in the selected area (Fig. **1**) by ArcGIS software. Using DIVA-GIS software, the point data of climate variables were extracted from the fishnet grid data. Principal component analysis of 19 climatic variables in the region was conducted by SPSS 13 software [16], highlighting that the main climatic factors may affect differentiation index of the wild barley populations in the region. The Ordinary Least Squares (OLS) in ArcGIS was used to construct the local regression model, to analyze the relationship between the differentiation coefficient of the 17 genes with major climatic factors.

4. EXPERIMENTAL RESULTS

4.1. Assessment of Molecular Data with GIS

In this paper, the differentiation indexes of 44 wild barley populations were used as the basis; the gene differentiation indexes in the research field were calculated through Ordinary Kriging Interpolation; the relationship between biological climatic factors and gene differentiation indexes was analyzed.

The distribution of wild barley accessions, number of variable sites in the DNA sequences and the differentiation indexes of 17 genes were visually represented through ArcGIS.

As shown in Fig. (**2**) s, *DHN4* was taken as an example. The geographical distribution of gene population, number of singleton and the different indexes. were graphically displayed. Throughout the analysis, the differentiation index of all genes, except *ADH1*, was observed to be much higher in the wild barley population distributed in Israel and Jordan, which may be greatly associated with the region's Canyon topography and climatic change [17].

4.2. Effect of Main Climate Factors in the Region

The climatic variable data in the research region was extracted using DIVA-GIS, and principal component analysis was conducted using SPSS software. Experimental results showed that cumulative contribution rate of 4 main components in the research area couldbe 91.44%, signifying biological climatic variable information.

The principal component 1 included information on Bio5, Bio10, Bio1 and Bio9, mainly reflected as the temperature in driest and warmest period; the principal component 2 included information on Bio6, Bio11 and Bio8, mainly reflected as the temperature in wettest and coldest period; the principal component 3 included information on Bio19, Bio15, Bio16 and Bio13, mainly reflected as the precipitation in the wettest and coldest period; principal component 4 included information on Bio4 and Bio7, reflected as changes in temperature differences. Based on the principal component analysis, Bio5, Bio6, Bio19 and Bio4 were selected to represent the 4 main components respectively. Meanwhile, the relationship between the gene differentiation indexes and 4 main climate factors was also studied. The principal component analysis of 19 biological climate factors in the research area is shown in Table **2**.

4.3. Spatial Dependence Between Genes and Climate Factors

Using OLS (ordinary least squares) regression method [18], the spatial correlation between the gene differentiation indexes and the main climatic factors is shown in Fig. (**3**).

In the past studies, it was reported that there was a direct relationship between the plant's DHN family and environmental stresses, such as drought and heat. Whereas, the alpha-amylase type B gene was not reported to have any significant relationship with drought or temperature.

According to the experimental results and analysis, the OLS regression result indicated a significant relationship between the DHN family and 4 main climatic factors, while there was less correlation between alpha-amylase type B gene and the main climate factor.

CONCLUSION

In this article, wild barley was used as the research object, and "Genetic diversity" and "Wild Barley" were chosen as key words; 17 genes and 44 wild barley populations were studied through Meta-analysis. The differentiation indexes of the same gene in different wild barley accessions were calculated, and the geographic distributions of these data were graphicallyrepresented. The information on 19 biological climatic factors was extracted, and principal component analysis was conducted using SPSS software. OLS was used to analyze the relationship between the main climatic factors and gene differentiation indexes, which determined the degree of gene evolution. The study found a significant spatial correlation between the 4 main

Table 2. The principal component analysis of 19 environment factors in the research area.

Factor	The Total Explained Variance		
	Initial Eigen Value		
	Total	Variance %	Total %
1	8.843	46.544	46.544
2	4.232	22.272	68.816
3	2.470	13.002	81.818
4	1.828	9.621	91.439

Environment	Factor			
	1	2	3	4
bio1	.856	.487	.040	.137
bio2	.593	-.599	.097	-.063
bio3	.591	.083	.015	-.635
bio4	.114	-.622	.113	.756
bio5	.889	.190	.105	.370
bio6	.667	.735	.008	.006
bio7	.208	-.808	.127	.477
bio8	.610	.557	-.266	.157
bio9	.856	.349	.084	.319
bio10	.876	.334	.074	.328
bio11	.777	.622	.015	-.029
bio12	-.786	.394	.411	.179
bio13	-.690	.376	.588	.100
bio14	-.655	.405	-.476	.220
bio15	.565	-.142	.628	-.230
bio16	-.693	.380	.591	.102
bio17	-.702	.398	-.449	.216
bio18	-.702	.348	-.453	.161
bio19	-.579	.421	.666	.065

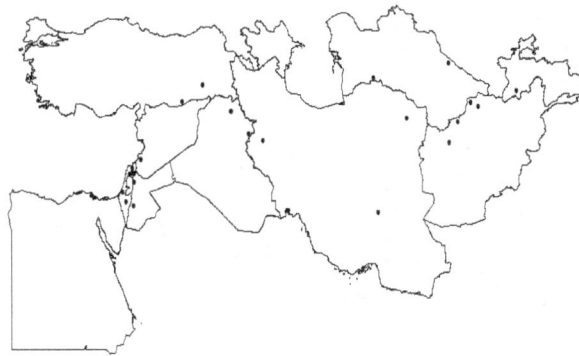

A. Distribution of DHN4 wild barley accessions

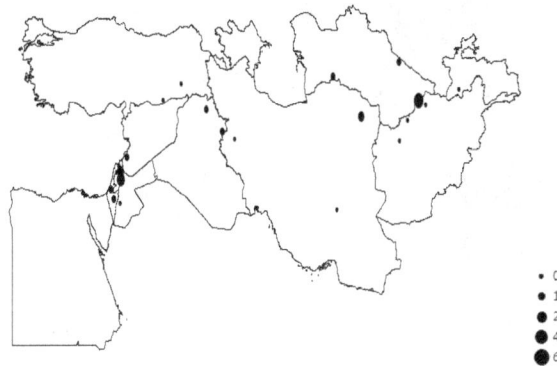

B. Distribution of singletons in DHN4 sequenced genomic region.

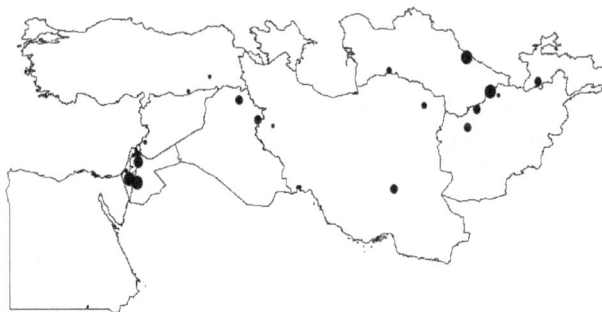

C. Distribution of differentiation index in DHN4 sequenced genomic region.

Fig. (2). The geographical distribution of DHN4 wild barley accessions, singleton and differentiation indexes. The size of the dots is **equivalen**t to the number of singletons/differentiation index.

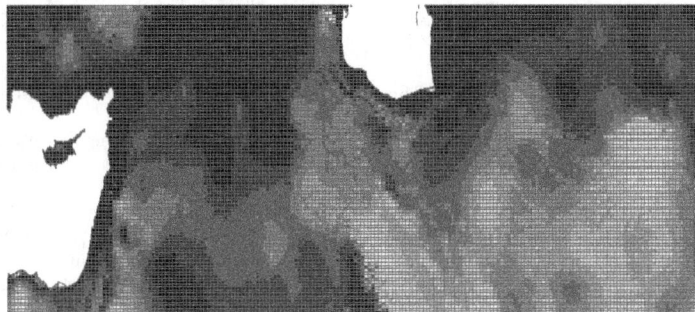

A. The spatial correlation between the DHN1 gene differentiation indexes and the major climatic factors.

Fig. (3). Contd...

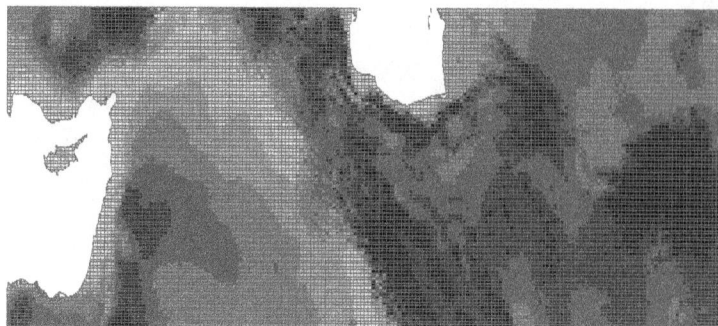

B. The spatial correlation between the DHN4 gene differentiation indexes and the major climatic factors.

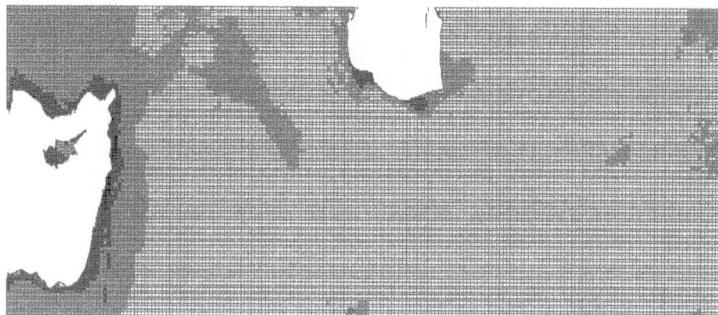

C. The spatial correlation between the alpha-amylase type B gene differentiation indexes and the major climatic factors.

Fig. (3). The spatial correlation between the gene differentiation indexes and the major climatic factors.

climatic factors with the differentiation indexes in stress-related genes, such as DHN family. Whereas, the genes reported not to be involved in drought or heat tolerance were found to have little correlation with main climatic factors. Thus, it can be seen that, Bio4, Bio5, Bio6 and Bio9 play important roles in gene evolution and in drought related stresses. In addition, this method could be used to predict the functions of unknown genes in the future.

REFERENCES

[1]   Olson MS, Levsen N, Soolanayakanahally RY, *et al.* The adaptive potential of Populus balsamifera L. to phenology requirements in a warmer global climate. Mol Ecol 2013; 22(5): 1214-30.

[2]   Zhou Y, Liu J, Savolainen O. Climatic adaptation and ecological divergence between two closely related pine species in Southeast China. Mol Ecol 2014; 23(14): 3504-22.

[3]   Wilczek AM, Cooper MD, Korves TM, Schmitt J. Lagging adaptation to warming climate in Arabidopsis thaliana. Proce Nat Acad Sci USA 2014; 111(22): 7906-13.

[4]   Cronin JK, Bundock PC, Henry RJ, Nevo E. Adaptive climatic molecular evolution in wild barley at the Isa defense locus. Proc Natl Acad Sci USA 2007; 104(8): 2773-8.

[5]   Strommer J. The plant ADH gene family. Plant J 2011; 66(1): 128-42.

[6]   de Bruxelles GL, Peacock WJ, Dennis ES, Dolferus R. Abscisic acid induces the alcohol dehydrogenase gene in Arabidopsis. Plant Physiol 1996; 111(2): 381-91.

[7]   Yang Y, He M, Zhu Z, *et al.* Identification of the dehydrin gene family from grapevine species and analysis of their responsiveness to various forms of abiotic and biotic stress. BMC Plant Biol 2012; 12: 140.

[8]   Tommasini L, Svensson JT, Rodriguez EM, *et al.* Dehydrin gene expression provides an indicator of low temperature and drought stress: transcriptome-based analysis of barley (Hordeum vulgare L.). Funct Integr Genom 2008; 8(4): 387-405.

[9]   Hoffmann MH, Glass AS, Tomiuk J, Schmuths H, Fritsch RM, Bachmann K. Analysis of molecular data of Arabidopsis thaliana (L.) Heynh (Brassicaceae) with Geographical Information Systems (GIS). Mol Ecol 2003; 12(4): 1007-19.

[10]   Fand BB, Kumar M, Kamble AL. Predicting the potential geographic distribution of cotton mealybug Phenacoccus solenopsis in India based on MAXENT ecological niche model. J Environ Biol 2014; 35(5): 973-82.

[11]   Elebead FM, Hamid A, Hilmi HS, Galal H. Mapping cancer disease using geographical information system (GIS) in Gezira State-Sudan. J Commun Health 2012; 37(4): 830-9.

[12]   Thompson JD, Gibson TJ, Plewniak F, Jeanmougin F, Higgins DG. The CLUSTAL_X windows interface: flexible strategies for multiple sequence alignment aided by quality analysis tools. Nucleic Acids Res 1997; 25(24): 4876-82.

[13]   Rozas J, Rozas R. DnaSP version 3: an integrated program for molecular population genetics and molecular evolution analysis. Bioinformatics 1999; 15(2): 174-5.

[14]   Cuervo PF, Rinaldi L, Cringoli G. Modeling the extrinsic incubation of Dirofilaria immitis in South America based on monthly and continuous climatic data. Vet Parasitol 2015; 209(1-2): 70-5.

[15]   Liu W, Du P, Wang D. Ensemble learning for spatial interpolation of soil potassium content based on environmental information. PLoS One 2015; 10(4): e0124383.

[16]   Valeri L, Vanderweele TJ. Mediation analysis allowing for exposure-mediator interactions and causal interpretation: theoretical assumptions and implementation with SAS and SPSS macros. Psychol Methods 2013; 18(2): 137-50.

[17]   Zhang T, Li GR, Yang ZJ, Nevo E. Adaptive evolution of duplicated hsp17 genes in wild barley from microclimatically divergent sites of Israel. Genet Mol Res 2014; 13(1): 1220-32.

30

The Growth and Plankton Changes of Intensive Ecological Aquaculture Model of *Pseudosciaena crocea*

Ruan Chengxu and Yuan Chonggui[*]

College of Biological Science and Technology, Fuzhou University, Fuzhou, Fujian 350116, China

Abstract: The growth of *Pseudosciaena crocea* and changes of plankton were studied under intensive ecological aquaculture model that does not involve any change in water. The results indicated that: the growth rate of *Pseudosciaena crocea* was significantly higher than that in cage culture ($p<0.05$), the essential amino acids, flavor amino acids and total amino acids of *Pseudosciaena crocea* were significantly higher than that in cage culture($p<0.05$), the PUFA and DHA、EPA of *Pseudosciaena crocea* were also significantly higher than that in cage culture($p<0.05$). Plankton biomass of aquaculture water increased with the cultivation time. The diversity of phytoplankton was better in the early stage, the diversity index and evenness index decreased in the later stage, and the diversity of phytoplankton was poor at the end of the experiment. The diversity of zooplankton was better along with a slightchange in its diversity index and evenness index with the zooplankton being in an equilibrium state.

Keywords: Ecological aquaculture, Growth, Plankton, *Pseudosciaena crocea*.

1. INTRODUCTION

Pseudosciaena crocea (Richardson) is a member of Perciformes, belonging to Sciaenidae and Pseudosciaena, commonly known as cucumber fish or yellow croaker [1]. It is one of the four main economic fishes and can be found in the list [2]. Recently, its culture model is represented as the cage culture model in seawater, which depends more on the natural weather condition and has a long growth cycle, easy to be affected by such natural disasters as typhoon and red tides and also from the pollution in seawater. All the problems mentioned above can not be ignored, as these cause difficulties for the culture of *Pseudosciaena crocea*. By constructing simple production facilities and regulating various ecological factors, this study proposes an indoor intensive culture for *Pseudosciaena crocea*. Different from the high-investment and high-cost industrial recirculating aquaculture, the intensive culture mode is used to intensively culture the yellow croaker without changing water and utilizes the complete biological structure to realize the purification of water and the prevention from diseases.

2. MATERIALS AND METHODS

2.1. Experimental Material

The Experimental fishes were bought from Kengyuan Seedling Culture Site in Ningde, Fujian, belonging to Dong

*Address correspondence to this author at the College of Biological Science and Technology, Fuzhou University, Fuzhou, Fujian 350116, China; E-mail: fdycg@126.com

Nationality's *Pseudosciaena crocea* that grew in Guanjingyang around the area of Fujian and Guangdong. They were spring seedlings having body length of 1 cm.

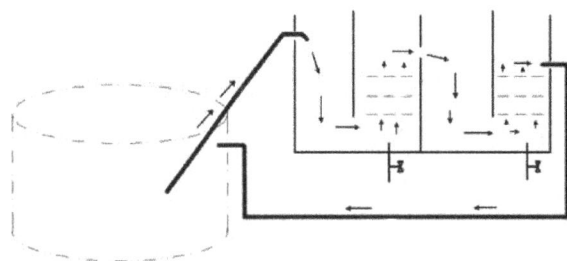

Fig. (1). Organic substance separation device of the breeding pool.

2.2. Experimental Design

The experiment was carried out in a plastic pool having a diameter of 4 m with 80cm depth of water. A parallel group, in each pool with about 2300 tails of fishes was set up with a density of 228.9 tail/m^3. The experimental seawater was the natural seawater with salinity of 25, while some fresh water was added every 10 days and after 2 months, the salinity decreased to 15. The pool was equipped with a microtubule aeration device that could aerate and provide oxygen for 24 hours to ensure the dissolved oxygen> 5.0mg/L, a water plumbing and heating device to ensure the temperature of water≥25°C and a self-designed organic substance separation device to separate the organic substance (see Fig. 1). In the device as shown in Fig. (1), the gas pipe was set up at the bottom of the pool and with the

Table 1. Evaluation standard for the biodiversity threshold.

Evaluation	Threshold	Level Description
I	<0.6	Poor
II	0.6-1.5	Commonly
III	1.6-2.5	Better
IV	2.6-3.5	Rich
V	>3.5	Very Rich

help of gas, the waste was absorbed in the precipitation tank, the inlet of which was higher than overflow hole and with the help of separation board, the water flowed into another precipitation tank along with the overflow hole and finally the water was sent back to the pool by the returned tube that was connected with this precipitation tank to complete the purification of water. Meanwhile, the precipitated waste could also be discharged out of the tank by the blow-down valve.

2.3. Culture Management

In the childhood phase, depending on their size, the fishes are fed with Hippo-brand slowly sinking feed with a combination of copepods and cladocerans for several times. When growing into the juvenile phase, they are fed Hippo-brand slowly sinking feed and at intervals are put into appropriate cladocerans for 4 times each day. When the body length of fishes grows to about 10cm, they are fully fed Xialin-brand extruded feed for 3 times a day and after they are finished, the rest of the feed is removed. During the whole phase of culture, in order to keep the dynamic equilibrium, the water should not be changed but the pH value was adjusted by lime to keep it in 7-8.

2.4. Experimental Project and Methods

After finishing the experiment, 20 tails were randomly selected from every pond to carry out statistical data of the growth pattern.

15 tails of *Pseudosciaena crocea* were randomly selected to study the quality of meat. After the experiment was finished, the Dong nationality of *Pseudosciaena crocea* in Fujian and Guangdong cultured by cage was bought in Guanjingyang of Ningde as a comparison. To determine the meat quality, the fish scales were first scraped, then the meat was removed from the spine and the fish bone was removed from the meat, finally the meat was kept into the blender to cut it into paste, which was stored in the refrigerator at 4°C for determination. The methods used in the experiment included the Kjeldahl method to determine the content of crude protein, the Soxhlet method to determine the content of crude fat, the Hitachi 835-50 type amino acid autoanalyzer to determine amino acids and the Shimadzu GC2010 gas chromatograph to determine the fatty acids after the effect of methyl esterification from benzene-petroleum ether lipids.

The samples of plankton were selected every 30 days for the study. The method utilized for selecting phytoplankton sampleinvolved relative selection of 500ml water sample

from the middle level of water in the middle of or around the experimental pool and after mixing that together, 500ml was selected to be fixed by 1.5% lugol. The method incorporated for the selection of zooplankton sample included selection of 1L water sample which was fixed by 5% formalin. In the laboratory, optical microscope was used to identify the species and to analyze the data.

Evaluation standard for the biodiversity threshold can be seen in Table 1 [3].

The content of NH_4^+-N can be determined by the Nessler's reagent colorimetric method [4] and the content of NO_2^--N can be determined by Naphthylethylenediamine spectrophotometry (GB17378.4-1998, 1999).

2.5. Data Statistics

The diversity index of plankton is calculated by Shannon-Weaver index equation [5].

The diversity index $H' = -\sum_{i=1}^{s} Pi\log_2 Pi$; $Pi = ni/N$

Where S represents the total number of species in sample, Pi is ni/N (ni the number of the i species, N the total number).

The evenness index of plankton is calculated by Pielou index equation [6].

The evenness index $J = H'/\log_2 S$.

S represents the total number of species in sample.

SPSS17.0 can be applied to the growth data of *Pseudosciaena crocea* to carry outsingle factor variance analysis. If the difference is significant(P<0.05), then the Duncan's multiple comparison can be utilized for further analysis.

3. RESULT AND ANALYSIS

3.1. The Growth Performance of *Pseudosciaena crocea*

The experiment was carried out from March 5 to August 28, stretching on to a total of 176 days and in this process, it did not break out any disease without any use of drugs and water change as described in Table 2. According to the survey, the growth performance of the same batch of spring seedling cultured in the cage of Sanduao, Ningde was as

Table 2. The growth performance of juvenile *Pseudosciaena crocea*.

Date (m.d)	Body Quality (g)	Body Length (cm)	Fatness (%)	Breeding Density (Tail/m³)	Feed Efficiency(%)	Survival Rate(%)
3.05	0.03±0.00	1.47±0.09	-	228.90±0.00	-	100.00±0.00
8.28	53.49±2.71	14.98±0.51	1.60±0.11	101.37±9.92	88.74±2.16	44.28±4.33

Table 3. Effects of ecological culture and cage culture on conventional nutrients of juvenile *Pseudosciaena crocea* (%).

Stem From	Water Content	Crude Protein	Crude Fat	Crude Ash
Ecological culture	73.78±0.38a	17.55±0.42a	7.36±0.29a	1.09±0.06a
Cage culture	75.21±0.43b	16.61±0.28b	6.88±0.17a	1.21±0.15a

follows: body quality: 36.58±3.27g, body length: 14.21±0.58cm and the fatness : 1.28±0.11%, which shows that its growth speed was significantly lower than the speed of those experimental fishes ($p<0.05$). Compared with some cage experimental results from other scholars [7, 8], the speed was faster and the feed efficiency was higher in this experiment (Tables **3-5**).

3.2. The Change of Plankton's Structure in Cultured Waters

As can be seen in Tables **6** and **7**, in the early stage of experiment, the biomass of phytoplankton was small, as it did not succeed to establish the stable ecological structure. When developing into the middle and later stage, its biomass increased, mainly concentrating on pyrrophyta, bacillario-phyta and chlorophyta. Especially in August, it mainly increased in pyrrophyta, which was due to the succession of water in this experiment. During this stage, owing to the culture implementation indoors and lack of enough light, the chlorophyta had no growth advantage and with the increase in feed, the NH^{4+}-N concentration rose, which proved beneficial for the growth of pyrrophyta.

The change in tendency of zooplankton was as follows: with the extension of culture time, the biomass of copepods increased and the rotifer decreased while the protozoan kept an inverse tendency for the rotifer. In the early stage of the experiment, the seedlings of zooplankton were fed with copepods, so the data at this time was not interfered. At this stage, the seedlings ate copepods, which resulted in no change in the biomass of copepods but in rotifer. The biomass of rotifer increased to a large number until May, which restricted the increase in protozoan. But *Pseudosciaena crocea* during its growth did not eat any more copeods in the middle and the latter of experiment, which then resulted in an increase in copepods and decrease in rotifer.

From Table **8**, it has been observed that the diversity of phytoplankton performs well in the early stage, for it cultures in natural seawater with complicated phytoplankton. Developing into the later stage, owing to the lack of light and the effect of high NH^{4+}-N, the diversity performs badly with the

decrease in diversity index and evenness index while the diversity of zooplankton is good, mainly because its culture system is a closure system without any invasion of alien species with the environmental ecological factors being stable.

3.3. The Change of Water Quality in Cultured Waters

From Fig. (**2**), it is found that in the first three months, the concentrations of $NH^{4+}-N$ and NO_2^--N increased slowly, for they had less feed and produced little pollution having a small body at this stage. After three months, with the increase in eating and pollution, the concentrations rose rapidly. Since in this experiment, the culture model without changing water was applied, therefore the pollution was severe. However, in the later stage, the increased speed slowed down and the concentrations of $NH^{4+}-N$ and NO_2^--N were under control, as the succession of water is suitable for the water environment with high $NH^{4+}-N$.

4. DISCUSSION

In this experiment, the method of intensive ecological model without changing water was applied to culture *Pseudosciaena crocea* resulting in a technical effect of culturing *Pseudosciaena crocea* at high density. Different from the crude ecological culture model, this experiment model is a high-density intensive culture and is different from the popular greenhouse intensive model in which 100% water needs to be changed, while the model in this experiment did not involve any change of water or drug usage. Meanwhile, by establishing the ecological equilibrium in water, it used dynamic equilibrium between microorganisms and cultured fishes to discompose the metabolic waste and let the water have the ability to purify itself and resist diseases, which in turn saved water and settled the problem of drug's residue from the source.

As a kind of poikilotherm, fishes' metabolic rate is determined by temperature [9, 10]. In this study, it implemented the culture by controlling and maintaining the temperature

Table 4. Effects of ecological culture and cage culture on amino acid compositions of muscle of juvenile *Pseudosciaena crocea* (% DW).

Amino Acid	Ecological Culture	Cage Culture
Lysine	5.45±0.09[a]	5.39±0.04[a]
Threonine	2.65±0.03[a]	2.62±0.05[a]
Serine	2.15±0.04[a]	2.30±0.08[b]
Glutamate	10.41±0.45[a]	9.92±0.31[a]
Glycine	2.95±0.04[a]	3.17±0.10[b]
Alanine	3.47±0.04[a]	3.32±0.04[b]
Cystine	0.36±0.03[a]	0.57±0.02[b]
Valine	2.84±0.05[a]	2.76±0.04[a]
Methionine	1.89±0.04[a]	1.65±0.03[b]
Isoleucine	3.75±0.03[a]	3.18±0.03[b]
Leucine	4.98±0.06[a]	4.68±0.09[b]
Tyrosine	2.75±0.06[a]	1.77±0.03[b]
Phenylalanine	2.41±0.04[a]	2.45±0.03[a]
Aspartic Acid	6.58±0.05[a]	5.58±0.09[b]
Histidine	1.22±0.05[a]	0.91±0.02[b]
Arginine	3.45±0.07[a]	3.33±0.05[a]
Proline	1.87±0.05[a]	1.16±0.05[b]
Essential Amino Acid	23.99±0.06[a]	22.75±0.02[b]
Flavor Amino Acid	26.86±0.31[a]	25.32±0.22[b]
Total Amino Acids	59.21±0.35[a]	54.78±0.28[b]

Note: Values in the same row with different superscript letters are significantly different($P<0.05$)

Table 5. Effects of ecological culture and cage culture on fatty acid compositions of muscle of juvenile *Pseudosciaena crocea* (%(DW).

Fatty Acid	Ecological Culture	Cage Culture
Myristic Acid (C14:0)	3.16±0.05[a]	3.86±0.08[b]
Palmitic Acid (C16:0)	27.31±0.46[a]	27.19±0.25[a]
Palmitic Acid (C16:1)	11.18±0.09[a]	11.17±0.11[a]
Stearic Acid (C18:0)	4.15±0.07[a]	5.21±0.18[b]
Oleic Acid (C18:1)	21.65±0.05[a]	23.58±0.17[b]
Linoleic Acid (C18:2)	1.15±0.02[a]	1.19±0.04[a]

Table 5. contd...

Fatty Acid	Ecological Culture	Cage Culture
Linolenic Acid (C18:3)	0.54 ± 0.03^a	0.52 ± 0.02^a
Parinaric Acid (C18:4)	0.63 ± 0.02^a	0.55 ± 0.02^b
Arachidonic Acid (C20:0)	0.62 ± 0.01^a	0.46 ± 0.02^b
Arachidonic Acid (C20:1)	0.96 ± 0.03^a	1.93 ± 0.05^b
EPA (C20:5)	5.30 ± 0.10^a	4.22 ± 0.08^b
Erucic Acid (C22:1)	1.19 ± 0.06^a	1.13 ± 0.02^a
DHA (C22:6)	17.07 ± 0.10^a	14.18 ± 0.06^a
Others	5.09 ± 0.76^a	4.81 ± 0.56^a
UFA	58.59 ± 0.23^a	58.47 ± 0.19^a
PUFA	24.69 ± 0.12^a	20.66 ± 0.17^b
DHA+EPA	22.37 ± 0.12^a	18.40 ± 0.10^b

Note: Values in the same row with different superscript letters are significantly different($P < 0.05$)

Table 6. Biomass of phytoplankton (104ind·L-1).

Date	Cyanophyta	Pyrrophyta	Bacillariophyta	Chlorophyta	Euglenophyta
March 15	0.21	0.33	0.54	0.42	0.15
April 15	0.12	0.58	0.78	0.45	0.00
May 15	0.00	3.66	2.42	1.59	0.19
June 15	0.12	8.61	4.76	3.87	0.45
July 15	0.15	11.07	1.98	4.03	0.00
August 15	0.00	40.86	0.72	3.24	0.06

Table 7. Biomass of zooplankton (ind·L-1).

Date	Copepods	Nauplius	Rotifer	Protozoan	Other
March 15	180	75	60	900	240
April 15	105	60	195	450	180
May 15	195	225	495	450	203
June 15	345	75	105	1050	263
July 15	435	165	0	750	338
August 15	375	105	45	450	188

Table 8. Species diversity index and evenness index of plankton.

Date	Phytoplankton		Zooplankton	
	Diversity Index	Evenness Index	Diversity Index	Evenness Index
March 15	2.19	0.94	1.64	0.71
April 15	1.79	0.77	2.01	0.87
May 15	1.63	0.70	2.20	0.95
June 15	1.68	0.72	1.90	0.82
July 15	1.32	0.57	1.86	0.80
August 15	0.51	0.22	2.07	0.89

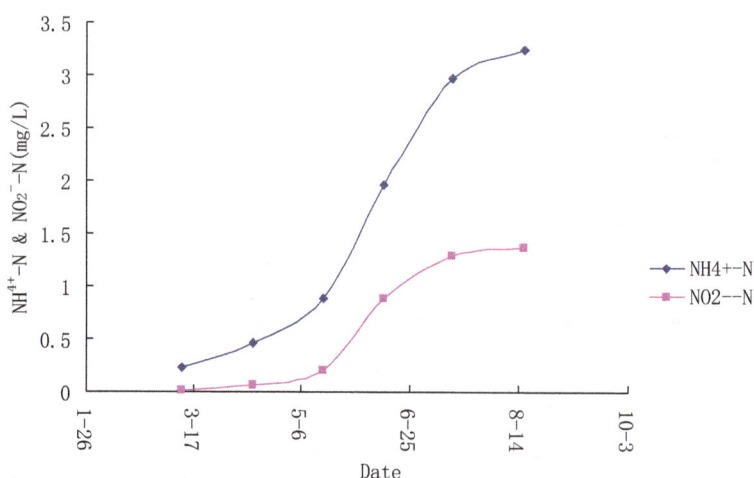

Fig. (2). The concentrations of NH^{4+}-N and NO_2-N.

above 25°C, making the growth rate higher than that cultured in cage. In general, the feeding rate, growth rate and feed efficiency rise along with the increase in temperature [11], which in turn influences the growth of fishes by impacting their feeding and feed utilization rate. The author proves that in his early study, the feeding amount, quality increase rate and specific growth rate rose and had significant differences along with the rise of temperature [12]. Therefore, the high growth rate and feed efficiency produced by the controlled temperature in the ecological culture can not be obtained in cage culture.

The composition of amino acids of meat is a key indicator for evaluating the nutritional value and flavor. Glutamate, aspartic acid, arginine, glycine and alanine are called flavor amino acids, the contents of which determine the flavor of the meat. In this study, essential amino acids and flavor amino acids of the muscles in intensive ecological culture are all higher than that in cage culture (P<0.05), illustrating that in intensive culture model, the *Pseudosciaena crocea* cultured by appropriate mixed feed has higher nutritional value and flavor, for the artificial mixed feed it uses more lipid contents. As from Campos's study, it is known that the content of amino acids in fish meat will rises along with the increase in lipids of the feed.

The composition of meat's fatty acid can be affected by the environment, feed and other external factors [13, 14]. Some studies have proved that the composition of fatty acids can reflect the fat contents of feed [15]. In this study, PUFA, DHA and EPA of *Pseudosciaena crocea* were all higher than that in cage culture model (P<0.05), mainly determined by the fats in artificial feed. DHA and EPA are beneficial for people's nutrition and health and can reduce the risk of suffering from coronary heart disease, asthma and inhibit the growth of tumor [16, 17], which shows that in intensive ecological model, the nutritional value cultured by appropriate feed is higher than that in conventional cage culture.

The plankton structure established in the intensive ecological model succeeds with the succession of culture process under stable condition in greenhouse and its variation also reflects the ecological characteristic of this model. In the early stage, the biomass of phytoplankton is small, reflecting good diversity while in the later stage, the biomass decreases, showing poor diversity, is similar to the variation of phytoplankton in the model of seawater culture [18]. The increase of phytoplankton is mainly due to the mixed water affected by fishes' activities and which propels the cycling of nutritional matters while the high-density culture promotes the increase of nutritional matters, especially the increase of N and P [19]. In other culture models in seawater, if there are filter feeding organisms, the biomass and diversity of phytoplankton are lowered [20]. In this study, however, culturing of *Pseudosciaena crocea*, does not belong to filter feeding organism, so the decrease in phytoplankton's diversity in later stage results from the environment. Considering the water quality, the rapid increase in the contents of NH^{4+}-N propels the cultivation of pyrrophyta, which may result in the lowering of diversity. But in the later stage, the water load turns terrible owing to the poor diversity and high contents of $NH^{4+}-N$, so the change of water or other measures is required to improve the environment for carrying out further culture.

The feeding of fishes to zooplankton makes the zooplankton structure more simpler and lowers its diversity. In the early stage of this experiment, *Pseudosciaena crocea* consumed such large zooplankton as copepods. Therefore, the copecods were restricted in the early stage but increased rapidly in the later stage. The biomass of phytoplankton was closely related to that of zooplankton, which increased along with the growth of phytoplankton in the middle and the later stage, but the relationship of catching and getting caught among zooplanktons made copepods, rotifers and protozoa grow reciprocally. In the later stage, the diversity of zooplanktons was not affected since no such planktons consumed them.

Contrary to the change regulation of plankton in this study, the structure of plankton cultured in the open sea is influenced by the water temperature. However, the biomass and diversity of plankton are not impacted by the water temperature owing to the fact that it applies the temperature control culture model.

After analyzing the growth effect of *Pseudosciaena crocea* and the evolution of biological structure in waters, it was found that the model uses the high-density culture without changing water. It develops the natural evolution to a stable dynamic equilibrium, suitable for this ecological environment by controlling the ecological condition, enabling the growth of *Pseudosciaena crocea* in an appropriate environment. Thus the growth rate, feed efficiency, amino acids and fatty acids in this model are obviously higher than that in cage culture. During the whole experimental process, it did not suffer from large-scale plant diseases and insect pests and did not make use of any drugs. Though, in the later stage, the phytoplankton had poor diversity owing to the simple environmental factor and the high $NH^{4+}-N$ of water. The diversity and evenness indexes of zooplankton performed well, which shows that the phytoplankton and zooplankton remained in an equilibrium state, giving a complete and stable structure of plankton. Therefore, the intensive ecological model can be used to establish stable water biological structure, by realizing the factory culture of *Pseudosciaena crocea* and by applying the technical indicators of water saving, energy saving and environment protecting, overall providing a new path to this model of culture.

ACKNOWLEDGEMENTS

This work is supported by The Scientific and Technological Project in Fuzhou University (0080822733).

REFERENCES

[1]   Yuanding Z, Hanlin W. Records of Fujian Fishes(II) Fuzhou: Fujian Science and Technology Press, 1985.

[2]   Cailan Z, Jiafu L, Yacui L. Analyzing the present condition and counter measure of cultured large yellow croaker *Pseudosciaena crocea* in fujian province. J Shanghai Ocean Univ 2002; 11: 77-83.

[3]   Qingchao C, Liangmin H, Jianqiang Y. Study on the diversity of zooplankton in waters around nansha islands. Beijing: Ocean Press 1994.

[4]   Jingping Z. Quickly determination of the ammonia-nitrogen in the seawater aquiculture by Nessler's reagent colorimetric method. Fujian Anal Test 2011; 20: 10-4.

[5]   Shannon CE, Weaver W. The mathematical theory of communication. Urbana IL: University of Illinois Press, Chicago, IL., 1998; pp. 144.

[6]   Pielou EC. An introduction to mathematical ecology, Wiley-Interscience, New York, 1969; pp. 294.

[7]   Chen H, Guowen L, Zhankun L. Study on the growth characteristic of *Pseudosciaena crocea* in cage culture. Mar Sci 2010; 34: 1-5.

[8]   Chengjin C. Observation on growth properties of cultured large yellow croaker, *Pseudosciaena crocea*. Modern Fish Inform 2011; 26: 24-9.

[9]   Xiaojun X, Ruyong S. Study on the main ecological factor that influencing the metabolizing of fishes. J Southwestern Normal University (Natural Science Edition) 1989; 14: 141-9.

[10]   Brett JR. Environmental factors and growth. New York: Academic Press 1979.

[11]   Shimeno S, Shikata T. Effects of acclimation temperature and feeding rate on carbohydrate-metabolizing enzyme activity and lipid content of common carp. Nippon Suisan Gakkaishi 1993; 59: 661-6.

[12]   Chengxu R, Defeng WU, Chonggui Y. Effects of temperature on the growth and hydrochemical state of juvenile *Pseudosciaena crocea*. J Guangzhou Univ 2013; 12: 36-9.

[13]   Campos P, Martino RC, Trugo LC. Amino acid composition of brazilian surubim fish (*Pseudoplatystoma coruscans*) fed diets with different levels and sources of fat. Food Chem 2006; 96: 126-30.

[14]   Bandarra NM, Batista I, Nunes ML, Christie WW. Seasonal changes in lipid composition of sardine (*Sardine pilchardus*). Food Sci 1997; 62: 40-2.

[15]   Kiessling A, Pickova J, Johansson L, *et al*. Changes in fatty acid composition in muscle and adipose tissue of farmed Rainbow trout (*Oncorhynchus mykiss*) in relation to ration and age. Food Chem 2001; 73: 271-84.

[16]   Chen IC, Chapman FA, Wei CI, Portier KM, O'Keefe SF. Differentiation of cultured and wild sturgeon (*Acipenser oxyrinchus* Desotoi) based on fatty acid composition. Food Sci 1995; 60: 631-5.

[17]   Justi KC, Hayashi C, Visentainer JV, Souza NE, Matsushita M.
        The influence of feed supply time on the fatty acid profile of *Nile
        tilapia* (oreochromis). Food Chem 2003; 80: 489-93.

[18]   Conner WE. Importance of n-3 fatty acids in health and disease.
        Nutr 2000; 17: 171-5.

[19]   Simopoulos AP. Omega-3 fatty acids in inflammation and autoim-
        mune diseases. Nutr 2002; 21: 495-505.

[20]   Pardini RS. Nutritional intervention with omega-3 fatty acids en-
        hances tumor response to anti-neoplastic agents. Chem Biol Inte-
        ract 2006; 162: 89-105.

31

High-Yield Laccase-Producing Strains Constructed by Protoplast Fusion Between Bacterium and Fungus

Lihong Zhao[*], Wenli Yin and Lele Wang

Civil & Architectural Engineering College, Liaoning University of Technology, Jinzhou, China

Abstract: This study is based on the construction of high-yield laccase-producing fusant achieved by inter-kingdom protoplast fusion between *Pleurotus ostreatus* and *Escherichia coli*. The optimized protoplasts formation and regeneration conditions were demonstrated with the presence of 1.5% cellulase +1.0% snailase and 0.6M mannitol at 30°C for 3h. The fusants were screened for different characteristics between two parental strains and further identified by laccase activity, offering one of the genetically stable fusants, Strain F. The fusant F produced the highest yield of laccase, being about 22% higher than that of the parental strain. The results suggest that the protoplast fusion technique can be considered as a promising technique for the control of environmental pollution.

Keywords: *Escherichia coli*, Laccase, *Pleurotus ostreatus*, Protoplast fusion, White rot fungi.

1. INTRODUCTION

Laccase, representing the largest subgroup of blue multicopper oxidases (MCO), uses distinctive redox ability of copper ions to catalyze the oxidation of a wide range of aromatic substrates concomitantly with the reduction of molecular oxygen in water [1]. Laccase has also been found in various basidiomycetous and ascomycetous fungi and, thus farfungal laccase has accounted for the most important group of MCOs with respect to the number and extent of characterization. Especially white-rot fungi produce laccase which efficiently degrades lignin. *Pleurotus ostreatus* is one of the unusual white rot fungi that it is able to mineralize the compound of native lignin. Moreover, it can also degrade almost all hazardous organic pollutants. Its potential would be enhanced, if genetic methods for producing strains with superior capacities were available. Therefore, laccase has important application value in the degradation of lignin, biological pulping, biobleaching and catalytic synthesis, biological elimination of toxic compounds [2-4], and has become a hotspot in the research of enzyme engineering and environmental protection. But due to its low yield and slow growth of white rot fungi, there is a problem to be solved for practical application of laccase in the industry.

Protoplast fusion has been an established technique for more than 30 years, but this traditional technique is still a powerful tool for the improvement of various microorganisms. Protoplast fusion has indicated broad applicability among microorganisms not only between intraspecies but also interspecies microbes and even between microbes from different kingdoms [5-9].

The aim of this work was to generate high-yield laccase-production and fast growth strains from *Pleurotus ostreatus* and *Escherichia coli* through the combination of protoplast fusion technique.

2. MATERIAL AND METHODS

2.1. Strains and Culture Medium

The fungal strain *P. ostreatus* was obtained from the microbiology laboratory, Civil and Architectural Engineering College, Liaoning University of Technology, Jinzhou, China. *E. coli* was obtained from the China Center of Industrial Culture Collection. *P. ostreatus* was maintained on potato-dextrose-agar (PDA) medium. *E. coli* was cultivated from Luria-Bertani (LB) medium (tryptone 10g/l, NaCl 10g/l, yeast extract 5g/l, pH7.6). Mycelia were cultured in PDA liquid medium under different conditions at 30°C for 6-7 days for *P. ostreatus* and *E. coli* was cultured in LB liquid medium under different conditions at 37°C for 12-13 hours for *E. coli* prior to protoplast isolation.

2.2. Formation of Protoplasts

Mycelia of *P. ostreatus* at different incubation time were harvested by suction filtration, carefullyto prevent the mycelium from drying extensively, and washed with corresponding osmotic stabilizer ($MgSO_4$, KCl, NaCl, mannitol). Cells (0.1 to 1.0 g) in 1 ml of stabilizer were treated with the cell-wall-degrading enzymes at 30°C with shaking for 3h [10]. The cell-wall-degrading enzymes such as snailase, lywallzyme and cellulase were tested in different combinations, so as to get the optimum digestion conditions. Remaining fragments were removed by filtration (coarse fritted glass funnel), and the protoplasts were sedimented by centrifugation at 4,000 rpm for 10 min and resuspended in the stabilizer as indicated. The number of protoplasts in suspension was determined with a hemacytometer.

E. coli strain was grown in LB medium to an optical density at 600 nm of 0.9 to 1.0 at 37°C. The cells were

*Address correspondence to this author at the Civil & Architectural Engineering College, Liaoning University of Technology, Jinzhou, China; E-mails: zhaolh05@163.com; 254297630@qq.com

harvested by centrifugation at 4°C at 10,000 rpm for 5 min. The bacteria were washed twice at 23°C with 10 mM tris(hydroxymethyl)aminomethane (Tris) buffer (pH 8.0). The pellet was suspended by pipetting. Suspension in 45 ml of 0.1 M Tris, with pH 8.0, containing 20% (w/w) sucrose was maintained at 37°C directly in the centrifuge tube. Cells were then transferred to a small flask, and the temperature was adjusted to 37°C. Within 1 min, 2.25 ml of a 2 mg/ml solution of lysozyme in doubledistilled water was added for a final concentration of 100 μg of lysozyme per ml. During the addition of lysozyme, the suspension of cells was stirred with a magnetic stirrer. After the addition, the temperature was adjusted to 37°C. Incubation was carried out bystirring for 12 min at 37°C, followed by addition of ethylenediamine-tetraacetic acid (EDTA) by slow dilution using 0.1 M dipotassium EDTA (pH 7.0) (1: 10, v/v, EDTA /cell) with prewarmed (37°C) EDTA, added slowly over 2.5 min with continuous stirring to avoid lysis. The temperature dropped during this addition, and was adjusted back to 37°C. Within 8 to 10 min, more than 99% of the cells became spherical [11]. Protoplasts were harvested by centrifugation at 4°C at 5,000 rpm for 10 min and suspended in SMM buffer (sucrose 0.5M, MgCl$_2$ 0.02M, sodium maleate 0.02M, pH 6.5).

2.3. Protoplast Fusion

Protoplasts from *P. ostreatus* and *E. coli* were prepared as described previously. After washing with osmotic stabilizer, protoplasts were mixed and centrifuged (4000 rpm, 10 min). The protoplasts were suspended in 1 ml of a pre-warmed (30°C) solution of 30%(w/v) polyethylene glycol (PEG) 6000 in 0.01M CaCl$_2$· 2H$_2$O and 0.05M glycine (pH 7.5) and incubated for 10 min at 30°C. The osmotic stabilizer was then added, and the protoplasts were resedimented by centrifugation and resuspended in stabilizer. Protoplasts in stabilizer were plated in 15ml of regeneration agar at 30°C. The regeneration agar consisted of solid PDA medium supplemented with osmotic stabilizer was adjusted to pH 6.0. Subsequently, the plates were incubated at 30°C for 3-4 days.

2.4. Analytical Method

Laccase activity was determined as described by Niku-Paavola *et al.* [12]. The protoplasts regeneration ratio was obtained as shown below.

$$\text{The protoplasts regeneration ratio } (\%) = \frac{C - B}{A - B} \times 100\% \quad (1)$$

Where A is total colony number of the solid PDA (the cells without enzyme-treatment were spread onto the solid PDA);

B is colony number of the solid PDA (the cells with enzyme-treatment were washed twice with distilled water to lyse protoplasts and spread onto the solid PDA) [13]; C is colony number of the regeneration medium plates (protoplasts from different populations were mixed and spread on the regeneration medium plates and were also incubated) [14].

3. RESULTS AND DISCUSSION

3.1. Effect of Different Enzyme Combinations on Protoplasts Formation and Regeneration of *P. ostreatus*

Considering the composition complexity of the cell walls of *P. ostreatus*, three kinds of hydrolytic enzymes were tested for protoplasts isolation. Comparison of different combinations showed that the enzyme mixture containing 1.5% cellulase and 1.0% snailase was more effective than cellulase, snailase, lywallzyme alone, or cellulase mixed with snailase and lywallzyme (Table 1). The results showed that the protoplasts formation and regeneration ratio of *P. ostreatus* were 5.01×108 and 7.31% in 1.5% cellulase and 1.0% snailase, respectively.

3.2. Effect of Incubation Time on Protoplast Formation and Regeneration of *P. ostreatus*

As illustrated in Fig. (1), the optimal incubation period for rapid protoplast formation and regeneration of *P. ostreatus* was 6 days. The incubation time showed a highly significant effect on the protoplast yield. As shown in Fig. (1), on the fifth and sixth day, the protoplast formation was efficient. The effect of enzymolysis is poor when the incubation time is too short or too long. Cell wall is the enzyme substrate. The fungi in different physiological state directly affect the cell wall structure, metabolism and the vitality of fungi. When the incubation time is 5-6 d, hypha growth is enough and easy to take off the wall.

3.3. The effect of Osmotic Stabilizer on Protoplast Formation and Regeneration of *P. ostreatus*

The osmotic stabilizer also plays an important role in protoplast isolation. Appropriate concentration of osmoticum can prevent the protoplast from bursting or shrinking. Though there is no uniform standard for the selection of a suitable osmotic stabilizer [10, 15], it was demonstrated in this study that mannitol was a good choice for *P. ostreatus*. The best results were achieved with the 0.6M mannitol osmotic stabilizer with protoplast formation and regeneration of *P. ostreatus* of 4.13×10[8] and 5.67%, respectively

Table 1.   Effect of different enzymes on protoplasts formation and regeneration of *P. ostreatus*.

Enzyme Solution	Protoplasts Formation (10[8])	Regeneration Ratio (%)
2% Cellulase	4.97	6.33
1.5% Snailase	4.35	6.73
1.0% Lywallzyme	3.59	4.91
1.5% Cellulase +1.0% Snailase	5.01	7.31
1.5% Cellulase +1.0% Snailase +0 .5% Lywallzyme	5.67	7.01

(Fig. **2**). The rest of the order was KCl> MgSO$_4$·7H$_2$O >NaCl.

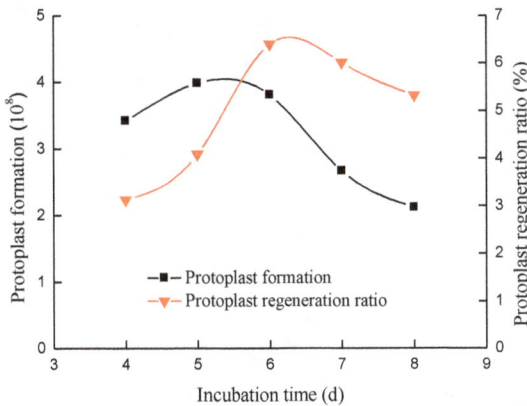

Fig. (1). Effect of incubation time on protoplast formation and regeneration ratio.

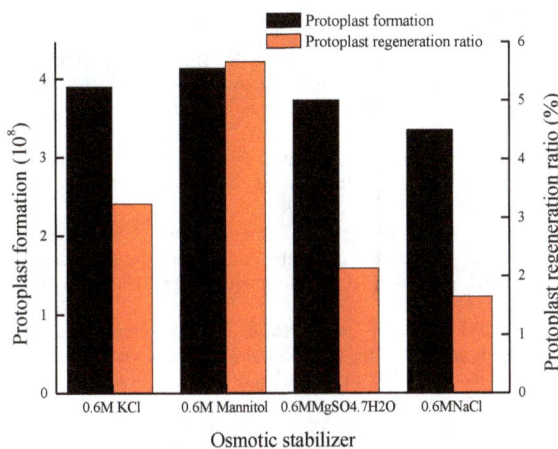

Fig. (2). Effect of osmotic stabilizer on protoplast formation and regeneration ratio.

3.4. Optimization of Protoplasts Formation and Regeneration Conditions

To optimize the protoplasts formation and regeneration conditions, a four-factor, three-level cross design test was conducted. The design and result are shown in Table **2**. Based on the results (k value) shown in Table **2**, the best combination for protoplasts formation and regeneration was 1.5%C+1.0%S, 3.0h, 30°C, 0.6M mannitol. Based on the value of r, the time of enzymolysis played the most important role. The rest of the order was enzyme combination> enzymolysis temperature> osmotic stabilizer.

3.5. Fusion of Protoplasts

A total of 6 fusants were selected for further screening. Since the aim of this study was to find a fusant with high laccase productivity, fusants based on the following criteria were screened: first, it must grow very well in selecting medium; second, it must have the highest laccase production. Based on these criteria, fusant F was selected for the study (Fig. **3**). The fusant F showed comparatively higher laccase activity to that of other fusants, and it had the fastest growth. In the present work, protoplast fusion of *P. ostreatus* and *E. coli* improved the laccase yield by 22%.

3.6. The Genetical Stability of The Fusant F

To ascertain whether fusant F is genetically stable, it was traced for 10 generations to measure the laccase activity of every other generation. All the generations demonstrated similar activity and productivity as the original F, suggesting that this fusant is genetically stable and suitable for industrial production. In addition, fermentation condition was also optimized for this strain (data not shown). The laccase activity reached 190.5U/ml.

Table 2.   Cross design test to optimize protoplasts formation and regeneration of *P. ostreatus*.

Experimental No.	Enzyme Combination	Enzymolysis Time	Enzymolysis Temperature	Osmotic Stabilizer	Protoplasts Formation (X)× Regeneration Ratio (Y) (10^6)
1	1.5%C	2.5h	28°C	0.6M KCl	8.96
2	1.5%C	3.0h	30°C	0.6M MgSO$_4$·7H$_2$O	27.33
3	1.5%C	3.5h	32°C	0.6M Mannitol	11.67
4	1.5%C+1.0%S+0.5%L	2.5h	30°C	0.6M Mannitol	10.12
5	1.5%C+1.0%S+0.5%L	3.0h	32°C	0.6M KCl	12.32
6	1.5%C+1.0%S+0.5%L	3.5h	28°C	0.6M MgSO$_4$·7H$_2$O	4.69
7	1.5%C+1.0%S	2.5h	32°C	0.6M MgSO$_4$·7H$_2$O	9.65
8	1.5%C+1.0%S	3.0h	28°C	0.6M Mannitol	25.09
9	1.5%C+1.0%S	3.5h	30°C	0.6M KCl	21.23
k$_1$	15.987	9.577	12.913	14.170	
k$_2$	9.043	21.580	19.560	13.890	
k$_3$	18.657	12.530	11.213	15.627	
r	9.641	12.003	8.347	1.737	

C, Cellulase; S, Snailase; L, Lywallzyme.

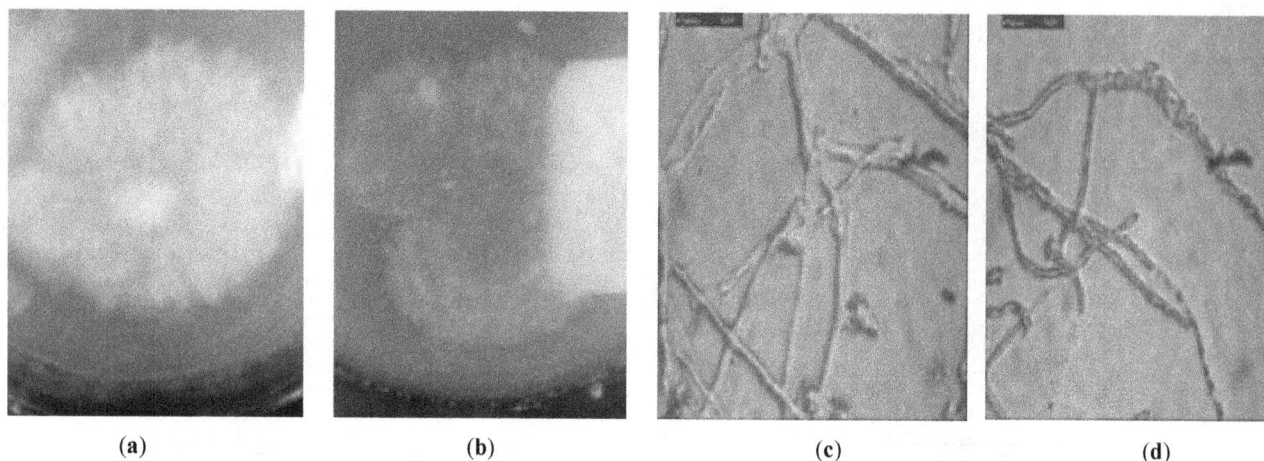

Fig. (3). The colonies and cell morphology of parental strains and fusant F. (**a**) The colonies of parental strain *P. ostreatus;* (**b**) The colonies of fusant F; (**c**) Cell morphology of parental strain *P. ostreatus;* (**d**) Cell morphology of fusant F.

CONCLUSION

Using the inter-kingdom protoplast fusion method, a fusant between *P. ostreatus* and *E. coli* was successfully produced. Results of the present work clearly confirmed the highly effective role of protoplast fusion in enhancing laccase production capacity of *P. ostreatus* and its rate of growth suggesting that this fusant is genetically stable. Protoplast fusion of *P. ostreatus* strains resulted in laccase yield increase by 22% compared to their parent strain, which is considered an excellent result. Thus, it is a very promising strain for industrial application. In conclusion, protoplast fusion proved its efficiency as a tool for constructing a second generation with much better characteristics for efficient and economical field applications.

ACKNOWLEDGEMENTS

This work is supported by the National Natural Science Foundation of China (No.51408290) and the Natural Science Foundation of Liaoning Province, China (No.2014020112).

REFERENCES

[1]   Giardina P, Faraco V, Pezzella C, Piscitelli A, Vanhulle S, Sannia G. Laccases: a never-ending story. Cell Mol Life Sci 2010; 67: 369-85.

[2]   Singh G, Kaur K, Puri S, Sharma P. Critical factors affecting laccase-mediated biobleaching of pulp in paper industry. Appl Microbiol Biotechnol 2015; 99: 155-64.

[3]   Mohajershojaei K, Khosravi A, Mahmoodi NM. Decolorization of dyes using immobilized laccase enzyme on zinc ferrite nanoparticle from single and binary systems. Fibers Polymers 2014; 15: 2139-45.

[4]   Kulshreshtha S, Mathur N, Bhatnagar P. Aerobic treatment of handmade paper industrial effluents by white rot fungi. Bioremed Biodeg 2012; 3: 1-7.

[5]   Imada C, Okanishi M, Okami Y. Intergenus protoplast fusion between *Streptomyces* and *Micromonospora* with reference to the distribution of parental characteristics in the fusants. J Biosci Bioeng 1999; 88: 143-7.

[6]   Verma V, Qazi GN, Parshad R. Intergeneric protoplast fusion between *Gluconobacter oxydans* and *Corynebacterium* species. J Biotechnol 1992; 26: 327-30.

[7]   Chen XY, Wei PL, Fan LM, Yang D, Zhu XC, Shen WH, Xu ZN, C PL. Generation of high-yield rapamycin-producing strains through protoplasts-related techniques. Appl Microbiol Biotechnol 2009; 83: 507-12.

[8]   Wei WL, Wu KL, Qin Y, Xie Z, Zhu XS. Intergeneric protoplast fusion between *Kluyveromyces* and *Saccharomyces cerevisiae* – to produce sorbitol from *Jerusalem artichokes*. Biotechnol Lett 2001; 23: 799-803.

[9]   Li M, Yi P, Liu Q, Pan Y, Qian GR. Biodegradation of benzoate by protoplast fusant *via* intergeneric protoplast fusion between *Pseudomonas putida* and *Bacillus subtili*. Int Biodeter Biodegr 2013; 85: 577-82.

[10]   Gold MH, Cheng TM, Alic M. Formation, fusion, and regeneration of protoplasts from wild-type and auxotrophic strains of the white rot basidiomycete *Phaonerochaete chrvsosporium*. Appl Environ Microbiol 1983; 46: 260-3.

[11]   Weiss RL. Protoplast formation in *Escherichia coli*. J Bacteriol 1976; 128: 668-70.

[12]   Niku-Paavola ML, Raaska L, Itävaara M. Detection of white rot fungi by a non-toxic stain. Mycol Res 1990; 94: 27-31.

[13]   John RP, Gangadharan D, Madhavan NK. Genome shuffling of *Lactobacillus delbrueckii* mutant and *Bacillus amyloliquefaciens* through protoplasmic fusion for 1-lactic acid production from starchy wastes. Bioresource Technol 2008; 99: 8008-15.

[14]   Zheng HJ, Gong JX, Chen T, Chen X, Zhao XM. Strain improvement of *Sporolactobacillus inulinus* ATCC 15538 for acid tolerance and production of D-lactic acid by genome shuffling. Appl Microbiol Biotechnol 2010; 85: 1541-49.

[15]   Sivakumar U, Kalaichelvan G, Ramasamy K. Protoplast fusion in *Streptomyces* sp. for increased production of laccase and associated ligninolytic enzymes. World J Microb Biot 2004; 20: 563-68.

32

On the Recognition, Measurement and Disclosure of Forest Biological Assets

Linfang Hou*

School of Economics and Management, Zhoukou Normal University, Zhoukou, 466001, Henan, China

Abstract: By analyzing the recognition, measurement and disclosure of forest biological assets, this paper aims to further enhance the recognition, measurement and disclosure levels of forest biological assets and also regulate the recognition, measurement and disclosure processes of forest biological assets. This paper both refers to the relevant provisions of the "Accounting Standards for Enterprises No. 5–Biological Assets" and uses cost measurement methods such as replacement cost method. The recognition, measurement and disclosure analysis of forest biological assets can help grasp the specific value of forest biological assets from another perspective. This will not only be able to improve the credibility of forestry enterprises, but also guarantee the reliability of forest biological assets.

Keywords: Forest biological asseta, Recognition, Measurement, Disclosure.

1. INTRODUCTION

With the increasingly rapid development of China's economy and society, all professions and trades have growing demand for forest biological assets [1-3]. At the same time, non-standard, unreasonable and undemanding recognition, measurement and disclosure of forest biological assets have gradually affected the statistical normalization and utilization of forest biological asset. Therefore, how to well recognize, measure and disclose forest biological assets has been a problem demanding prompt solution for local competent departments [4-7]. In this case, this paper makes a comprehensive analysis of recognition, measurement and disclosure of forest biological assets while conducting targeted studies of the concrete recognition, measurement and disclosure of different forest biological assets.

2. RECOGNITION OF FOREST BIOLOGICAL ASSETS

In order to well recognize the forest biological assets, the first thing to do is understand the specific concept of forest biological assets. In accordance with relevant accounting theory, forest biological assets can be recognized as forest resources if an enterprise possesses or controls the biological asset as a result of past transaction or event [8-10], and the economic benefits or service potential concerning this biological asset are likely to flow into the enterprise.

As can be seen from the above definition, whether a certain class of forest resources can be recognized as forest

biological assets is mainly depending on the following standards:

First of all, forest resources that have been recognized should be biological assets. The so-called biological assets can be simply understood as living forest resources grown in woodlands. To be identified as forest biological assets, all types of forest resources should meet the above basic requirements.

In the second place, the forest resources that have been recognized must exist currently, rather than those that existed in the past or will possibly exist in the future. In addition, the forest resources should still be possessed or controlled by an enterprise as a result of past transaction.

In the third place, the forest resources that have been recognized should be able to bring real benefits to the enterprise. Only by meeting this condition can the forest resources be able to be called forest biological assets. For bringing real benefits to enterprises, it incorporates two situations: First, the enterprise that owns the forest biological assets can directly earn some benefits wherefrom; second, although the forest biological assets are not possessed by the enterprise, it can effectively control it in order to gain certain benefits.

Finally, after meeting the above three requirements, forest resources should also meet one more requirement if to be recognized as forest biological assets: the forest resources as a kind of forest biological assets must be able to be indeed profitable for the enterprise. Under normal circumstances, this possibility should be greater than 50%.

After confirming a certain type of forest resources as forest biological assets, before carrying out measurement and disclosure work, people also need to simply classify forest biological assets. More often than not, the classification criteria of forest biological assets should conform to business

---
*Address correspondence to this author at the School of Economics and Management, Zhoukou Normal University, Zhoukou, 466001, Henan, China;

Table 1. Measurement Methods and Attributes of Purchased Forest Biological Assets.

	Types of forest biological assets	Method of measurement	Measurement attributes
Forest biological assets purchased	Consumption of forest biological assets	Monetary measurement + physical measurement	According to the measurement of historical cost as obtained
	Production of forest biological assets		
	The public welfare forest biological assets		

purposes. In accordance with this criterion, forest biological assets can be classified into three categories, namely, consumptive forest biological assets, productive forest biological assets and public welfare forest biological assets. Just as the names suggest, this classification standard is generally deemed as "object classification".

3. MEASUREMENT OF FOREST BIOLOGICAL ASSETS

3.1. Initial Measurement

As per the relevant provisions of the "Accounting Standards for Enterprises No.5–Biological Assets", forest biological assets should be well initially measured. The so-called "initial measurement" is to take the cost of forest biological assets as the measurement standard. Generally speaking, the acquisition methods of forest biological assets are divided into three categories: sourced from the nature, self-cultivating and purchased.

3.1.1. Purchased Forest Biological Assets

To take purchased forest biological assets as an example, its specific measurement method is:

Purchased forest biological assets = monetary measurement + physical measurement (1)

In Equation (1), purchased forest biological assets can be divided into three types as aforementioned, which are consumptive purchased forest biological assets, productive purchased forest biological assets, and purchased public welfare forest biological assets. When measuring the specific assets, all the measurement attributes are the historical costs acquired by the forest biological assets (See Table 1).

3.1.2. Self-Cultivating Forest Biological Assets

The cost of self-cultivating forest biological assets is ascertained in a way quite different from that of purchased forest biological assets. Among them, the biggest difference is it introduces the concept of "canopy closure". The so-called "canopy closure" can be understood as the degree of forest stand density, which in short can be summarized as:

Canopy closure (forest stand density) = the ratio between the vertical projection area of the forest canopy and the area of the forest land (2)

In Equation (2), if the canopy closure (forest stand density) is 1, then the forest area calculated is completely covered by forest resources; if the canopy closure (forest stand density) is ≥20%, the forest biological asset is usually seen as a closed forest. Conversely, if the canopy closure (forest stand density) is ≤20%, then the forest biological asset is not closed.

For self-created forest biological assets, the key to assets measurement lies in the cost of its assets. Firstly, it is necessary to capitalize forest biological assets. Secondly, expense forest biological assets in the management phase. Finally, when chopping down trees, capitalize forest biological assets and then include the calculation results into the total cost of the asset.

Cost of self-created forest biological assets = cost of forest culture and management production stage + management cost +cost of chopping down (3)

In Equation (3), the specific cost accounting objects are usually tree species or production unit. Moreover, according to the different operation purposes of forest biological assets, cost accounting needs to take into account different factors, typically including these types of factors: government financial subsidies, planting, tending, facilities, soil preparation and after-replacement (See Table 2, Table 3).

3.1.3. Forest Biological Assets Sourced from the Nature

For forest biological assets sourced from the nature, its measurement method is different from those of purchased forest biological assets and self-cultivating forest biological assets. The cost of a forest biological asset sourced from the nature is ascertained in accordance with its nominal amount. The so-called "nominal amount" can be determined by RMB 1 yuan, and only when there is conclusive evidence that the asset is possessed by the enterprise.

To take state administration transfer as an example, if the state transfers a forest biological asset sourced from the nature to a certain enterprise, then the enterprise should take the forest biological asset transferred as government subsidies in accordance with 1 yuan, and should also include it into the current profits or losses.

This cost accounting method is controversial among enterprises. First, although a forest biological asset is sourced from the nature, the government transfers it to an enterprise and the enterprise should be responsible for its management, such as pest control, and these activities require cost input.

Table 2. Self-Cultivating Forest Biological Assets.

Forest Biological Assets of Each Stage	The Occurrence Time	Charge	The Cost of the Occurrence of Events
Forest production stage	Before canopy closure	Capitalization	Preparation - on various types of land consolidation
			Planting - refers to the consumption when the planting of seeds and saplings
			Afforestation - refers to the forest canopy in front of the clay
			Tending - refers to the forest canopy before weed fertilization, mowing and pest control operation etc.
Stage management	After canopy closure	The cost	Maintenance - including forest protection personnel wages, related repair costs, pest control fee etc.
			Facilities - including fire anti-theft devices, lookout and other simple facilities etc.
			Other relevant protection cost - in addition to the above except
Harvesting stage	After canopy closure	Capitalization	Replanting - including the replant and necessary preparation fee

Table 3. Cost of Productive Forest Biological Assets.

	The stage	Charge	Cost accounting object
To create or plant production of forest biological assets	Immature production of forest biological assets	Capitalization	Tree or production unit
	Mature productive forest biological assets	The cost	

Secondly, a forest biological asset sourced from the nature can be converted to a consumptive forest biological asset under certain conditions. Then, the enterprise will face relatively less assets evaluation. Therefore, the author believes that a more reasonable accounting method can be taken to measure forest biological assets sourced from the nature.

Cost of forest biological assets sourced from the nature = consumable biological assets + productive biological assets + public welfare biological assets (4)

In Equation (4), the commonly used accounting measurement method is the replacement cost method, or the amount that fair value subtracts expected disposal costs. The above measurement methods can not only properly reflect the current profits and losses, but also more rationally, completely and accurately embody the value of forest biological assets sourced from the nature.

3.2. Subsequent Measurement

Subsequent measurement of forest biological assets mainly aims to adjust forest biological assets after initial measurement. Specifically, subsequent measurement of forest biological assets includes: subsequent expenditures of accounting treatment, depreciated accounting treatment,

accounting treatment that provides for diminution in value, and harvested and disposed accounting treatment. For length and simplicity reasons, the paper focuses on analyzing the common subsequent expenditure accounting treatment in subsequent measurement of forest biological assets. A forest biological asset is characterized by long life cycle and high input costs; particularly the subsequent input cost is more likely to exceed the initial investment. Therefore, subsequent measurement will directly affect current profits and losses for a business. Subsequent expenditure accounting treatment, in a nutshell, is to capitalize and expense the subsequent investment, which can not only accurately calculate the current profits of an enterprise, but also be greatly beneficial to the calculation of its current profits or losses.

To sum up, the subsequent expenses of forest biological assets shall be capitalized by the enterprise before canopy closure and included in the total cost of the asset; and its subsequent expenses shall be expensed by the enterprise after canopy closure. This is mainly because there are little or no changes in the appearance of forest biological assets after canopy closure. Therefore, the enterprise shall expense the subsequent expenses and additionally set a secondary subject under management expenses, thus enabling it to better clarify its current profits and losses.

Table 4. Basic information disclosure of forest biological assets.

	Category	Numbers (MU)	Planting	Canopy Density	The Government Subsidies
Consumption of forest biological assets	Timber forest				-
Production of forest biological assets	Bamboo forest				-
	Fruit tree				-
The public welfare forest biological assets	The ecological public welfare forest				

4. DISCLOSURE OF FOREST BIOLOGICAL ASSETS

4.1. Important Events Before the Financial Report of Forestry Enterprises

After a long term of development, disclosure of forest biological asset has gradually shifted from the traditional accounting information disclosure to a combination of non-financial information and accounting information disclosure. Disclosure of forest biological assets should be more accurate and vivid on the basis of traditional assets disclosure. At the same time, the disclosure of forest biological assets should also have some pictures and videos, thereby displaying related situation of forest biological assets in an all-round, three-dimensional way. Therefore, a forestry enterprise shall disclose the information concerning the forest biological assets as follows:

(1) The categories of forest biological assets. The so-called categories of forest biological assets not only refer to the division of its operation purpose, but also include various forest biological assets, namely quantities of physical output and book value of immature and mature forest biological assets.

(2) The natural condition reports, including plant diseases and insect pests in various forest biological assets and whether there are natural disasters.

(3) The appearance of massive updates of forest biological assets, especially improvements of biotechnology results in a substantial change in biological assets.

(4) The categories, obtainment methods and quantities of physical goods of the forest biological assets sourced from the nature (productive natural forests, consumptive natural forests or public welfare natural forests).

4.2. The Proposed Content and Format for Financial Statements of Forestry Enterprises

For forestry enterprises, forest biological assets are disclosed by income statements, cash flow statements and balance sheets. Since forest biological assets have the particularity of canopy closure, the measurement of historical cost is put a priority in the disclosure of assets. Meanwhile, the fair value shall be disclosed in the notes. The specific measure of

including forest biological assets and the earnings in the financial statements is to increase the appropriate items in the statements, and separately disclose the financial condition and economic performance indicators regarding forest biological assets. This paper analyzes the financial statements required to disclose forest biological assets, in the hope of promoting the standardization of forest biological assets disclosure, as well as enhancing its scientificity and reasonability (See Table 4, Table 5).

4.2.1. Income Statement

Concerning the income statement of forest biological assets, in actual disclosure of assets, "the income of biological assets" can be added according to the actual situation. In addition, for public welfare forest biological assets, government subsidies and other factors should be taken into account, and then correspondingly, in the income statement, "government subsidies", "productive subsidies "and "welfare benefits", among many others, should be added.

4.2.2. Cash Flow Statement

For the cash flow statement of forest biological assets, in actual disclosure of assets, "cash inflow" and "net flow" and more can be added according to the actual situation. To take as an example disclosure of public welfare forest biological assets, if the cash flow generated by the government's ecological compensation to forestry enterprises, then the cash flow statement is appropriately more perfect. Similar cases encompass the addition of cash resulted from selling forest biological assets in operation activity cash flow.

4.2.3. Balance Sheet

When forestry enterprises disclose forest biological assets, the balance sheet shall be involved inevitably. In the forestry enterprise's balance sheets, it is necessary to list respectively productive biological assets and public welfare biological assets. Meanwhile, the consumptive biological assets subject shall be added in the secondary subject of stock. However, this operation method has obvious shortcomings in the actual disclosure of assets. In particular, incorporating all biological assets under the subject of biological assets leads to a lack of rationality. In actual operations,

Table 5. Detailed Statement of Forest Biological Assets.

	At the Beginning of the Book Balance	The Increase Amount	Decrease for the Period	The Final Book Balance	Market Value
Consumption of forest biological assets					
Production of forest biological assets					
The public welfare forest biological assets					

accounting personnel prefer another method, which is to appropriately increase subjects in the balance sheet based on the above methods. For example, add the "consumptive forest biological assets" under the subject of "consumptive biological assets". This can more accurately and objectively reflect the disclosure requirements of forest biological assets.

CONCLUSION

In summary, with the increase in forest biological assets in our country, the recognition, measurement and disclosure of assets have been increasingly difficult. In real asset recognition, measurement and disclosure work, relevant personnel should carry out work flexibly as per the actual situation. Depending on the different operation purposes of forest biological assets, well measure the assets while taking the canopy closure as an important standard to rationally and scientifically capitalize and expense forest biological assets. In addition, relevant personnel should disclose forest biological assets. Especially for public welfare forest biological assets, scientific and reasonable government subsidy disclosure plays a very important role for both the reliability of forest biological assets and the credibility of all forestry enterprises.

ACKNOWLEDGEMENTS

Declared none.

REFERENCES

[1] Herbohn, Kathleen, and John Herbohn. "International Accounting Standard (IAS) 41: what are the implications for reporting forest assets?." *Small-scale Forest Economics, Management and Policy*, 5(2), pp. 175-189, 2006.

[2] Argilés, Josep M., Josep Garcia-Blandon, and Teresa Monllau. "Fair Value versus historical cost-based valuation for biological assets: Predictability of financial information." *Revista de Contabilidad*, 14(2), pp. 87-113, 2011.

[3] Morse, Wayde C., et al. "Consequences of environmental service payments for forest retention and recruitment in a Costa Rican biological corridor." Ecology and Society, 14(1), pp.23, 2009.

[4] Ali, Tanvir, *et al.* "Impact of participatory forest management on financial assets of rural communities in Northwest Pakistan." *Ecological Economics*, 63(2), pp. 588-593, 2007.

[5] Bray, David Barton, Camille Antinori, and Juan Manuel Torres-Rojo. "The Mexican model of community forest management: The role of agrarian policy, forest policy and entrepreneurial organization." *Forest Policy and Economics*, 8(4), pp. 470-484, 2006.

[6] Sayer, Jeffrey, et al. "The implications for biodiversity conservation of decentralized forest resources management." *The Politics of Decentralization: Forests, Power and People* ,pp. 121-137, 2005.

[7] Anderson, Jon, et al. "Forests, poverty and equity in Africa: new perspectives on policy and practice." International Forestry Review, 8(1) ,pp. 44-53, 2006.

[8] Mamo, Getachew, Espen Sjaastad, and Pål Vedeld. "Economic dependence on forest resources: A case from Dendi District, Ethiopia." *Forest Policy and Economics*, 9(8),pp. 916-927, 2007.

[9] Gibson, Clark C. "Forest resources: Institutions for local governance in Guatemala." *Protecting the commons: A framework for resource management in the Americas*,pp. 71-89, 2001.

[10] Rudel, Thomas K. "Changing agents of deforestation: from state-initiated to enterprise driven processes, 1970–2000." *Land use policy*, 24(1),pp. 35-41, 2007.

33

The Status and Countermeasure of Nitrate Pollution Under Double Crop-ping Systems in China

Xiujuan Ren[*], Sumei Yao, Runqing Wang, Dafu Wu and Shilin Chen

Henan Institute of Science and Technology, Xinxiang City, 453003, China

Abstract: The multiple cropping is the main agricultural production pattern, including double cropping of rice in south China and double cropping of wheat and corn in north China. Household investigation, typical farmland survey and county statistics inspection were performed to analyze the status and countermeasure of nitrate pollution in China. The result showed that the nitrate rate of underground water was slowly increased, but it was not above the standard of WHO. Under the multiple cropping of winter wheat-summer corn, the nitrate level of irrigation well water had raised year after year since 1991. Some countermeasures should be taken to avoid the nitrate content of underground water, such as increasing the fertilizer use efficiency by taking agronomic measures, according to the local conditions, a reasonable allocation and application of fertilizer and others.

Keywords: Fertilizer, multiple cropping, nitrate pollution.

1. INTRODUCION

With the increased agricultural intensification degree, especially the rising Nitrogen fertilizer application rate, nitrates from agricultural fertilizer could continue to leach into underground water for at least 80 years after the initial use [1]. In China, as early as the 1960s, the problem of nitrate pollution has been a matter of concern, which now has become one of the hotspots. The Chinese farmland loss of nitrogen comes from chemical nitrogen fertilizer entered into the environment accounting for about 4.93 million tons in 2004, including 284 thousand tons of nitrogen to form N_2O and 2.84 million tons of nitrogen to translate into NH_3 emission into the atmosphere, but the nitrogen loss of 1.29 million tons gets into the surface water, and 517 thousand tons enter into the underground water [2].

Nitrate is concentrated rapidly in the surface water and groundwater, continuously expanding the hazards of drinking water resources. Once the nitrate contents in drinking water increased over a certain level, it brings about the direct threat to human health, by damaging the ecosystems directly or indirectly. There were many reports on nitrate pollution of water bodies [3-11]. Meanwhile, the relationship was positive between groundwater nitrate concentrate of grain field and nitrogen fertilizer use level in Beijing suburban [7].

*Address correspondence to this author at the Henan Henan Institute of Science and Technology, Xinxiang, Henan Province, China;
E-mail: singrule@163.com

2. MATERIALS AND METHODS

The research paper focussed on the nitrate pollution of underwater in the multiple cropping regions. Many research methods were adopted to gain the status and countermeasure of nitrate pollution in China, such as household investigation, typical farmland survey, and county statistics inspection. Three planting patterns of winter wheat-summer corn, winter wheat-rice, and double cropping of rice were chosen. Three regions were chosen, which were Yujiang County, Xinxiang County, and Jing County, respectively. Investigation was carried out in 2011. Water samples were selected from the three counties, information was obtained on well depth, crop rotation, and yield and fertilizer application.

The advanced water quality laboratory series HACH-DR/EL 5 was used for the analysis of NO3-N at programmed wavelengths. Data were analyzed using regression analysis.

3. RESULTS

3.1. Nitrate Pollution Caused by Different Planting Patterns

3.1.1. Nitrate Pollution Caused by Double Cropping of Winter Wheat-Summer Corn in Jing County

According to the household investigation results, the water samples of irrigation wells were analyzed; the nitrate rates of irrigation water are shown in Table **1**. Under the conditions of double cropping of winter wheat-summer corn, the nitrate level was between 0.18 and 3.12 mg per liter

Table 1. The changes in nitrate rate of different irrigation well.

Sampling Location	Yield Level	Sampling Date (Month/Year)	Nitrate Content (mg/l)	Well Depth (m)
Village 1	low	5/2011	0.18	40
	high	5/2011	1.01	60
Village 2	low	5/2011	0.77	60
	high	5/2011	1.08	30
Village 3	low	5/2011	0.18	50
	high	5/2011	0.18	50
Village 4	low	5/2011	0.18	30
	high	5/2011	0.63	50
Village 5	low	5/2011	0.75	50
	high	5/2011	0.96	50
Village 3	high	12/2011	3.12	50
		12/2011	1.14	50

Table 2. The surface water nitrate concentration of point paddy fields during the rice growth period.

No.	Nitrogen Fertilizer Rate (kg/ha)	Before Rice Seedling Planting (mg/L)	Seedling Stage (mg/L)	Heading Stage (mg/L)	Filling Stage (mg/L)	Average (mg/L)
1	600	9.95	2.30	10.89	7.52	7.65
2	550	12.69	3.06	6.30	7.83	7.47
3	500	10.80	2.93	6.80	7.25	6.93
4	470	10.17	2.93	6.89	6.89	6.71
5	420	10.26	3.29	8.33	6.57	7.07
6	370	10.31	2.21	7.61	5.94	6.53
7	300	11.07	3.02	6.53	5.94	6.62

(Table 1). The nitrate levels had great variation for irrigation water by twice testing in a year (first time was May 29, 2011; second time was December 4, 2011). The nitrate rate did not surpass the standards of WHO.

3.1.2. Nitrate Pollution Caused by Double Cropping of Winter Wheat-Rice in Xinxiang County

The nitrate concentration of surface water of paddy-field cause impact on double cropping of winter wheat-rice. After the winter wheat is harvested, all wheat straws were returned to field. The arable land was first irrigated, then plowed and harrowed by tractors. The rice seedlings were planted in the end of June or early July. During the rice growth season, the paddy fields of seven households were chosen, and pointed,

and the surface water of paddy field was taken by regular intervals. The time of drawing the water sample was before rice seedling planting, seedling stage, heading stage, and filling stage. The nitrate level is shown in Table **2**.

The results showed that whether it is the different growth stage or the whole growth period of rice, the maximum surface water nitrate contents was only 12.69 mg/L; it conformed wth the WHO standard.

According to the results of nitrate test and the nitrogen fertilizer use level, the model was established as follows (1):

$$y=0.0035x+5.3691 \quad R2=0.7298 \quad r=0.8543** \tag{1}$$

**y is the nitrate concentration, and x is the fertilizer use level; there was a significant correlationship between them.

Table 3. The water nitrate rate of irrigation well variation in 2011.

Date (M-D)	7-10	7-31	8-10	8-31	9-15	9-30	12-20	4-10	5-20
Nitrate (mg/L)	6.76	5.62	10.29	7.63	6.29	7.17	3.21	7.03	5.04

Table 4. The nitrate rate of different water bodies in Yujiang County (mg/L).

Village	BRSP	HSER	MSER	HSLR	MSLR
Hongyuanwu	0.45	2.15	0.57	0.58	0.02
Panjia	0.31	0.46	0.31	0.06	0.43
Tangjing	0.94	0.65	1.67	0.04	0.59
Xinhelin	1.37	2.86	1.40	0.59	2.80
Huangni	0.54	0.55	1.32	0.55	0.44
Wujia	1.99	0.29	0.41	1.38	0.60

The water nitrate rate of irrigation well.

The investigation of irrigation well for water nitrate contents level began with the wheat planting, and finished by the next year after the wheat harvesting is completed. The water depth of irrigation well was between 40 and 50 meters. The testing nitrate levels are shown in the Table 3. The results showed that, within one year, the maximum nitrate rate reached only 10.29 mg/L; the other numbers were fluctuated between 3.21 and 7.63 mg/L.

3.1.3. Nitrate Pollution Caused by Double Cropping Rice in Yujiang County

The surface water samples were drawn from a large area. The time of water sample was drawn before rice seedling planting (BRSP), heading stage of early rice (HSER), maturing stage of early rice (MSER), heading stage of late rice (HSLR), and maturing stage of late rice (MSLR). The nitrate testing results are shown in Table 4. During the growth period of double cropping rice, the nitrate rate in different water bodies was below the standard, but the nitrate levels varied in different locations.

By comparng the three cropping patterns, winter wheat-rice in Xinxiang County was more likely to cause a rapid increase in the nitrate content of groundwater.

3.2. Nitrate Pollution Prevention and Countermeasures

China is facing the challenge of feeding its large sized and increasing population from a limited and decreasing area of cultivated land while striving to achieve a clean and safe environment [12]. The fertilizer-based intensive agriculture brings serious negative effects on the environment [13]. Nitrogen is the basic element of all life on Earth as we know. The need to provide sufficient nitrogen to grow food and support agriculture has governed the rise and fall of civilizations [14]. On the other hand, nitrogen is an important factor of growing pollution; nitrogen is different from other pollutants as a special object [5].

Today, there are many factors that affect the nitrate pollution. The most important factor is the nitrogen fertilizer application level. The nitrogen fertilizers have been heavily used recently, thereby increasing the nitrate rate of water body. Some research showed that the nitrate level of water body has a positive relationship with the nitrogen fertilizer use level [5, 6, 10, and 11]. The application and management of nitrogen fertilizer is imminent.

3.2.1. According to the Local Conditions, a Reasonable Allocation and Application of Fertilizer, Especially Nitrogen Fertilizer That is a Limited Agro-Resource in China

According to the China statistics yearbook of 2002, the fertilizer consumpton reached 42.54 million tons; every hectare of cultivated land used 327.13 kilograms in the same year (Table 5). In terms of arable land, the fertilizer amount used was the highest in Fujian province; the No. was 818.29 kilograms per hectare. Tibet is the example of lowest use, where the fertilizer application level was only one tenth of Fujian province, which was 82.74 kg/ha.

The planting index (PI) was considered. Different regions had different PI value (Table 6). Therefore, the fertilizer application levels were figured out by sown areas for every province in 2002 (Table 7).

Considering the actual agriculture production, there was a difference between FURc and FURs. The FURs of Shanghai municipality is the largest, it was 502.96 kg/ha, whereas the FURs of Heilongjiang province was the lowest; the latter was only about 1/4 of the former.

The model was built based on nitrogen fertilizer application amount and nitrate level of groundwater, for which data came from the investigation of both Wen County, Henan province, Huantai County, and Shandong province, where the yield of double cropping of winter wheat-summer corn was more than 15 tons per hectare. The model is as follows (2):

Table 5. The fertilizer use rate per hectare of cultivated land (FURc) in different Chinese provinces (kg/ha).

Province	FURc	Province	FURc
nationwide	327.13	Beijing	456.53
Shanxi [1]	184.98	Shanghai	644.24
Heilongjiang	104.68	Neimenggu	96.70
Anhui	470.12	Fujian	818.29
Henan	544.62	Hubei	495.61
Guangxi	381.36	Hainan	354.28
Yunnan	186.87	Tibet	82.74
Qinghai	103.92	Ningxia	193.49
Province	FURc	Province	FURc
Tianjin	356.26	Hebei	397.19
Liaoning	263.01	Jilin	204.54
Jiangsu	667.76	Zhejiang	424.88
Jiangxi	366.47	Shandong	557.40
Hunan	466.23	Guangdong	596.23
Sichuan	231.21	Guizhou	142.76
Shanxi [2]	254.94	Gansu	131.55
Xinjiang	209.00		

Note. [1] the capital is Taiyuan City; [2] the capital is Xi'an City.

y=0.0851x-23.283 R2=0.5212 r=0.7219* r0.01=0.765

r0.05=0.632   (2)

*y is nitrate concentration, x is fertilizer use level.

If y=0, then x=273.6 kg/ha

Thereby, once the nitrogen fertilizer application level is over 273.6 kg/ha, the nitrate will start accumulating in the groundwater.

If y=50, then x=861.1

If there is no nitrate pollution, the nitrogen fertilizer use amount must not exceed 861.1 kg/ha.

3.2.2. Rational Fertilization, Depending on the Rules of Different Crop Varieties, Demand Different Fertilizer Kinds

To enhance and improve Chinese people' living standards, the crop types and scopes have been expanded. The fertilization areas could be synchronously enlarged. Considering different soil types, crop varieties, climate, water resources, and agricultural production conditions, the reasonable amount of fertilization must be recommended in order to reduce the negative effects on the environment. However, this work was done in 1980s in China; by that time the application amount of NPK was suitable for the main food crops (Table 8).

Although the project has already been formulated for about 30 years, the situation has tremendously changed now. The project must therefore, be modified. Firstly, the most soil did not lack potash in China during 1980s. Therefore, the proposal of fertilization project did not include the potash fertilizer. But now that the 1/3 of arable land lack potash nutrient today [15], with the growing crop yields, this trend must be taken seriously; the potash fertilizer use must be increased. Secondly, the crop yield was between 4500 and 6000 kg/ha in 1980s, the main staple crop (wheat, rice, and corn) yield is as much as 7500 kg/ha now. Under the double cropping condition, the annual crop output could reach 15 t/ha. In the high output farming system, the input must be kept at the same pace with output. Thirdly, with the policy reforming and opening of aid, most rural Young men got into the factories of towns and cities and the left behind were women, older man and children. The organic fertilizer had almost not been used in the grain field besides the crop straws returning to the field. So, according to the newly changing production situation, the new project of fertilization should be re-enacted.

The fertilizer usage increased efficiency to some extent on the basis of a reasonable ratio of NPK. The investigation of household in the three double cropping region for the high crops yield of households, came up with the concluson that phosphorus and potash fertilizer account for high percentage

Table 6. The planting index in different provinces of China.

Province	PI (%)	Province	PI (%)
nationwide	119.57	Beijing	92.47
Shanxi [1]	82.71	Shanghai	128.09
Heilongjiang	85.65	Neimenggu	75.79
Anhui	153.60	Fujian	172.95
Henan	171.67	Hubei	147.07
Guangxi	147.22	Hainan	102.1
Yunnan	94.27	Tibet	64.81
Qinghai	69.29	Ningxia	86.64
Province	PI (%)	Province	PI (%)
Tianjin	102.84	Hebei	127.64
Liaoning	90.94	Jilin	88.81
Jiangsu	150.96	Zhejiang	133.53
Jiangxi	175.46	Shandong	139.62
Hunan	201.81	Guangdong	147.16
Sichuan	140.96	Guizhou	97.97
Shanxi [2]	81.74	Gansu	74.15
Xinjiang	93.61		

Note. [1] the capital is Taiyuan City; [2] the capital is Xi'an City.

of the chemical fertilizer used in the households leading to high yields; the ratio of NPK was reasonable.

For instance, in the households for double cropping of rice region, the nitrogen, phosphorus, and potassium fertilizer application level were ranging from 45.0 to 375.0, from 0.0 to 230.0, and from 0.0 to 210.0 kg/ha, respectively. There were 4 households without application of phosphorus fertilizer. Another 4 households did not use the potash fertilizer during double cropping rice production. As for the ratio of NPK, in combination with the yield level, the household was divided into the low, medium, and high yield; the ratio of NPK is shown in Table **9**.

Among 30 households in the region of double cropping of winter wheat-summer corn in Jing County, the principle fertilizer used was based on nitrogen, phosphorus and potassium. The nitrogen peak was more than 1.9 times that of the lowest. The highest amount of phosphate was more than 3 times that of the lowest. There was 1 household that did not use potash fertilizer for the wheat and corn production. In the other 29 households, the maximum potash fertilizer application used was more than 6.6 times that of the lowest. The production of nitrogen, phosphorus, and potash fertilizer is shown in the Table **10**.

Among 30 households in the region of double cropping of wheat-rice, there were 3 households that did not use phos-

phorous fertilizer. There were 5 households without the use of potash fertilizer. However, the yield of wheat and rice was not low. The wheat straws must be returned to the fields. To a certain extent, the wheat straws containing NPK nutrients could make up for the nutrient elements depletion during the crop production.

If the ratio of NPK is not reasonable, not only does this lead to the waste of nitrogen fertilizer and increased nitrogen loss, but also caused the water body polluted with nitrate. The experimental results lasting over 18 years (1978-1996) on Malan farm, Xinji City, Hebei province, showed that the nitrate rate of soil was higher in the single nitrogen fertilizer application treatments. On the other hand, if the organic manure use overloaded, the soil contained high rate of nitrate, thereby exerting the same effects on the groundwater [10].

3.2.3. Increasing the Fertilizer Use Efficiency by Agronomic Measures

Optimization of the fertilization methods, such as using ammonium bicarbonate and urea as the basal fertilizer in the rice production, when they were topdressing, gave the nitrogen use efficiency to be 17 % and 28%, respectively. At the same time, the nitrogen loss could reach from 47 % to 70% respectively. Moreover, if they were buried material fertilizers or mixed with organic fertilizer, the nitrogen use

Table 7. The fertilizer use rate per hectare of sown area (FURs) in different provinces of China (kg/ha).

Province	FURs	Province	FURs
nationwide	273.59	Beijing	493.71
Shanxi[1]	223.65	Shanghai	502.96
Heilongjiang	122.22	Neimenggu	127.59
Anhui	306.07	Fujian	473.14
Henan	317.25	Hubei	336.99
Guangxi	259.04	Hainan	346.99
Yunnan	198.23	Tibet	127.67
Qinghai	149.98	Ningxia	223.33
Province	FURs	Province	FURs
Tianjin	346.42	Hebei	311.17
Liaoning	289.21	Jilin	230.31
Jiangsu	442.34	Zhejiang	318.19
Jiangxi	208.86	Shandong	399.23
Hunan	231.02	Guangdong	405.16
Sichuan	164.03	Guizhou	145.72
Shanxi[2]	311.89	Gansu	177.41
Xinjiang	223.27		

Notes.[1] the capital is Taiyuan City; [2] the capital is Xi'an City.

Table 8. The suitable use amount of NPK for main food crops in China (1981-1983).

Item	Rice	Wheat	Maize
N (kg/ha)	108.3	104.9	108.4
P_2O_5 (kg/ha)	36.8	66.3	68.3
K_2O (kg/ha)	37.9	0	0
Total (kg/ha)	183	171.2	176.7

Note: Data from the reference [15].

Table 9. The production of fertilizer in different households in the double cropping of rice region (kg/ha).

Yield Level	Actual Yield	Nitrogen Rate	Phosphorous use Rate	Potash Rate	N : P : K	Household %
<6750	5693.7	205.9	94.7	165.7	1 : 0.46 : 0.80	37.61
6751-9000	7474.1	191.2	90.5	140.0	1 : 0.47 : 0.73	53.85
>9001	10429	226.5	67.6	164.5	1 : 0.30 : 0.73	8.54

Table 10. The production of fertilizer in different households in the double cropping of winter wheat-summer corn region (kg/ha).

Yield Level	Actual Yield	Nitrogen Rate	Phosphorous Rate	Potash Rate	N : P : K	Household %
<11250	10346.3	499.3	219.8	105.9	1 : 0.44 : 0.21	26.7
11251-13500	12708.5	593.7	246.0	141.5	1 : 0.41 : 0.24	56.7
>13501	14118.0	589.8	235.5	137.1	1 : 0.40 : 0.23	16.6

efficiency was 26 % and 38%, respectively, while the nitrogen loss could reach from 38 % to 51%.

During corn production, if the nitrogen use efficiency was the maximum, the nitrogen fertilizer use times should be increased, and the amount of application times should be decreased; at the same time, the irrigation times should be increased, the growth speed must be promoted, and the amount of nitrogen absorption should be increased.

In short, as long as nitrogen use efficiency is improved, the crop yield should be increased, the nitrogen loss must be reduced, the nitrate accumulation in the environment must be cut down, and resultantly the nitrate pollution could be prevented to some degree.

CONCLUSION

The nitrate pollution of underground water is a hot topic worldwide. However, China being the largest population size in the world, is facing serious issues in terms of food security problem that is brought to focus worldwide through reserach. The total yield of food grain has increased since 2004, and in this connection, the multiple cropping of winter wheat-summer corn, winter wheat-rice, and double cropping rice play important roles. Besides, the chemical fertilizers have contributed invaluably. With the increased chemical fertilizer application gradually or even overload in some regions of China, the nitrate level of underground water raised or even went over the standard level of WHO. It gave rise to the nitrate pollution of the water body.

The amount of chemical fertilizer application at household, farmland and county levels was obtained. Careful study data came from three levels (county level, household level, and farmland level), three counties (Yujiang county, Xinxiang county, and Jing county), and three models of multiple cropping (double cropping rice, Winter wheat- rice, and winter wheat-summer corn). The results showed that the chemical fertilizer played an important role in increasing crop yields. There was difference between counties and households in that the amount of fertilizer application was gradually increased. The nitrate rate of underground water was slowly increased, but it did not rise over the standard of WHO. Under the multiple cropping of winter wheat-summer corn, the nitrate level of irrigation well water had raised year after year since 1991.

In order to avoid the nitrate pollution of underground water, some countermeasures should be taken, such as: increasing the fertilizer use efficiency by taking agronomic measures; the fertilizer use efficiency increases to some extent on the basis of the reasonable ratio of NPK; according to the local conditions, a reasonable allocation and application of fertilizer, especially nitrogen fertilizer that is one of the limited agro-resources in China; and depending on the nature of different crop varieties demand different fertilizer types.

ACKNOWLEDGEMENTS

This work was funded by the "12th Five Year Plan" Circular Agriculture Science and Technology Project Central Plains Economic Zone Agricultural Circular Technologies Integration and Demonstration, Project No. 2012BAD14 B08-1.

REFERENCES

[1] Boxall B. Nitrate pollution continues for decades after fertilizer use. Fertil Nitrate Water Pollut 2013; 10: .

[2] Zhu LZ. Research on soil nitrogen in China. Acta Pedologica Sinica 2008; 45(5); 778-83.

[3] Zhang FD. The trends and countermeasures of chemical fertilizers. Environ Sci 1985; 6: 54-9.

[4] Zhu JC, Tian YB. The chemical nitrogen fertilizer and groundwater pollution. Hydrogeol, Eng Geol 1986; 5: 60-5.

[5] Zhu ZL, Wen QX. Nitrogen in soil of China. Jiangsu Science and Technology Press: Nanjing 1992.

[6] Zhang WL, Tian ZX, Zhang N, Li XQ. Investigation of nitrate pollution in groundwater due to nitrogen fertilizer in agriculture in north China. Plant Nutrit Fertiliz Sci 1995; 1(2): 80-7.

[7] Zhu JG. The hazards and research prospects of NO3-—N pollution. Acta Pedologica Sinica1995; 32: 60-5.

[8] Zhang WL, Wu SX, Ji HJ, Kolbe H. Estimate of agricultural non-point pollution in China and the alleviating strategies. Scientia Agricultura Sinica 2004; 37(7): 1008-17.

[9] Zhao TK, Zhang CJ, Du LF, Liu BC, An Z. Investigation on nitrate concentration in groundwater in seven provinces (city) surrounding the bo-hai sea. J Agro-Environ Sci 2007; 26(2): 779-83.

[10] Wu DF, Chen HW. Effect of different planting patterns of food crop on nitrate content in underground water. Res Agric Modern 2007; 28(1): 107-109-3.

[11]   Wu DF, Zhang W, Sun XY, Li DF, Ren XJ. Nitrogen fertilizer applications having effects on the nitrate contents of groundwater in the double cropping of wheat-maize regions. J Henan Agric Sci 2008; 10: 63-7.

[12]   Brown LR. Who will feed China? World Watch 1994; 7(5): 66-76.

[13]   Zhu ZL, Norse D, Sun B. Policy for reducing non-point pollution from crop production in China. Beijing: China Environmental Science Press 2006.

[14]   Leigh GJ. World's great fix: a history of nitrogen and agriculture. Cary. NC, USA: Press Incorporated 2004.

[15]   Shen SM. Soil fertility of China. Beijing: China Agricultural Press 1998.

34

MEKC Technique-Based Dairy Product Biochemical Analysis

Wei Wang[*]

Baotou Light Industry Vocational Technical College, Baotou, Inner Mongolia 014035, China

Abstract: The paper applies MEKC technique to simultaneously carry out quantitative analysis on dairy product's melamine (Mel) and 5-hydroxymethyl furfural (HMF) bi-component. The quantization analytic experiment starts with a systematic research on operating voltage, electrophoretic buffer solution gravity and optimum pH value, which further influences additives' concentration as well as other factors, and detects 10 kinds of practical samples on the basis of defined optimum conditions. Detection result indicates that both Mel and HMF showed peaks in 16min, simultaneously acquiring two indicator parameters in one analytical test, This test verifies applicability and scientific validity of MEKC technical approach in dairy product's Mel and HMF contents detection, which can play an effective role in providing rapid and reliable daily monitoring support for dairy industry.

Keywords: 5-hydroxymethyl furfural, capillary electrophoresis, dairy product, melamine, standard extract.

1. INTRODUCTION

Dairy product contains rich amino acids, which possess many benefits for consumers. If dairy products exceed the concentration of Mel, it can generate negative effects on consumers [1]. Generally, dairy product manufacturers apply ultra high temperature sterilization procession on products in order to extend shelf-life of the goods, but raw milk generates a series of unhealthy physicochemical and biological changes during the heating process, from which lactose and protein occur. HMF is an important intermediate product of Mailard's reaction, and if HMF exceeds the standard limit, it also generates negative effects on consumers [2]. Therefore, it is necessary to detect and monitor Mel content and HMF contents in dairy products. This paper thus presents one detection approach based on micelle electro kinetic capillary electrophoresis (MEKC) technique for the purpose of providing a convenient method for dairy product quantization detection analysis.

MEKC technique has been widely applicable in a variety of researches. Among which, Han Le *et al.* [3] (2013) applied MEKC-DAD approach simultaneously determining different batches of Scutellaria laterifolia and its processed products such as baicalin, scutellaria laterifolia element, wogonin, oroxylin A and apigenin's contents. It was pointed out that the approachwas simple and accuratewith good repeatability. Hu Jiang-Tao *et al.* [4] (2014) set up MEKC analytic approaches which simultaneously detected chilli powder, chilli oil and chilli sauce's Sudan Red I, II, III, IV, G. Yin Wan-Lu *et al.* (2014) [5] used MEKC analytic approach to detect vaccenic acid content in milk.

Zhang Li-Chun *et al.* [6] (2015) made use of a seasoning sorbic acid and benzoic acid contents MEKC measurement approach, tested and investigated detection wavelength, buffer solution selection, buffer solution pH values,

separation voltage and column temperature as well as effects of other factors on separation. Luo Yi-Yuan *et al.* [7, 8] established an MEKC-DAD approach, simultaneously determining different origins and commodities' polygonum multiflorum's stilbene glucoside, emodin, aloe-emodin, rhein, physcion, and catechinic acid contents. Thereby, MEKC technique application feature is able to simultaneously detect the contents of various kinds of materials as compared to the current widely used gas chromatography-mass spectrometry and liquid chromatography tandem mass spectrometry. Moreover, it is capable of overcoming high experiment expending, complicated operational process, long time-consuming as well as other drawbacks. Therefore, this research paper introduced MEKC technique for detecting Mel and HMF contents in dairy products in the hope of making an initial effort in the improvement and reformation of dairy product monitoring technique.

2. EXPERIMENTAL DESIGN

Capillary electrophoresis apparatus : Beckman P/ACE MDQ capillary electrophoresis system (Fullerton, CA) was used. Beckman company 32Karat[TM] 8.0 version software workstation was applied. The equipment included a photodiode array detector which could scan at a scope of 190-600nm. Apparatus diagram is shown in Fig. (1) As can be seen, the structure consisted of a high voltage power supply, a capillary tube, a detector and two buffer solution storage bottles and sample cells that could support insertion at the two ends of the capillary tube and connection with power supply.

In electrolyte solution, charged particle transfer by electric field shifts from the opposite direction of its carried electrical charge at different speeds, is called electrophoresis. Capillary electrophoresis has a unified term of a kind of liquid phase separation technique that regards high voltage electric field as a driving force [8], uses capillary tube as a separation channel, and implements separation according to the samples' component mobility and allocation behavior

*Address correspondence to this author at the Baotou Light Industry Vocational Technical College, Baotou, Inner Mongolia 014035, China; E-mail: 13674728328@163.com

differences. Capillary electrophoresis separation is a typical differential motion process. All factors that affect migration rate would also affect separation result, which provides an evidence for controlling and optimizing separation. Meanwhile, it also points out the need for developing a new mode for capillary electrophoresis separation. The new mode applied for separation in this research was micellar electrokinetic chromatography (MEKC).

Non-coating quartz capillary column of 75μm i.d., 375μm o.d., was provided by Polymicro Technologies, Phoenix, AZ, USA. Two kinds of electronic scales were selected and used, one was the Hunan instrument scale instrument limited company produced instrument with the scale measurement accuracy being 0.01g; while the other being Mettler Toledo instrument (Shanghai) limited company produced instrument with the scale measurement accuracy of 0.0001g.

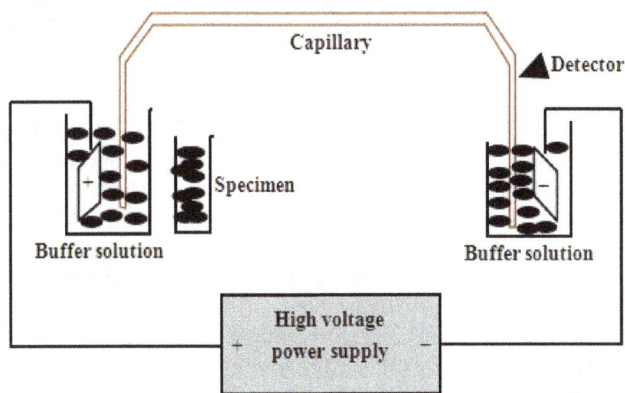

Fig. (1). Capillary electrophoresis apparatus configuration sketch.

Other instruments included ultrasonic cleaner, Centrifuge5415D typed high speed centrifuge, XW-80A mini-sized vortex mixer, SG2typed pH meter and 0.45μm microfiltration membrane. Manufacturers of these instruments Tianjin

Automatic Science Instrument limited company, Eppendorf company, Shanghai Huxi Analysis Instrument limited company, Mettler-Toledo instrument (Shanghai) and Shanghai Bandao industrial limited company purification instrument factory.

Table **1** shows basic information of each experimental reagent. In Table **1**, Mel manufacturer is Shanghai Sinopharm Chemical Reagent Limited Company, with its concentration above 99%; HMF and Tris are purchased from Sigma-Aldrich (St.Louis, MO). Above all, reagents adopt ultrapure water preparation.

Standard stock solution preparation included dissolving 5mg Mel and HMF into 1% TCA solution so that 1mg/mL Mel and HMF standard stock solution was achieved. The prepared 1mg/mL standard stock solution was kept in the refrigerator at 4 degree centigrade, and it was made sure that these two solutions were be used in one month. TCA solution having 1% concentration was used to dilute 1mg/mL Mel and HMF solutions, further diluting them into 0.05μg/ml, 0.10μg/ml, 0.20μg/ml, 0.50μg/ml, 2.0μg/ml, 5.0μg/ml, 20.0μg/ml and 100μg/ml standard stock solutions.

In electrophoresis buffer solution preparation, the solution was composed of 15mM SDP, 15mM DSP and 80mM SDS. The pH value of the solution was adjusted to 6.85, and then 0.45μm filter membrane was used for filtration, carrying out ultrasonic 10min bubble removal treatment on solution after filtration. The solution was then maintained at room temperature, ensuring its availability in two weeks.

For liquid milk sample treatment, 1mL liquid milk and 1mL 4% TCA solution were blended into 50mL centrifuge tube, and 2mL ultrapure water was then added. Subsequently, vortex shaking (2min) and ultrasound (10min) were carried out. Interlayer clear solution was drawn and filtration was conducted in 0.45μm filter membrane. Filter liquor could directly implement the sample detection.

Table 1. Experiment reagent basic information list.

Reagent Name (Chinese)	Reagent Name (English)	Chemical Formula	Abbreviation
ShiErWanJiLiuSuanNa	Sodium Dodecyl Sulfate	$C_{12}H_{25}\text{-}OSO_3Na$	SDS
Linsuan	Phosphoric Acid	H_3PO_4	PC
BengSunNa	Sodium Tetraborate	$Na_2B_4O_7.10H_2O$	ST
LinSuanErQingNa	Sodium Dihydrogen Phosphate	NaH_2PO_4	SDP
LinSuanQingErNa	Dibasic Sodium Phosphate	Na_2HPO_4	DSP
SanLuYiSuan	Trichloroa Cetic Acid	Cl_3CCOOH	TCA
QingYangHuaNa	Sodium Hydroxide	$NaOH$	SH
YanSuan	Hydrochloric Acid	HCl	HA
SanJuQingAn	Melamine	$C_3N_3(NH_2)_3$	Mel
JingJiaJiKangQuan	Hydroxy Methyl Furfural	$C_6H_6O_3$	HMF
SanJingJiaJiAnJiJia Wan	Tris (Hydroxymethyl) Methyl Aminomethane	$NH_2C(CH_2OH)_3$	THAM

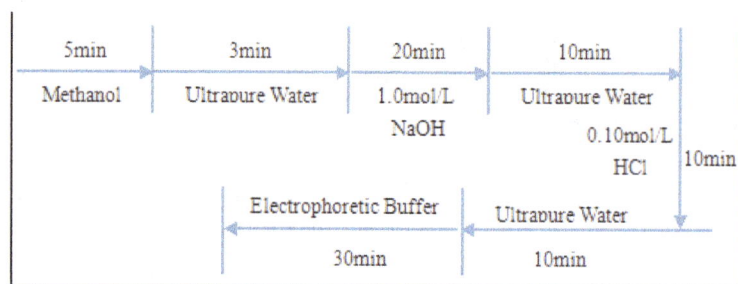

Fig. (2). New capillary tube handling and balancing flow.

Solid milk powder sample treatment involved dissolving of 2g milk powder into 6mL 1% TCA solution, successively carrying out vortex shaking for 2min and ultrasound for 10min, from which 1% TCA solution had a constant volume of 10mL. 12000rpm centrifugal treatment was then undertaken for 4min, drawing an interlayer clear solution. 12000rpm centrifugal treatment was repeated on clear solution for 4min, and then again interlayer clear solution was drawn after centrifugation. Finally, filtration was performed on clear solution in a 0.45μm filter membrane. Filter liquor could directly implement the sample detection.

Before starting capillary electrophoresis, new capillary tube carried out handling and balancing as given in Fig. (2):

In case there are same applied samples to be detected, it should use electrophoresis buffer solution to wash them for 5 minutes ahead of time.

1) The experiment used capillary tube having a non-coated quartz, with an inner diameter of 75μm, total length of 58.5cm, and valid length of 50cm.

2) 214nm and 280nm double –wavelengths were adopted to detect Mel and 5-HMF respectively.

3) Electrophoresis buffer solution pH value was 6.85, which was compounded by 30mmol/L PC and 80mmol/L SDS.

The sample was introduced with an air pressure (0.5psi×5s) and 15kV operating voltage. Capillary card case temperature was controlled at 25 degree centigrade.

4) Detected data formed drawing in 32Karat^{TM}8.0 chromatographic work station after signal collection.

3. RESULT ANALYSIS

For optimization of Mel and HMF detection wavelengths, capillary electrophoresis apparatus diode array detector was adopted to carry out full wavelength scanning at a scope of 190-400nm, by obtaining Mel and HMF ultraviolet absorption spectrograms. By comparing, Mel and HMF maximum absorption wavelengths were defined as 202nm and 280nm respectively. The research, however, selected 214nm and 280nm respectively as Mel and HMF detection wavelengths. Fig. (3) illustrates that the two wavelengths detected electrophoresis images and two kinds of standard stock solution ultraviolet absorption spectrograms.

In Fig. (3), vertical axis represents absorbance, with the unit being mAU. As shown, standard stock solution was 5.0μg/mL Mel standard substance and 5.0μg/mL HMF standard substance mixed solution. Electrophoresis buffer

solution was 15mM SDP-15mM DSP and 80mM SDS with pH value of 6.85. Also, electrophoresis voltage wasis 15kV with 0.5psix5s sample introduction .

By Fig. (3), it is clear that capillary electrophoresis separation-DAD detection can well separate and distinguish Mel and HMF, apply its migration time and ultraviolet absorption characteristic spectrum to determine the nature of Mel and HMF, and also use peak area to carry out quantitative analysis.

1) Electrophoresis buffer solution type optimization: Buffer solution type experience directly affected working current, samples' separation degree, and peak shape and detection sensitivity and so on. By investigating THAM-PC buffer solution, ST buffer solution and SDP-DSP buffer solution of the three types, the paper respectively optimized the three different types of buffer systems' different concentrations and different pH values, finally adopting 30mM SDP-DSP buffer solution as an electrophoresis buffer system.

Electrophoresis buffer solution pH value optimization : Buffer solution pH value has a great impact on electro osmotic flow and target molecular carried charge, and exerts effects on Mel and HMF separation degree. By investigating SDP-DSP buffer solution with pH value of 6.70 to 7.00, variation process impacts on target molecular carried charge and electro osmotic flow. The paper selected SDP-DSP buffer solution pH value as 6.85.

Additive concentration optimization: The paper defines SDS with 60mmol/L-100mmol/L concentration gradient to investigate additive influences during capillary electrophoresis separation process. During separation process, SDS composed micelle moves towards positive pole in capillary, and interacts with analyte through electrostatic function and formed hydrogen bond. Comprehensive analysis considered that SDS concentration impacts on separation degree, operating time, sensitivity, finally defining that 80mmol/L is additive to SDS optimum concentration.

2) Separation voltage optimization: Operating voltage has important impacts on migration time of samples, working current and samples' each component separation degree. When column length is defined, operating voltage increases, and electro osmotic flow absolute value also increases, while migration time shortens. When voltage increases and goes beyond pole, joule heat influence intensifies, current gets excessively higher, and base line is not stable, which further reduces samples' separation de-

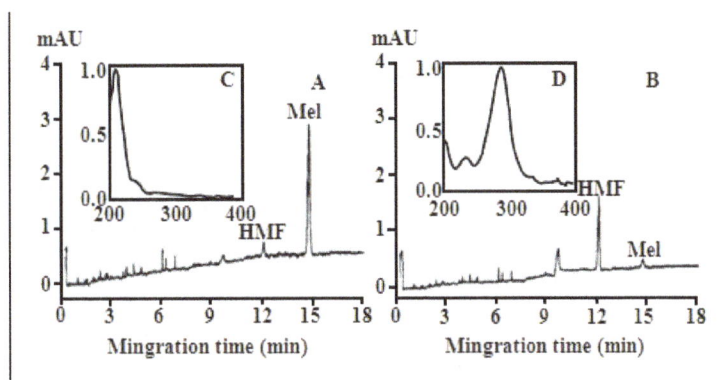

Fig. (3). Electrophoretic image (A: 214nm wavelength: B: 280nm wavelength) and ultraviolet absorption spectrum (C: Mel: D: HMF).

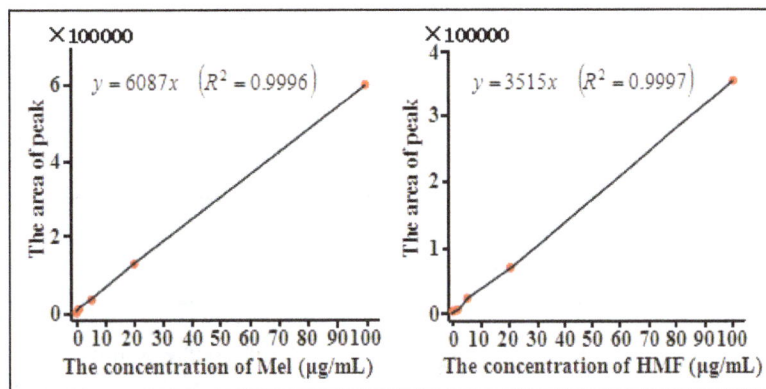

Fig. (4). Standard curve chart (Left: Mel: Right: HMF).

Table 2.　Mel and HMF LOD, LOQ, LR and CC measurement result list.

Compound	LOD (µg/mL)	LOQ (µg/mL)	LR (µg/mL)	CC (R^2)
Mel	0.047	0.05	0.05-100	0.9996
HMF	0.067	0.10	0.10-100	0.9997

gree. In order to separate Mel and HMF as well as other complicated ground substances in a short span of time, the research finally selected 15kV as the operating voltage.

To sum up, detection wave was applied at a 214nm wavelength to detect Mel, and at a 280nm wavelength to detect HMF. Optimum pH value was defined as 6.85. Optimum SDP-DSP buffer solution concentration was 30mmol/L. Optimum SDS additive concentration was 80mmol/L. Optimum separation voltage was 15kV.

0.05µg/ml,0.10µg/ml,0.20µg/ml,0.50µg/ml,2.0µg/ ml,5.0µg/ml,20.0µg/ml and 100µg/ml Mel and HMF standard solutions were selected to carry out capillary electrophoresis analysis, while taking Mel and HMF peak area as vertical coordinate, and regarding sample concentration as horizontal coordinate. Fig. (4) shows the standard curve chart.

By Fig. (4), it is clear that Mel had good linear relations at the concentration scope of 0.05-100µg/mL and HMF had good linear relations at a concentration scope of 0.10-100 µg/mL.

Table 2 shows Mel and HMF limits of detection LOD, limits of quantification LOQ, linear ranges LR and correlation coefficients CC. Among them, LOD value is 10 times blank sample introduction evaluation signal values plus 3 times' standard deviation; LOQ value is 10 times blank sample introduction evaluation signal values plus 10 times' standard deviation. The paper applies capillary electrophoresis –PDA which detects that Mel LOD is 0.047µg/mL, LOQ is 0.05µg/mL, while international reported milk Mel maximum residue limit is 1.0-2.5mg/kg. Therefore, the approach fully meets the demands of daily work's milk Mel detection. Besides, the approach also fully meets the demands of daily work's HMF detection.

Reproducibility research contains Mel and HMF forwarding time and peak area intraday relative standard deviation and interday relative standard deviation measurement. Among these, intraday deviation measures the same sample five times for sample pretreatment in the same day and undertakes capillary electrophoresis detection; inter-day deviation is carried out by continuously measuring the same sample for five days pretreatment and taking capillary electrophoresis detection. The purpose is to measure the accuracy

Table 3. Mel and HMF's MT and PA reproducibility research result table.

Sample		MT (%)	PA (%)
Intraday (n=5)	Mel	2.25	2.58
	HMF	1.69	3.53
Interday (n=5)	Mel	4.30	4.61
	HMF	4.23	5.42

Table 4. Milk Mel and HMF recovery research results table.

Compound	SC (μg/mL)	AR (%)	RSD (%) (n=4)
Mel	1.0	97.02	3.09
	5.0	94.85	4.20
HMF	1.0	99.20	3.11
	5.0	95.26	2.85

of the evaluation approach. Table **3** shows the Mel and HMF migration time MF and peak area PA relative standard deviation. Table **3** data shows that the approach has good accuracy.

Approach recovery evaluation selected Mel and HMF with 1.0μg/mL and 5.0μg/mL, two kinds ofspiked concentration (SC) to add a milk practical sample before sample treatment, for conducting a test according to the experimental design sample pretreatment approach and capillary electrophoresis detection approach, and for obtaining average recovery (AR) and relative standard deviation (RSD). Table **4** shows milk Mel and HMF recovery conditions.

By Table **4**, it is clear that Mel average recovery is 95.95%, HMF average recovery is 97.23%, and the two components relative standard deviations are less than 5%. Therefore, the approach recovery condition can be considered as good.

According to the experimental design, 10 kinds of dairy products are selected.Among them, seven kinds are liquid milk, and three are milk powder. The experiment optimization approach is adopted to carry out analytic detection on samples.

At first, TCA extracting solution is used to dilute, and then successively vortex shaking , ultrasonic extraction and centrifugal filtration are carried out, finally implementing capillary electrophoresis detection on samples.

As Fig. (**5**) shows typical milk practical samples extract liquor and extract liquor capillary electrophoretic image after respectively adding 5μg/mL Mel and HMF. In practical samples, Mel and HMF concentrations are quantized through Mel and HMF peak areas. As far as liquid milk is concerned [9], concentration unit finally is conversed into μg/mL ; as far as milk powder is concerned [10], concentration unit finally is conversed into μg/g. As Table **5** shows, capillary electrophoresis measured two kinds of different samples' Mel and HMF concentrations, from which comparison was carried out between Mel concentration and high performance

liquid chromatography. High performance liquid chromatography approach measured data obtained through inspection and quarantine bureau handling with milk practical samples according to "People's Republic of China national standard GB/T22388-2008 raw milk and dairy product melamine detection approach" of Fujian province Quanzhou city [11] stipulating sample pretreatment procedures and then carrying out detection by high performance liquid chromatography. Carrying out comparison on (Table **5**) CE-PDA and HPLC-UV measuring ten kinds of samples' Mel concentrations, it can be found that CE-PDA measured concentration was slightly higher than that of HPLC-UV measurement. Data correlation analysis carried out on concentration, highlighted that CE-PDA measured concentration and HPLC-UV measured concentration data correlation coefficient was 0.99485, which shows that two groups of data differences obtained by two approaches were mainly from system error. Two groups of data differences may be caused by two approaches' sample pretreatment differences. In national standard recommended raw milk and dairy product Mel detection approach, sample pretreatment also contains sample detection after carrying out solid phase extraction on extracting solution, as compared to capillary electrophoresis approach. However, extra solid phase extraction handling procedure may lead to samples' Mel recovery declination. The data provided by Fujian province Quanzhou city inspection and quarantine bureau shows that average recovery after solid phase extraction treatment is around 80%.

By Table **5**, following conclusions can be drawn:

1) Sample-1 is a kind of children drunk milk beverage, and use of capillary electrophoresis approach and high performance liquid chromatography approach did not detect Mel.

Addition of Sample-2 1010μg/mL Mel into Sample-1, capillary electrophoresis detection result was 10.26 μg/mL, which shows good accuracy and recovery of the approach to a certain extent.

Table 5.   Practical samples' Mel and HMF contents measurement result table.

Sample	Mel		HMF
	CE-DAD (μg/mL)	HPLC (μg/mL)	CE-DAD (μg/mL)
1 — Milk drink for kids	/	/	0.59
2 — #1 spiked with 10μg/mL Mel	10.26	8.32	0.54
3 — Whole milk powder	/	/	/
4 — Ordinary milk, same brand as #3	33.84	26.99	1.59
5 — Baby milk powder	20.10	16.95	2.25
6 — Baby milk powder, same brand as #5, different batch	23.63	18.18	1.27
7 — Imported baby milk powder	/	/	0.60
8 — Baby milk powder	/	/	2.10
9 — Milk powder	1.32	1.02	/
10 — Soy milk powder	31.73	26.80	0.91

HPLC: The HPLC results of Mel were provided by Quanzhou Inspection and Quarantine Bureau of China, the mean of recoveries was 80%. /: Not detected.

2) Sample-3 and Sample-4 are whole milk powder and milk beverage of the same brands', however Sample-4 contains higher concentration of Mel, as detected by capillary electrophoresis approach and high performance liquid chromatography.

Sample-5 and Sample-6 respectively are milk powders of same brands belonging to different batches, and both were detected with Mel.

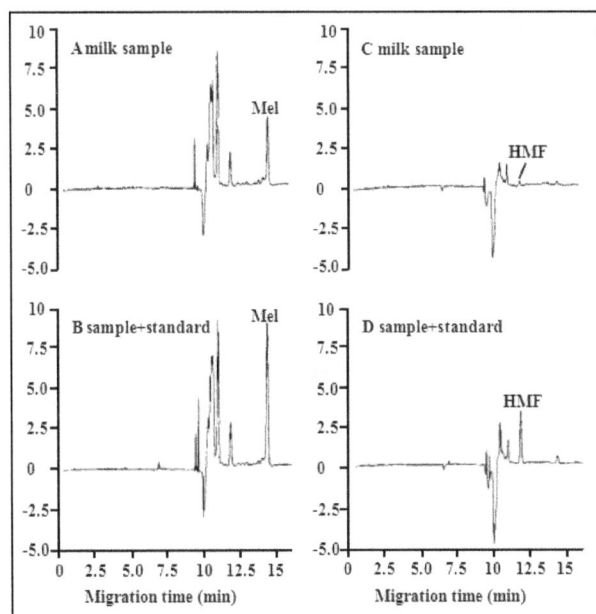

Fig. (5). Electrophoretic image of extracting solution that typical milk practical samples and same samples add standard substance (**A** and **B** is 214nm wavelength detected spectrogram; **C** and **D** is 280nm wavelength detected spectrogram).

3) Sample-7 and Sample-8 are imported infant milk powders. Mel is not detected in these two samples.

4) Sample-9 detects Mel with 1.32μg/g concentration, which gets closer to national standard restrictive Mel maximum residue limit (1.0-2.5mg/kg).

5) Sample-10 is a soybean milk powder, from which very high content of Mel is detected.

6) Sample-3 and Sample-9, two milk powders' HMF concentration was lower than 0.50μg/g, however the concentration cannot be accurately quantized. For other eight kinds of samples, HMF content concentration, was between 0.54-2.25μg/g, from which infant milk powder's HMF concentration was between 0.60-2.25μg/g, as a result conforming to document literature [10] liquid milk and milk powder HMF contents.

To sum up, MEKC detection technique is a kind of effective approach that can simultaneously analyze milk samples' Mel and HMF contents, and can provide reliable daily monitoring support for dairy industry.

CONCLUSION

The paper applies micellar electrokinetic capillary chromatography (MEKC) technique to detect dairy product melamine (Mel) and 5-hydroxymethyl furfural (HMF) contents. By detection results, it is clear that the technique approach has very strong applicability and scientific validity. Based on MEKC technique, dairy product biochemical analysis has the following three features:

1) Mel and HMF concentrations show peak values in 16min and it meets food inspection rapidly gaining high efficiency demands.

2) The approach is simple and easy to operate, and can effectively save experiment cost.

3) It consumes little reagent equipment, and is a kind of economic and environmental friendly technical method.

CONFLICT OF INTEREST

The author confirms that this article content has no conflict of interest.

ACKNOWLEDGEMENTS

Declared none.

REFERENCES

[1]   Z. Ping-Zhen, and S. Qi, "Food melamine attribute, harm and common inspection approach,"*Scientific Practice*, vol. 15, no. 3, pp. 282-283, 2014.

[2]   L. Wen, Z. Xue-Mei, and L. Yang, " Low lactose milk browning suppression technique," *Chinese Dairy Industry*, vol. 22, no. 1, pp. 20-22, 2004.

[3]   H. Le, C. Qiao-Xia, and L.Xun-Hong, "MEKC-DAD simultaneously scutellaria laterifolia six flavonoid constituent contents." *Medicine Analysis Magazine*, vol. 33, no. 9, pp. 1525-1531, 2013.

[4]   H. Jiang-Tao, Z.Wei, and C.Lu-Ying, "Micellar electrokinetic capillary chromatography analyzes chilli products' five kinds of Sudan Red,"*Analysis Test Room*, vol. 33, no. 12, pp. 1408-1412, 2014.

[5]   Z. Li-Chun, G. Ke, and W. Lian, "Micellar electrokinetic chromatography approach rapid measure seasoning sorb and benzoic acid," *Chinese Health Inspection Magazine*, vol. 15, no. 10, pp. 1510-1512, 2015.

[6]   L. Yi, L. Yi-Ping, and B. Yu, "Micellar electrokinetic chromatography simultaneously polygonum multiflorum's seven kinds of indicators constituents contents research, "*Chinese Pharmaceutical Journal*, vol. 50. no. 9, pp. 802-807, 2015.

[7]   L. Yi, L. Yi-Ping, and B. Yu, "Capillary tube zone electrophoresis high performance separation and high sensitivity detection α - whey protein and β - milk globulin," *Research Bulletin of Analytical Chemistry*, vol. 41, no. 10, pp. 1597-1600, 2013.

[8]   L. Ly, M. Qing, and Z. Li-Jia, "Food common additives analysis and detection approach research development," *Journal of Zunyi Normal College*, vol. 16, no. 6, pp. 69-73, 2014,

35

Experimental and Numerical Analysis of Drip Tape Layout for Irrigation of Sugarcane in Latosol

Kai Huang[1,2,*], Desuo Cai[1], Jinchuang Guo[2] and Wei Pan[2]

[1]College of Civil Engineering and Architecture, Guangxi University, Nanning 530004, China; [2]Guangxi Institute of Hydraulic Research, Nanning 530023, China

Abstract: A laboratory soil column experiment was first conducted to analyze water movement in latosol of sugarcane field under drip irrigation from single-point source at different emitter discharge rates. Next, a mathematical model of soil water movement under drip irrigation from single-point source was built using Hydrus-3D, which could accurately simulate the shape of the wetted soil volume and the distribution of volumetric water content in the experiment. Further, a Hydrus-3D model of soil water movement under drip irrigation from double-point source was built and then used to analyze the effects of critical parameters on irrigation uniformity. Results showed that emitter spacing affected irrigation uniformity greatly, but emitter discharge rate did not. According to the irrigation uniformity, project cost and operational management patterns, appropriate drip tape parameters for irrigation of sugarcane in latosol were determined: emitter discharge rate 1.38 L/h, emitter spacing 30 cm, and single-emitter irrigation volume 9.0 L.

Keywords: Drip tape, hydrus-3D, latosol, sugarcane field.

1. INTRODUCTION

Drip irrigation is a water-saving irrigation technique that uses emitters to directly deliver water or a mixture of water and fertilizer to the crop roots in a uniform and accurate way [1, 2]. The benefit of drip irrigation and the efficiency of water use are directly affected by the characteristics of the wetted soil volume, while the wetted soil volume is controlled by a variety of factors such as soil type, initial soil water content, emitter discharge rate, emitter spacing and irrigation volume [3, 4]. Thus far, relevant experimental [5-7] and numerical studies [8-10] have been conducted extensively, which provide important references for appropriate design of drip irrigation systems.

Latosol is one of the major soil types in sugarcane fields of Guangxi Province and widely occurs in the coastal areas of southern Guangxi, China. No studies have investigated the selection of drip tape layout parameters for appropriate irrigation of sugarcane in latosol. This situation leads to some degree of arbitrariness in the design of drip irrigation systems and negatively affects the benefits of drip irrigation project in Guangxi. Therefore, to find out the effects of drip tape layout parameters on the wetted soil volume in latosol under sugarcane has great implications for the design of sugarcane irrigation project in the vast regions of Guangxi.

In this study, a representative soil was selected to perform soil column experiment of drip irrigation from single-point source. Soil water movement was analyzed under drip irrigation at three emitter discharge rates according to the

emitter discharge levels in common use. A mathematical model of soil water movement under drip irrigation from double-point source was built using Hydrus-3D and then used to systematically analyze the effects of critical parameters on water distribution uniformity of the wetted soil volume. Appropriate emitter spacing and discharge rate of drip tapes were selected for irrigation of sugarcane in latosol. Further, preferred irrigation volume and time were determined for drip irrigation of sugarcane at different growth stages in accordance with the patterns of sugarcane cropping, root distribution and project management.

2. MATERIAL AND METHODS

2.1. Experimental Soil

The experimental soil was collected from a sugarcane field in the pilot demonstration base for efficient water-saving irrigation technology in Wulangjiang Village of Xichang Town in Hepu County, southwestern Guangxi Province, China. Eight tons of soil was taken from the plow layer of 10–40 cm. The soil was air-dried, crushed and passed through a 2-mm sieve before use. When preparing for the experiment, the soil was loaded into boxes by layers (5-cm thick each) and compacted with a flat plate to ensure the uniformity. The contact surface between layers was roughened to prevent stratification. The initial volumetric soil water content and dry density were measured immediately after the soil was filled into the boxes. The physical properties of the test soil were shown in Table **1**.

2.2. Experimental Device

The experimental system was comprised of a water supply, a soil box and a soil-water monitor. The water supply

*Address correspondence to this author at the Guangxi University, Nanning 530004, China
E-mail: gxhuangkai@126.com

Table 1.   Physical properties of experimental soil.

Soil Classification	Mechanical Composition/%			Water Content/(cm³·cm⁻³)	Dry Density/ (g·cm⁻³)
	<0.002 mm	0.002–0.05 mm	>0.05 mm		
Latosol	1.9	40.5	57.6	0.065	1.53

was a peristaltic pumpcapable of adjusting the flow rate to simulate different emitter discharge rates. The soil box was a rectangular plexiglass box (100 cm long × 60 cm wide × 85 cm high). An emitter was fixed to the middle of the long side of the glass box and kept 2 cm away from the glass wall to avoid the impact of wall. Four probes of an AZS-2 soil water sensor were buried at different depths (5, 15, 25 and 35 cm, respectively), with 15 cm horizontal distance from the emitter. Data were recorded by the probes every 10 min.

When preparing for the experiment, the soil was loaded into boxes by layers (5-cm thick each) and compacted with a flat plate to ensure the uniformity. The contact surface between layers was roughened to prevent stratification.

2.3. Experimental Design

A laboratory soil column experiment of water movement under surface drip irrigation was conducted in common scenarios of emitter discharge rate (1.38, 2.20 and 2.80 L/h). The irrigation volume was set to 18 L. Each experiment was repeated three times and the results were expressed as the mean values. During the experiment, the soil surface was covered with a plastic film to prevent evaporation. Data were recorded every 30 min, including the surface ponding radius, the wetted surface radius and the vertical wetted depth. At the end of irrigation, soil samples were taken at 5-cm intervals by mesh-stratified sampling and used drying method for measuring soil moisture.

3. NUMERICAL MODEL BUILDING

3.1. Principle

It is assumed that the soil is an isotropic and homogeneous medium. Then soil water movement can be described using Richard's equation as follows:

$$\frac{\partial \theta}{\partial t} = \frac{\partial}{\partial x}\left[K(h)\frac{\partial h}{\partial x}\right] + \frac{\partial}{\partial y}\left[K(h)\frac{\partial h}{\partial y}\right] + \frac{\partial}{\partial z}\left[K(h)\frac{\partial h}{\partial z}\right] + \frac{\partial K(h)}{\partial z} \quad (1)$$

Where θ is volumetric soil water content (cm³/cm³) ; h is negative hydraulic head (cm); t is time (min); and $K(h)$ is unsaturated hydraulic conductivity (cm/min).

The parameters of soil water movement were determined using pedo-transfer functions [11]. The van Genuchten model was used to describe Hydrus-3D as follows:

$$\theta(h) = \begin{cases} \theta_r + \dfrac{\theta_s - \theta_r}{[1 + |\alpha h|^n]^m}, h < 0 \\ \theta_s, h \geq 0 \end{cases} \quad (2)$$

$$K(h) = \begin{cases} K_s S_e^l [1 - (1 - S_e^{1/m})^m]^2, h < 0 \\ K_s, h \geq 0 \end{cases} \quad (3)$$

in which

$$S_e = \frac{\theta - \theta_r}{\theta_s - \theta_r} \quad (4)$$

$$m = 1 - 1/n, \; n > 1 \quad (5)$$

where S_e is effective soil water saturation (cm³/cm³); θ_s is saturated water content (cm³/cm³); θ_r is residual water content (cm³/cm³); K_s is saturated hydraulic conductivity (cm/min); l is pore connectivity factor ($l = 0.5$); m and n are shape factors; and α is air intake factor (cm⁻¹).

3.2. Simulation Area

For simulating soil water movement under drip irrigation from single-point source, the coordinate origin was set at the emitter and the lengths of 40, 40 and 50 cm were taken along the horizontal x-, horizontal y- and vertical z-directions. For simulating soil water movement under drip irrigation from double-point source, the length of emitter spacing was taken along the horizontal x-direction, with one emitter located at the coordinate origin and the other at the end of the length along x-direction; the lengths of 40 and 50 cm were taken along the horizontal y- and vertical z-directions.

3.3. Boundary Conditions

During the experiment, the ponding area at the soil surface changed over time, which could be regarded as a dynamic-head boundary. Since Hydrus-3D can not simulate the dynamic-head boundary, this condition was regarded as a constant-head boundary according to previous studies [12, 13]. Additionally, the constant-head radius Rs was set as the ponding radius when surface ponding generally stabilized in the laboratory soil column experiment. Then the above boundary can be described as follows:

$$h = 0 \quad 0 \leq \sqrt{x^2 + y^2} \leq R_s, \; z = 0, \; 0 \leq t \leq T \quad (6)$$

The other boundaries were set as confining boundaries.

3.4. Initial Conditions

The initial soil water content measured before the start of the experiment was taken as the initial condition of water movement. This condition can be described as follows:

$$\theta(x,y,z,0) = \theta_0 \quad 0 \leq x \leq X, \; 0 \leq y \leq Y, \; Z \leq z \leq 0, \; t = 0 \quad (7)$$

3.5. Model Parameter Determination

Pedo-transfer functions were used to generate the hydraulic parameters of the van Genuchten model and obtain the preliminary parameters of the model. Then the basic parameters of the model were obtained by repeated calibration in accordance with experimental records were shown in Table

Table 2.   v-G model parameters of the experimental soil.

Soil Classification	θ_s / (cm³/cm³)	θ_r / (cm³/cm³)	α / (1/cm)	n	K_s / (cm/min)
Latosol	0.3362	0.0274	0.0332	1.4006	1.96

2. According to the experimental observations, the infiltration radius were 6.5 cm, 8 cm, 9 cm at the emitter discharge rates of 1.38, 2.20 and 2.80 L/h, respectively.

4. RESULTS

4.1. Experimental Data Analysis

Fig. (1) illustrated the relationship between wetting front migration and irrigation time under drip irrigation at the emitter discharge rates of 1.38, 2.20 and 2.80 L/h (irrigation volume 18 L). Clearly, emitter discharge rate affected the wetted depth greatly, but had little effect on the wetted radius. During the same irrigation time, the wetted depth increased with increasing emitter discharge rate to a greater degree than did wetted radius. Under irrigation of the same volume, emitters with a greater discharge rate resulted in slightly shorter wetted radius and depth than did emitters with a slightly smaller discharge rate immediately after the completion of irrigation. However, the differences were not significant since the total irrigation time was shorter for emitters with a greater discharge rate. During the same irrigation time, the wetted depth was 1.3–1.5-fold the wetted radius, indicating more significant infiltration in the vertical direction. The wetted radius and depth exhibited a power function relationship with irrigation time.

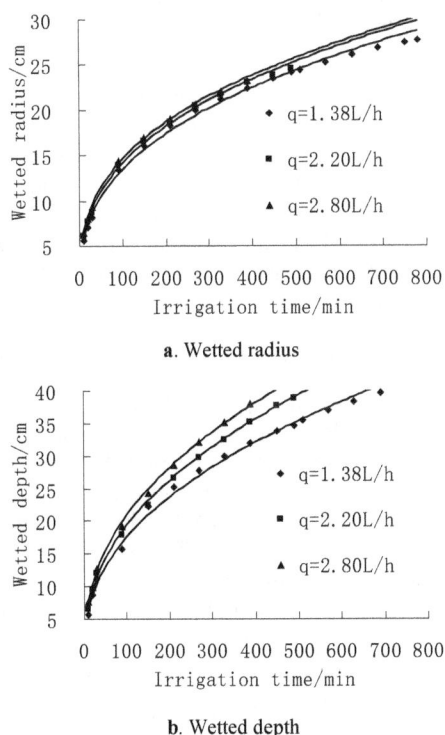

a. Wetted radius

b. Wetted depth

Fig. (1). Exponent relationship between wetting front migration and drip irrigation time at different emitter discharge rates (L/h).

4.2. Model Accuracy Analysis

First, we compared wetting front migration between the simulated and measured results. Fig. (2). presented the measured and simulated distances of soil wetting front from the point source at different irrigation time (discharge rate 1.38 L/h, volume 18 L). There was clearly a gap between the measured and simulated values of the wetting front at the initial stage of irrigation. The simulated wetting front moved faster than the measured one at 30, 90 min. Especially, the simulated values of the wetted radius were 2–3 cm greater than the measured values at the same irrigation time. At the late stage of irrigation, the gap between the measured and simulated values was gradually diminished with increasing irrigation time and almost disappeared at the irrigation time of 150 min.

The model overestimated wetting front migration mainly because Hydrus-3D could not simulate the moving-head boundary condition and thus regarded it as a constant-head boundary condition in the simulation. Additionally, the infiltration radius was set as the ponding radius when surface ponding generally stabilized in the laboratory soil column experiment. As a matter of fact, the ponding radius gradually changed before reaching the stable state, and the actual ponding radius was smaller than the simulated values before the stable state. Therefore, the measured values of the wetting front were smaller than the simulated values at the initial stage of irrigation, which accounted for the shorter distance of wetting front migration in the measurements than in the simulations. As the time elapsed, the difference between the measured and simulated values was gradually diminished and almost disappeared. Therefore, the simulated results of wetting front migration could better reflect the actual situation at the irrigation time more than 150 min.

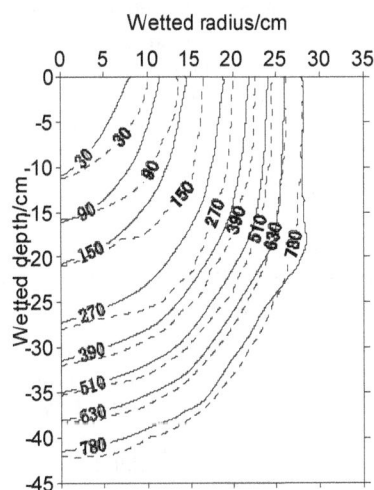

Fig. (2). Measured (solid) and simulated (dashed) shapes of the wetting front at different irrigation time (min).

a. Probes buried at 5 cm depth

b. Probes buried at 15 cm depth

c. Probes buried at 25 cm depth

d. Probes buried at 35 cm depth

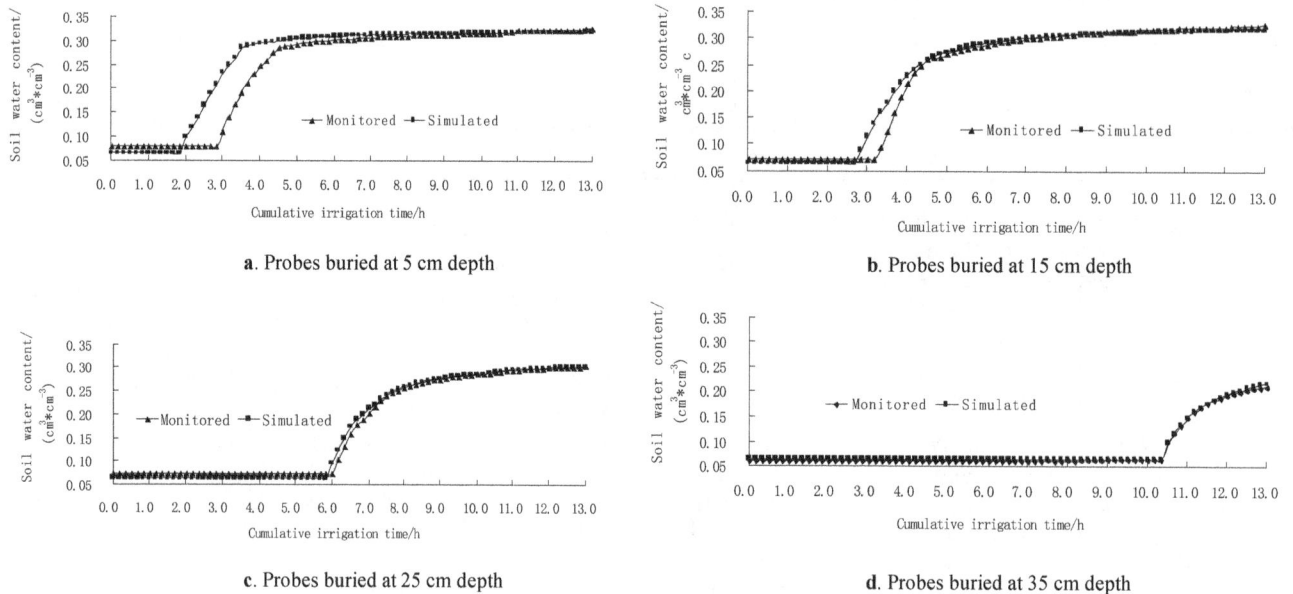

Fig. (3). Monitored and simulated water contents of the wetted soil volume at different irrigation time (h).

Next, we compared the simulated and actual irrigation volumes and examined the difference in soil water content at four monitoring points between the simulated and measured results. According to statistics, the irrigation volume during model calculation was 4.48 L. Since the simulation area accounted for 1/4 of the wetted soil volume, the simulated single-emitter irrigation volume was 17.92 L. The experimental irrigation volume was 9.0 L. Since the experimental area accounted for 1/2 of the wetted soil volume, the actual single-emitter irrigation volume was 18.0 L. There was a minor difference of 0.44% between the simulated actual irrigation volumes, which met the accuracy requirement.

Fig. (3) showed the changes in the monitored and simulated soil water contents over time at 5, 15, 25 and 35 cm depths with 15 cm horizontal distance from the emitter (single-emitter irrigation volume 18 L). On balance, the monitored and simulated values of soil water content followed similar trends at the four monitoring points. However, the time of the starting point of the changes showed difference between the monitored and simulated values. Such difference was gradually diminished with increasing depth of the monitoring point. With 15 cm horizontal distance from the emitter, the arrival time of the simulated wetting front was earlier than the than the monitored result by 1 h at 5 cm depth were shown in Fig. (3a), by 0.5 h at 15 cm depth were shown in Fig. (3b) and by 0.17 h at 25 cm depth were shown in Fig. (3c); the arrival time of the simulated wetting front was generally consistent with the monitored result at 35 cm depth were shown in Fig. (3d).

As mentioned earlier, the dynamic-head boundary condition was regarded as a constant-head boundary condition in model simulation. Thus, the simulated values of wetting front migration were greater than the measured values at the initial stage of irrigation. The simulated wetting front arrived earlier than the monitored wetting front at the point closer to the emitter, but this time difference was diminished gradually with increasing irrigation time. Once soil water content

entered a stable state, the errors between the simulated and measured results were 1.7% at 5 cm depth, 2.6% at 15 cm depth, 0.5 % at 25 cm depth and 3.0% at 35 cm depth. The accuracy of the model simulation completely met the requirements of irrigation decision, and the simulated results of soil water content could exactly reflect the changes in the actual soil water content.

In summary, the simulation results of the proposed model could accurately reflect the actual situation of wetting front migration and soil water content changes when drip irrigation lasted more than 270 min.

Additionally, sugarcane is a crop characterized by drill sowing, shallow roots, and close-planting. The cropping pattern of wide-narrow rows (wide row spacing 1.2–1.3 m; narrow row spacing 0.4–0.5 m) is commonly used in sugarcane fields with drip irrigation. It has been reported that sugarcane roots are mainly distributed in the 0–20 cm (62%) and 20–40 cm soil layers (23.4%); the optimal irrigation depth and width for sugarcane of the vigorous growth stage are approximately 30 and 40 cm, respectively.

4.3. Effect of Emitter Spacing on Irrigation Uniformity

Emitter spacing is an important parameter of drip irrigation system design. We used Hydrus-3D to build a model of water movement under drip irrigation from double-point source. The irrigation conditions were as follows: emitter discharge rate 1.38 L/h, single-emitter irrigation volume 5.5 L, and emitter spacing 30, 40 and 50 cm. Fig. (4). showed a simulated water content distribution in the wetted soil volume at different emitter spacing by the end of irrigation.

The results showed that emitter spacing was an important factor affecting irrigation uniformity for drip irrigation in latosol of sugarcane field. At the emitter spacing of 30 cm, the wetted depth was 33.5 cm and the irrigation uniformity appeared good; the wetted soil volume formed a wetting zone with consistent wetted depth and uniform water distri-

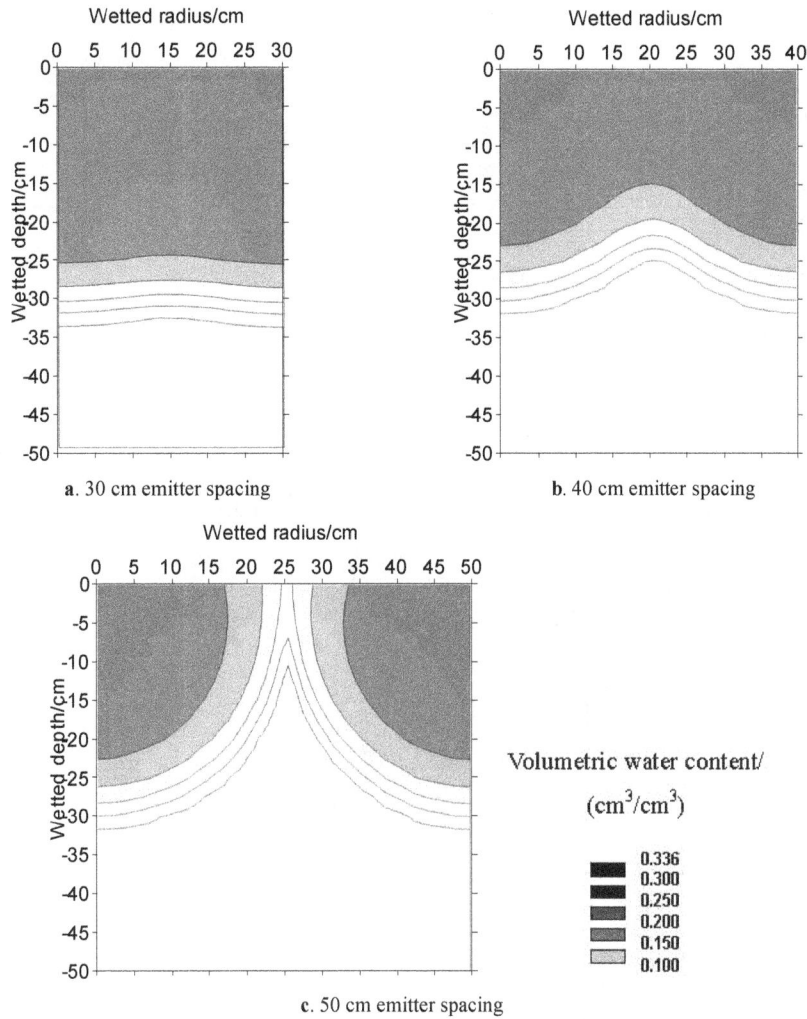

Fig. (4). Simulated water content distribution in the wetted soil volume under drip irrigation at different emitter spacing.

bution were shown in Fig. (**4a**). At the emitter spacing of 40 cm, the wetted depth directly below the emitter was 31.5 cm, while that directly below the midpoint of two emitters was 25 cm, with a 6.5 cm difference; the change in the shape of the wetting front was 20.6%, indicating the poor irrigation uniformity were shown in Fig. (**4b**). At the emitter spacing of 50 cm, the wetted depth directly below the emitter was 31.5 cm, while that directly below the midpoint of two emitters was only 10 cm, with a 21.5 cm difference and 68.3% change in the shape of the wetting front; water content in the core area of the wetted soil volume directly below the midpoint of two emitters was 0.153, significantly lower than that directly below the emitter which was 0.332; these results were indicative of the worse irrigation uniformity were shown in Fig. (**4c**). Since the cropping pattern of sugarcane (drill sowing and close planting) has higher demand for irrigation uniformity, we recommend the emitter spacing of 30 cm for drip irrigation at the emitter discharge rate of 1.38 L/h in latosol of sugarcane field.

4.4. Effect of Emitter Discharge Rate on Irrigation Uniformity

Emitter discharge rate is another important parameter of drip irrigation system design. We used Hydrus-3D to build a model of water movement under double-point source of drip irrigation with the following parameters: emitter spacing 40 cm, single-emitter irrigation volume 9 L, and emitter discharge rate 1.38, 2.20 and 2.80 L/h. Fig. (**5**). Showed a simulated water content distribution in the wetted soil volume at different emitter discharge rates by the end of irrigation.

The results showed that emitter discharge rate had little effect on irrigation uniformity in latosol of sugarcane field. Under irrigation of the same volume, emitters with a greater discharge rate had a shorter irrigation time and thus resulted in a smaller wetted volume, slightly increasing the irrigation uniformity. At the emitter discharge rate of 1.38 L/h, the wetted depth directly below the emitter was 31.5 cm, while that below the midpoint of two emitters was 25 cm; the 6.5 cm difference accounted for 20.6% change in the shape of the wetting front were shown in Fig. (**5a**). At the emitter discharge rate of 2.20 L/h, the wetted depth directly below the emitter was 31.0 cm, while that below the midpoint of two emitters was 24 cm; the 6.0 cm difference accounted for 19.4% change in the shape of the wetting front were shown in Fig. (**5b**). At the emitter discharge rate of 2.80 L/h, the wetted depth directly below the emitter was 29.0 cm, while that below the midpoint of two emitters was 23.5 cm; the 5.5 cm difference accounted for 19.0% change in the shape of

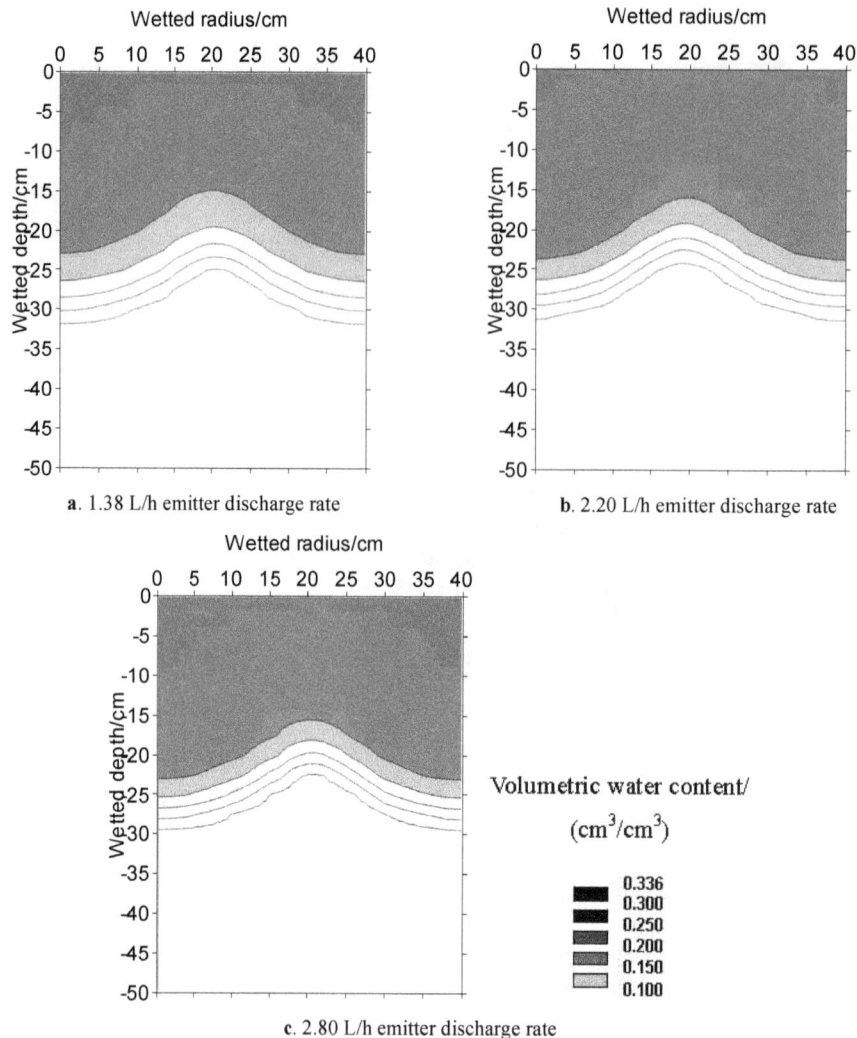

a. 1.38 L/h emitter discharge rate

b. 2.20 L/h emitter discharge rate

c. 2.80 L/h emitter discharge rate

Fig. (5). Simulated water content distribution in the wetted soil volume under drip irrigation at different emitter discharge rates.

the wetting front were shown in Fig. (**5c**). Therefore, it is infeasible to improve irrigation uniformity by a greater emitter discharge rate in latosol of sugarcane field.

From the perspective of project cost, using a smaller emitter discharge rate at the same emitter spacing can significantly reduce project investment. However, the irrigation time will need to be increased to obtain the same irrigation volume. Hence, both the project cost and operational management should be taken into consideration in order to select the appropriate emitter discharge rate.

4.5. Reasonable Parameter Analysis of Drip Tapes in Latosol of Sugarcane Field

Based on the above analysis, three combinations of emitter discharge rate and spacing (1.38, 2.20 or 2.80 L/h and 30 cm) enable common drip tapes to meet the requirement of irrigation uniformity in latosol of sugarcane field in Guangxi. The combination of lower emitter discharge rate (1.38 L/h) and 30 cm emitter spacing is most cost effective, while the combination of higher emitter discharge rate (2.80 L/h) and 30 cm emitter spacing is most expensive and time efficient, and thus is the easiest to manage. Given that the single-emitter irrigation volume desired for sugarcane of the

growing period is 9 L in latosol, then one rotational irrigation group needs the irrigation time of 6.5 h at the emitter discharge rate of 1.38 L/h, 4.1 h at the emitter discharge rate of 2.20 L/h, and 3.2 h at the emitter discharge rate of 2.80 L/h.

According to the existing management system for sugarcane fields, two rotational irrigation groups are completed in one day. The operator needs to manually control the valve in the field for the changing over between rotational irrigation groups, and the pre-set control time is at least 0.5 h. Hence, the selection of emitter discharge rate at 1.38, 2.20 or 2.80 L/h leaves enough time for field control of the valve by the operator. The corresponding 1-d cumulative running time was 13, 8.2 and 6.4 h. Since rainfall is abundant in Guangxi, sugarcane fields are generally irrigated at a relatively low frequency (10–12 times a year). If using the emitter discharge rate of 1.38 L/h, the 1-d cumulative running time of the irrigation system is estimated to be 13 h. This strategy can be recommended as the preferential option as it not only complies with the regulatory requirements but also fits the work schedule of sugarcane farmers and meets general requirements of irrigation management. For sugarcane fields with special requirement for 1-d cumulative running time of the irrigation system, the appropriate emitter discharge rate should be determined in accordance with the single-emitter irrigation volume and running time.

Taking into consideration the project cost and operational management pattern, we recommend the following reasonable design parameters of common drip tapes for use in latosol of sugarcane fields in Guangxi: emitter discharge rate 1.38 L/h, emitter spacing 30 cm, and single-emitter irrigation volume 9 L.

CONCLUSION

Emitter spacing strongly affected irrigation uniformity in latosol of sugarcane field. An emitter spacing of 30 cm ensured irrigation uniformity, while the emitter spacing of 40 or 50 cm resulted in poor irrigation uniformity in the experimental soil. Hence, drip tapes with 30-cm emitter spacing are recommended for irrigation of sugarcane cropped in the drill sowing and close planting pattern.

Emitter discharge rate had little effect on irrigation uniformity in latosol of sugarcane field. The appropriate emitter discharge rate should be selected by taking into consideration the project cost and operational management requirements.

In accordance with the project cost, operational management pattern, and water demand of sugarcane, the appropriate design parameters of drip tapes for irrigation of latosol under sugarcane were determined: emitter discharge rate 1.38 L/h, emitter spacing 30 cm, and single-emitter irrigation volume 9 L.

CONFLICT OF INTEREST

The authors confirm that this article content has no conflict of interest.

ACKNOWLEDGEMENTS

This work is supported by the Agricultural Sci-Tech Achievements Transformation Fund of Guangxi, China (No.14125004-4), the Non-profit Industry Financial Program of Ministry of Water Resources, China (No.201301013).

REFERENCES

[1]   Charles MB. Rapid field evaluation of drip and micro spray distribution uniformity. Irrigat Drainage Syst 2004; 18: 275-9.

[2]   Barragn J, Bralts V, Wu IP. Assessment of emission uniformity for micro-irrigation design. Biosyst Eng 2006; 93: 89-97.

[3]   Sun H, Li MS, Ding H, Wang YX, Cui WM. Experiments on effect of dripper discharge on cotton-root distribution. Transact Chin Soci Agricult Eng 2009; 25: 13-8.

[4]   Subbaiah R. A review of models for predicting soil water dynamics during trickle irrigation. Irrigat Sci 2013; 31: 225-58.

[5]   Wang ZR, Wang WY, Wang QJ, Zhang JF. Experimental study on soil water movement from a point source. J Hydraul Eng 2000; 6: 39-44.

[6]   Cote CM, Bristow KL, Charlesworth PB, Cook FJ, Thorburn PJ. Analysis of soil wetting and solute transport in subsurface trickle irrigation. Irrigat Sci 2003; 22: 143-56.

[7]   Bhatnagar PR, Chauhan HS. Soil water movement under a single surface trickle source. Agricult Water Manag 2008; 95: 799-808.

[8]   Provenzano G. Using HYDRUS-2D simulation model to evaluate wetted soil volume in subsurface drip irrigation systems. J Irrigat Drainage Eng 2007; 133: 342-9.

[9]   Maziar M, Kandelous, Jiri Simunek. Comparison of numerical, analytical, and empirical models to estimate wetting patterns for surface and subsurface drip irrigation. Irrigat Sci 2010; 28: 435-44.

[10]   Dabach S, Lazarovitch N, Šimůnek J, Shani U. Numerical investigation of irrigation scheduling based on soil water status. Irrigat Sci 2013; 31: 27-36.

[11]   Liu JL, Xu SH, Liu H. A review of development in estimating soil water retention characteristics from soil data. J Hydraul Eng 2004; 2: 68-76.

[12]   Li MS, Kang SZ, Sun HY. Relationships between dripper discharge and soil wetting pattern for drip irrigation. Transact Chin Soci Agricult Eng 2006; 22: 32-5.

[13]   Zhao Y, Li MS. Analysis of soil surface ponding radius movement model under point source drip irrigation. Water Saving Irrigat 2014; 12: 16-22.

PERMISSIONS

LIST OF CONTRIBUTORS

Utpal Mohan, Shubhangi Kaushik and Uttam Chand Banerjee
Biocatalysis and Protein Engineering Group, Department of Pharmaceutical Technology, National Institute of Pharmaceutical Education and Research, Sector 67, S.A.S. Nagar-160062, Punjab, India

Wenjing Huang, Yanjie Tong, Wangxiang Huang, Ke Wang, Qiming Chen, Yuanxin Wu and Shengdong Zhu
Key Laboratory for Green Chemical Process of Ministry of Education, Hubei Key Laboratory of Novel Chemical Reactor and Green Chemical Technology, School of Chemical Engineering and Pharmacy, Wuhan Institute of Technology, Wuhan 430073, P.R. China

Jane yuxia Qin, Yan Chen, Dianshuai Gao and Ye Xiong
Department of Neurobiology, Xuzhou Medical College, Jiangsu Province, 221994, China

Sun Jingjing, Zhu Mijia, Zhang Chi and Yao Jun
School of Civil & Environmental Engineering and National "International Cooperation Based on Environment and Energy" and Key Laboratory of "Metal and Mine Efficiently Exploiting and Safety" Ministry of Education, University of Science and Technology Beijing, Beijing 100083, P.R. China

Yang Xiaoqia
College of Resources and Environment Science, Agricultural University of Hebei, Baoding, Hebei, 071001, P.R. China

Shang Shi Yu and Zhao Yi Jun
Key Laboratory of Ecological Remediation of Lakes and Rivers and Algal Utilization of Hubei Province, College of Resources and Environmental Engineering, Hubei University of Technology, Wuhan 430068, China

Gao Ying
Department of Life Sciences, Huazhong Normal University, Wuhan 430079, China

Cheng Kai
Key Laboratory of Ecological Remediation of Lakes and Rivers and Algal Utilization of Hubei Province, College of Resources and Environmental Engineering, Hubei University of Technology, Wuhan 430068, China
Department of Life Sciences, Huazhong Normal University, Wuhan 430079, China

Huang Z. Guang
South China Institute of Environmental Sciences, Guangzhou 510655, China

Yuen Ling Ng and Yi Yang Kuek
Department of Chemical and Environmental Engineering, University of Nottingham, University Park, Nottingham NG7 2RD, United Kingdom

Wang Yanhua, Wu Fuhua, Guo Zhaohan, Peng Mingxing, Xia Min, Pang Zhenling, Wang Xiaoli, Liang Zian and Zhang Naiqun
Life Science and Technology College, Nanyang Normal University, Henan Province, China

Geng Sha and Su Wei
School of Agriculture, Ningxia University, Yinchuan, Ningxia, 750021, P.R. China

Shi Yun and Mi Wenbao
School of Resources and Environment, Ningxia University, Yinchuan, Ningxia, 750021, P.R. China

Li Yuqiu, Tan Hua, Li Da, Li Zhoulin, Chi Yanping, Jiang Yuanyuan, Liu Xiangying, Wang Jinghui and Li Qiyun
Center of Agro-food Technology, Jilin Academy of Agricultural Sciences, Changchun, Jilin, 130033, P.R. China

Lijie Shan
Food Safety Research Base of Jiangsu Province, School of Business, Jiangnan University, Wuxi 214122, China

Hua Li
Synergetic Innovation Center of Food Safety and Nutrition, Wuxi 214122, China

Linhai Wu
Food Safety Research Base of Jiangsu Province, School of Business, Jiangnan University, Wuxi 214122, China
Synergetic Innovation Center of Food Safety and Nutrition, Wuxi 214122, China

Ruixin Liu
Food Safety Research Base of Jiangsu Province, School of Business, Jiangnan University, Wuxi 214122, China
School of Tourism, Yangzhou University, Yangzhou 225127, China

Yafan Bi, Mingke Lei, Yezi Lv, Yangyang Liu, Jiali Liu, Lili Lu and Yali Ma
Key Laboratory for Green Chemical Process of Ministry of Education, Hubei Key Laboratory of Novel Chemical Reactor and Green Chemical Technology, School of Chemical Engineering and Pharmacy, Wuhan Institute of Technology, Wuhan 430073, P.R. China

Qijun Wang
Key Laboratory for Green Chemical Process of Ministry of Education, Hubei Key Laboratory of Novel Chemical Reactor and Green Chemical Technology, School of Chemical Engineering and Pharmacy, Wuhan Institute of Technology, Wuhan 430073, P.R. China
College of Horticulture and Landscape Architecture, Key Laboratory of Horticulture Science for Southern Mountainous Regions, Ministry of Education, Southwest University, Chongqing 400715, P.R. China

Tang Yongzheng, Chu Shaohua, Lu Zhicheng, Yu Yongqiang and Li Xuemeng
Ocean School, Yantai University, Yantai 264005, PR China

Tapobrata Panda
Biochemical Engineering Laboratory, Department of Chemical Engineering, Indian Institute of Technology, Madras, Chennai 600036, Tamil Nadu, India

A. Seenivasan
Biochemical Engineering Laboratory, Department of Chemical Engineering, Indian Institute of Technology, Madras, Chennai 600036, Tamil Nadu, India
Department of Chemical Engineering, SSN College of Engineering, Chennai-603110, Tamil Nadu, India

Sathyanarayana N. Gummadi
Department of Biotechnology, Indian Institute of Technology, Madras, Chennai – 600036, Tamil Nadu, India

Thomas Théodore
Department of Chemical Engineering, Siddaganga Institute of Technology, Tumkur - 572103, Karnataka, India

Ren Yu
Key Laboratory of Ecology and Biological Resources in Yarkand Oasis at Universities under the Education Department of Xinjiang Uygur Autonomous Region, Department of Kashi Normal College, No.29. Xueyuan Road, Kashi Prefecture, Xinjiang Uygur Autonomous Region, China

Yaojun Bo
College of Life Sciences, Yulin University, Yulin, Shaanxi, 719000, China

College of Water and Soil Conservation, Beijing Forestry University, Beijing 100083, China

Qingke Zhu
College of Water and Soil Conservation, Beijing Forestry University, Beijing 100083, China

Weijun Zhao
Key Laboratory of Tourism and Resources Environment in Colleges and Universities of Shandong Province, Taishan University, Taian, Shandong 271021, China

Wei Li
Computer School, China West Normal University, Nanchong, Sichuan, 637002 China

Qinghua Yang
Medical Imaging Department, North Sichuan Medical College, Nanchong, Sichuan, 637000 China

Xudong Jiang, YaoLing Liao and GuiXi Lu
School of Medicine, Guangxi University of Science and Technology, Liuzhou, 545005, China

Zhike Xiao
Liuzhou WanYou Pest Control Research Institute, Liuzhou, 545616, China

Hong Zhang, Wenhui Xing and Huiping Chang
Department of Life Science, Henan Normal of Education, Zhengzhou 450046, China

Wuling Chen
College of Life Science, Northwest University, Xi'an 710069, China

Ruimin Fu
Department of Life Science, Henan Normal of Education, Zhengzhou 450046, China
College of Life Science, Northwest University, Xi'an 710069, China

Fang Lin
Department of Life Science and Technology, Xinxiang University, Xinxiang, 453003, China

Feng Zou
Hunan Provincial Cooperative Innovation Center of Non-wood Forest Cultivation and Utilization, Central South University of Forestry and Technology, Changsha, Hunan, 410004, China

Shixin Xiao
Key Laboratory of Cultivation and Protection for Non-Wood Forest Trees (Central South University of Forestry and Technology), Ministry of Education, Changsha, Hunan, 410004, China

Jun Yuan, Deyi Yuan and Xiaofeng Tan
Hunan Provincial Cooperative Innovation Center of Non-wood Forest Cultivation and Utilization, Central South University of Forestry and Technology, Changsha, Hunan, 410004, China
Key Laboratory of Cultivation and Protection for Non-Wood Forest Trees (Central South University of Forestry and Technology), Ministry of Education, Changsha, Hunan, 410004, China

Ma Jiaheng, Wu Xiaoying and Wang Cheng
School of Civil & Environmental Engineering and National "International Cooperation Based on Environment and Energy" and Key Laboratory of "Metal and Mine Efficiently Exploiting and Safety" Ministry of Education, University of Science and Technology Beijing, Beijing 100083, P.R. China

Zhang Dakun, Song Guozhi and Huang Cui
School of Computer Science and Software Engineering, Tianjin Polytechnic University, Tianjin 300387, China

Chuang Lu, Bo Wang, Xiu-Yuan Peng, Xiao-Lei Hou, Bing Bai and Chun-Meng Wang
Liaoning Academy of Agricultural Sciences, Shenyang, Liaoning, China

Jin Jia, Xiujuan Wang, Junli Lv, Shan Gao and Guoze Wang
School of Mathematics, Physics and Biological Engineering, Inner Mongolia University of Science and Technology, no. 7, Aerding Street, Kundulun District, Baotou, CN-014010 Inner Mongolia, China

Cheng Yang, Xuguang Zhao, Weiwei Chen and Sifa Zhang
School of Computer Science, China University of Geosciences (Wuhan), Wuhan Hubei 430074, P.R. China

Yaping Wang
State Key Laboratory of Freshwater Ecology and Biotechnology, Institute of Hydrobiology, Chinese Academy of Sciences, Wuhan Hubei 430072, P.R. China

Xiaojun Kang
State Key Laboratory of Biogeology and Environmental Geology, Wuhan Hubei 430074, P.R. China
School of Computer Science, China University of Geosciences (Wuhan), Wuhan Hubei 430074, P.R. China
State Key Laboratory of Freshwater Ecology and Biotechnology, Institute of Hydrobiology, Chinese Academy of Sciences, Wuhan Hubei 430072, P.R. China

Zhewen-Zhao, Jingfeng-Huang, Zhuokun-Pan and Yuanyuan-Chen
Institute of Remote Sensing and Information Application, Zhejiang University, Hangzhou, 310058, China
Key Laboratory of Polluted Environment Remediation and Ecological Health, Ministry of Education, College of Natural Resources and Environmental Science, Zhejiang University, Hangzhou, 310058, China
Key Laboratory of Agricultural Remote Sensing and Information System of Zhejiang Province, Hangzhou, 310058, China

Liu Zhong-Hua, Ge Hong-Lian, Luo Rui-Ling and Zhao Jin-Hui
College of Life Science and Agronomy, Zhoukou Normal University, Zhoukou, Henan, 466000, China

Zhilong Xiu
College of Life Science and Technology, Dalian University of Technology, Dalian, 116021, China

Chenchen Cao, Gang Wang and Yibing Zhou
Key Laboratory of Marine Bio-resources Restoration and Habitat Reparation in Liaoning Province, Dalian Ocean University, Dalian,116023, China
Key Laboratory of North Mariculture, Ministry of Agriculture, Dalian Ocean University, Dalian 116023, China

Dazuo Yanga
College of Life Science and Technology, Dalian University of Technology, Dalian, 116021, China
Key Laboratory of Marine Bio-resources Restoration and Habitat Reparation in Liaoning Province, Dalian Ocean University, Dalian,116023, China
Key Laboratory of North Mariculture, Ministry of Agriculture, Dalian Ocean University, Dalian 116023, China

Jing Fan, Jinhua Liao and Mingyuan Huang
College of Life Science, Leshan Normal University, Leshan, 614004, P.R. China

Jianping Hu
College of Chemistry, Leshan Normal University, Leshan, 614004, P.R. China

Lanyang Gao
Sichuan Academy of Botanical Engineering, Chengdu, 641200, P.R. China

Lina Zhao, Tao Pei, Gege Yang and Chenghu Zhou
State Key Laboratory of Resources and Environmental Information System, Institute of Geographical Sciences and Natural Resources Research, Chinese Academy of Sciences, China

Ruan Chengxu and Yuan Chonggui
College of Biological Science and Technology, Fuzhou
University, Fuzhou, Fujian 350116, China

Lihong Zhao, Wenli Yin and Lele Wang
Civil & Architectural Engineering College, Liaoning
University of Technology, Jinzhou, China

Linfang Hou
School of Economics and Management, Zhoukou
Normal University, Zhoukou, 466001, Henan, China

**Xiujuan Ren, Sumei Yao, Runqing Wang, Dafu Wu
and Shilin Chen**
Henan Institute of Science and Technology, Xinxiang
City, 453003, China

Wei Wang
Baotou Light Industry Vocational Technical College,
Baotou, Inner Mongolia 014035, China

Desuo Cai
College of Civil Engineering and Architecture, Guangxi
University, Nanning 530004, China

Jinchuang Guo and Wei Pan
Guangxi Institute of Hydraulic Research, Nanning
530023, China

Kai Huang
College of Civil Engineering and Architecture, Guangxi
University, Nanning 530004, China
Guangxi Institute of Hydraulic Research, Nanning
530023, China

Index

www.ingramcontent.com/pod-product-compliance
Lightning Source LLC
Chambersburg PA
CBHW080524200326
41458CB00012B/4324